Lecture Notes in Mechanical Engineering

Series Editors

Fakher Chaari, National School of Engineers, University of Sfax, Sfax, Tunisia

Francesco Gherardini ⓘ, Dipartimento di Ingegneria "Enzo Ferrari", Università di Modena e Reggio Emilia, Modena, Italy

Vitalii Ivanov, Department of Manufacturing Engineering, Machines and Tools, Sumy State University, Sumy, Ukraine

Mohamed Haddar, National School of Engineers of Sfax (ENIS), Sfax, Tunisia

Editorial Board

Francisco Cavas-Martínez ⓘ, Departamento de Estructuras, Construcción y Expresión Gráfica Universidad Politécnica de Cartagena, Cartagena, Spain

Francesca di Mare, Institute of Energy Technology, Ruhr-Universität Bochum, Bochum, Germany

Young W. Kwon, Department of Manufacturing Engineering and Aerospace Engineering, Graduate School of Engineering and Applied Science, Monterey, USA

Tullio A. M. Tolio, Department of Mechanical Engineering, Politecnico di Milano, Milano, Italy

Justyna Trojanowska, Poznan University of Technology, Poznan, Poland

Robert Schmitt, RWTH Aachen University, Aachen, Germany

Jinyang Xu, School of Mechanical Engineering, Shanghai Jiao Tong University, Shanghai, China

Lecture Notes in Mechanical Engineering (LNME) publishes the latest developments in Mechanical Engineering—quickly, informally and with high quality. Original research or contributions reported in proceedings and post-proceedings represents the core of LNME. Volumes published in LNME embrace all aspects, subfields and new challenges of mechanical engineering.

To submit a proposal or request further information, please contact the Springer Editor of your location:

Europe, USA, Africa: Leontina Di Cecco at Leontina.dicecco@springer.com
China: Ella Zhang at ella.zhang@cn.springernature.com
India, Rest of Asia, Australia, New Zealand: Swati Meherishi at swati.meherishi@springer.com

Topics in the series include:

- Engineering Design
- Machinery and Machine Elements
- Mechanical Structures and Stress Analysis
- Automotive Engineering
- Engine Technology
- Aerospace Technology and Astronautics
- Nanotechnology and Microengineering
- Control, Robotics, Mechatronics
- MEMS
- Theoretical and Applied Mechanics
- Dynamical Systems, Control
- Fluid Mechanics
- Engineering Thermodynamics, Heat and Mass Transfer
- Manufacturing Engineering and Smart Manufacturing
- Precision Engineering, Instrumentation, Measurement
- Materials Engineering
- Tribology and Surface Technology

Indexed by SCOPUS, EI Compendex, and INSPEC.

All books published in the series are evaluated by Web of Science for the Conference Proceedings Citation Index (CPCI).

To submit a proposal for a monograph, please check our Springer Tracts in Mechanical Engineering at https://link.springer.com/bookseries/11693.

Liselott Ericson · Petter Krus
Editors

Advancements in Fluid Power Technology: Sustainability, Electrification, and Digitalization

Proceedings of the Global Fluid Power Society PhD Symposium 2024

Editors
Liselott Ericson
Division of Fluid and Mechatronic Systems
(FLUMES), Department of Management
and Engineering
Linköping University
Linköping, Sweden

Petter Krus
Division of Fluid and Mechatronic Systems
(FLUMES), Department of Management
and Engineering
Linköping University
Linköping, Sweden

ISSN 2195-4356   ISSN 2195-4364 (electronic)
Lecture Notes in Mechanical Engineering
ISBN 978-3-031-84504-8   ISBN 978-3-031-84505-5 (eBook)
https://doi.org/10.1007/978-3-031-84505-5

© The Editor(s) (if applicable) and The Author(s) 2025. This book is an open access publication.

Open Access This book is licensed under the terms of the Creative Commons Attribution 4.0 International License (http://creativecommons.org/licenses/by/4.0/), which permits use, sharing, adaptation, distribution and reproduction in any medium or format, as long as you give appropriate credit to the original author(s) and the source, provide a link to the Creative Commons license and indicate if changes were made.
The images or other third party material in this book are included in the book's Creative Commons license, unless indicated otherwise in a credit line to the material. If material is not included in the book's Creative Commons license and your intended use is not permitted by statutory regulation or exceeds the permitted use, you will need to obtain permission directly from the copyright holder.
The use of general descriptive names, registered names, trademarks, service marks, etc. in this publication does not imply, even in the absence of a specific statement, that such names are exempt from the relevant protective laws and regulations and therefore free for general use.
The publisher, the authors and the editors are safe to assume that the advice and information in this book are believed to be true and accurate at the date of publication. Neither the publisher nor the authors or the editors give a warranty, expressed or implied, with respect to the material contained herein or for any errors or omissions that may have been made. The publisher remains neutral with regard to jurisdictional claims in published maps and institutional affiliations.

This Springer imprint is published by the registered company Springer Nature Switzerland AG
The registered company address is: Gewerbestrasse 11, 6330 Cham, Switzerland

If disposing of this product, please recycle the paper.

Organization

Program Committee Chairs

Liselott Ericson
Petter Krus
Samuel Kärnell

Preface

The Division of Fluid and Mechatronic Systems (Flumes) at Linköping University and the Hudiksvall Hydraulics Cluster (HHK), Sweden, proudly hosted the Global Fluid Power Society (GFPS) Ph.D. Symposium 2024, held from June 17–20 in Hudiksvall. The GFPS Symposium, organized every two years by the Global Fluid Power Society, serves as a platform for young researchers to present and discuss their work with senior researchers and industry professionals from around the world. This book, titled Advancements in Fluid Power Technology: Sustainability, Electrification, and Digitalization, compiles the 23 peer-reviewed papers presented at the symposium. These papers explore a range of research topics within the fields of fluid power, with a special emphasis on the themes of sustainability, electrification, and digitalization-critical areas driving the future of fluid power technology. In addition to the technical papers' presentations, the symposium also included insights from keynote speeches delivered by experts, offering broader perspectives on the trends and future directions of the industry. While the papers in this book vary in scope and depth, they collectively contribute to the ongoing dialogue in the fluid power community. We hope that this collection will provide valuable insights and inspiration for both current researchers and future investigations.

Linköping, Sweden

Liselott Ericson
Petter Krus
Samuel Kärnell

Contents

Downstream Throttled Pneumatic Drives with Time-Controlled Pneumatic-Mechanical Quick-Exhaust Valve 1
Olivier Reinertz, Christian Reese, and Katharina Schmitz

Subsystem-Based Learning Control of Hydraulically Driven Nonlinear Rotary Actuators with Unknown Input Backlash 15
Mahdi Hejrati and Jouni Mattila

Evaluation of a Simple Method to Estimate the Shaft Torque in a Gerotor Pump 31
Giuseppe Totaro, Barbara Zardin, and Massimo Borghi

Optimization-Based Energy Efficient Power Transmission Design Methodology Applied to a Compact Excavator 49
Grégory Tardy, Éric Bideaux, Christophe Gostomski, and Armando Fonseca

Analysis of the Operating Point Method for Dimensioning of Pneumatic Drives Under Variable Loading Conditions 63
Vinícius Vigolo, Antonio Carlos Valdiero, and Victor Juliano De Negri

Experimental Analysis of Friction Forces of Hydraulic Rod Seals—Effect of Pressure, Sliding Speed, Sealing Type, and Different Rod Coatings 83
Kivi Knuuti and Olof Calonius

Digital Twin-Based Classification of Hydraulic Excavator Duty Cycles in Road Construction 99
Johannes Sprink, Bernhard Sender, and Katharina Schmitz

Design of a Cartesian Hybrid Force-Position Controller for a Hydraulic Manipulator 109
Lukas Bachmann, Paul Remde, and Jürgen Weber

**Parameter Identification for Optimized Simulation Models
in Mobile Hydraulic Applications** 127
Bernhard Sender, Johannes Sprink, and Katharina Schmitz

**Intelligent Approach to Enhance Redundancy in Novel
Steer-by-Wire for Heavy Earth Moving Machinery** 141
Vinay Partap Singh, Abid Abdul Azeez, and Tatiana Minav

**Safety Function-Failure Mode and Effect Analysis a Novel
Approach of FMEA for Safety Application in Mobile Working
Machinery** 157
Christa Maria Düsing and Frank Will

**Improvement of Mobile Crusher Energy Efficiency Through
Hybridization and Electrification** 183
Jesse Backman and Tatiana Minav

**Effect of Electrification on the Energy Efficiency of Boom
Trajectories of Semi-Autonomous Mobile Cranes** 199
Timofei Komarov, Victor Zhidchenko, and Heikki Handroos

**Study of Cavitation Conditions Inside a Proportional Spool Valve
by Means of Modal Analysis on Sound Pressure Level** 211
Luca Romagnuolo, Emma Frosina, Carmela Galdi,
Maurizio De Bisceglie, and Adolfo Senatore

**Evaluating the Performance of Semi-autonomous Kinematically
Redundant Loader Crane Operation** 225
Amy Rankka, Marcus Rösth, and Alessandro Dell'Amico

**Design of Hydraulic Power Take-Offs for Wave-Powered Reverse
Osmosis Desalination: Meeting Constraints on Pressure Variation** 239
Jeremy Simmons and James Van de Ven

Experimental Bladder Accumulator-Based Passive Resonator 259
Ville Närvänen

**Investigation of the Heat Conduction in Axial Piston Pumps
by Direct Measurement and Simulation** 271
Roman Ivantysyn, Ahmed Shorbagy, Amey Vedpathak,
and Jürgen Weber

**Numerical Analysis of a High-Power Piezoelectric Pump using
Computational Fluid Dynamics (CFD) Simulations** 289
Francesco Sciatti, Vincenzo Di Domenico, Paolo Tamburrano,
Nathan Sell, Andrew R. Plummer, Elia Distaso, Giovanni Caramia,
and Riccardo Amirante

**Power Analysis of an ePump Applied to the Linear Functions
of an Agricultural Planter** 305
Jacob Lengacher, Ryan Jenkins, Peng Li, Michael Conboy,
and Andrea Vacca

**Analysis of Opportunities for Integrated Thermal Management
on Battery Powered Mobile Machines** 323
Fabian Lagerstedt, Samuel Kärnell, Marcus Rösth, and Liselott Ericson

**A Hydraulic Architecture Based on Multi-common Pressure Rail
Principle Using Multi-chamber Cylinders for Excavators** 339
Zihao Xu, Mateus Bertolin, Andrea Vacca, and Jan Nilsson

Harmonic Characterisation of Electrically Driven Pumps 357
Thomas Heeger, Martin West, and Liselott Ericson

Downstream Throttled Pneumatic Drives with Time-Controlled Pneumatic-Mechanical Quick-Exhaust Valve

Olivier Reinertz, Christian Reese, and Katharina Schmitz

1 Introduction and State of the Art

Pneumatic drives are widely used in industrial production since they are robust, allow for an easy realization of point-to-point motion tasks, have low initial costs, and high-power density. To control the cycle time of pneumatic drives, downstream throttling is the most common method. This involves driving the movement with the active chamber of the cylinder utilizing the full supply pressure, while the speed is regulated by the counter-acting chamber through adjusting the downstream throttle, thus creating a pressure that opposes the movement (referred to as back pressure). This system is however often regarded as having high energy saving potential, since the maximum amount of compressed air for the movement is always used, independently from the actual load.

Climate change stands out as the challenge of the century. Coupled with the rise in energy prices, this has intensified the focus on energy efficiency. Given the highly energy-demanding process involved in compressed air production, the development of more efficient pneumatic drive circuits has been the subject of numerous research projects. Many circuits aim to harness the expansion energy, which can be accomplished by an early supply shut-off in mid-stroke. This allows the compressed air already supplied to expand and complete the movement without further air consumption [1]. A very promising design option for such systems is the use of four 2/2-way valves arranged as a full bridge [2–5]. These systems provide full flexibility for independent metering and not only allow control of the air supply but also the exhaust. Rager describes an optimal control sequence with such a system, which includes an early supply shut-off and a braking phase to slow down the cylinder before it reaches the end-cushion [5]. This braking phase is achieved by closing the exhaust toward the

---

O. Reinertz (✉) · C. Reese · K. Schmitz
Institute for Fluid Power Drives and Systems, RWTH Aachen University, Aachen, Germany
e-mail: olivier.reinertz@ifas.rwth-aachen.de

end of the stroke, leading to great performance of the cylinder without overloading the end-cushion.

Such circuits do allow for very efficient control strategies due to their high flexibility. However, they suffer from higher costs associated with the necessary components and increased efforts in commissioning and control implementation. The focus of this paper lies, therefore, on a novel simple to implement switching scheme, which controls the exhaust process to achieve better performance. The system does not control the air supply and will therefore not benefit from using the expansion energy. However, it consists of a simple pneumatic—mechanical solution, which might reduce its threshold for broad applicability.

2 Cylinder Sizing Parameters Interactions

The efficiency of pneumatic drives is closely tied on accurately sizing cylinders for their intended tasks, given that compressed air consumption is dependent on cylinder chamber volumes and pressure levels. Common approaches to sizing pneumatic actuators involve using empirical formulas and computer-based tools provided by manufacturers [6, 7]. Simulation-based sizing, recommended by various researchers, offers an alternative and can be integrated with optimization schemes to achieve optimal sizing [8, 9]. However, these "black box" approaches do not provide any insights into the interactions between the different parameters relevant to the sizing task.

Doll proposed a heuristic method for the dimensioning of downstream throttled drives for motion tasks [10, 11], a widely adopted approach in academia [12–14]. The method relies on a dimensionless number called the 'pneumatic frequency ratio,' denoted as Ω. This ratio relates the actuator dynamics, determined by an estimation of its eigenfrequency ω_0, to the required dynamics for the motion task, calculated by the reciprocal of the transfer time ω_f. Where c corresponds to the cylinder stiffness midstroke and M to the moving mass. A well-dimensioned cylinder typically exhibits Ω values between 1.1 and 1.7.

$$\Omega = \frac{\omega_0}{\omega_f} = \frac{t_f}{2\pi}\sqrt{\frac{c}{M}} \qquad (1)$$

Furthermore, Doll proposes to approximate the stiffness by considering a rodless cylinder with length L and piston area A. Furthermore, it is assumed that both chambers and dead volumes in the lines are pressurized with the supply pressure p_s.

$$c = \frac{4 \cdot A \cdot p_s}{L} \qquad (2)$$

The primary advantage of this approach is that it results in a simple algebraic equation that encapsulates all the relevant parameters for the sizing task. By employing this equation and incorporating the air consumption V_n, Doll obtains the following relationship between compressed air demand and transfer time t_f.

$$V_n = A \cdot L \cdot \frac{p_s}{p_0} = \frac{\pi^2 \cdot \Omega^2 \cdot M \cdot L^2}{p_0 \cdot t_f^2} \tag{3}$$

It becomes evident that air consumption depends quadratically on the transition time. Therefore, by reducing the transfer time, energy consumption will decrease disproportionately on a well-dimensioned downstream throttled drive. This insight will be applied in the approach proposed in the next section, aiming to enhance system dynamics compared to conventional downstream throttled drives. The goal of this paper is to reduce the size of the cylinder using the novel system while ensuring comparable cycle times and end cushion performance to an optimally sized downstream throttled drive.

3 Time Controlled Quick-Exhaust Valve

The use of quick-exhaust valves is a well-known measure to increase the speed and reduce the cycle time of pneumatic drives, especially with longer tube lengths between valve and cylinder. However, it is not advisable to arbitrarily reduce the downstream resistance without taking end-cushion into consideration. One possibility to increase damping capacity at the end of the stroke is to use external hydraulic shock absorbers. However, these are additional components that increase cost and installation space. Furthermore, they also have a limited service-life and, therefore, need to be replaced at regular intervals. On the other hand, the loading capacity of the internal pneumatic end-cushion is limited, and an increase in cylinder speed beyond the allowed limit will also reduce lifetime of the drive and eventually result in mechanical failure.

The main concept proposed in this paper is illustrated in Fig. 1 and differs only in the exhaust process from conventional downstream throttling. The driving chamber remains connected to the full supply pressure in both systems. Unlike the conventional downstream throttled system, the cylinder speed is initially increased using a switchable quick exhaust valve. The quick exhaust valve should only remain active during the initial and mid-part of the stroke and be later switched off, allowing the exhaust air only to flow through a downstream throttle arranged in parallel. The resistance of the downstream throttle should be adjusted significantly higher than the quick exhaust valve, thereby reducing the cylinder speed before it reaches the end-cushion. As described by Rager, the desired outcome of the braking phase is to achieve lower cycle times through increased cylinder speeds in the preceding movement section without overloading the pneumatic end-cushion [5].

The right side of Fig. 2 depicts the novel circuit with switchable quick exhaust valves (SQE) investigated in this paper. The conventional downstream throttling

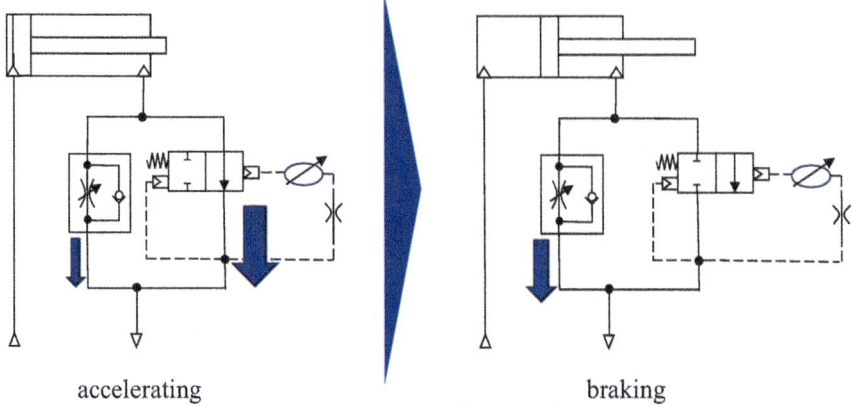

Fig. 1 Schematic representation of the concept with a switchable quick exhaust

(DT) is shown on the left side, which will be used as a reference to evaluate the novel system's performance.

To achieve the functionality described in Fig. 1, a pneumatic-mechanical approach is followed. This has the advantage of reducing system complexity and implementation effort. A possible design implementation can be seen in Fig. 3. The key element to achieve the desired control is a pneumatic time delay circuit, similar to the ones used in commercially available pneumatic time valves [15, 16]. This mechanism causes a delay in the pressure changes inside the chamber p_c compared to the acting pressure signal p_1. The dynamics of the chamber pressure p_c are defined by the

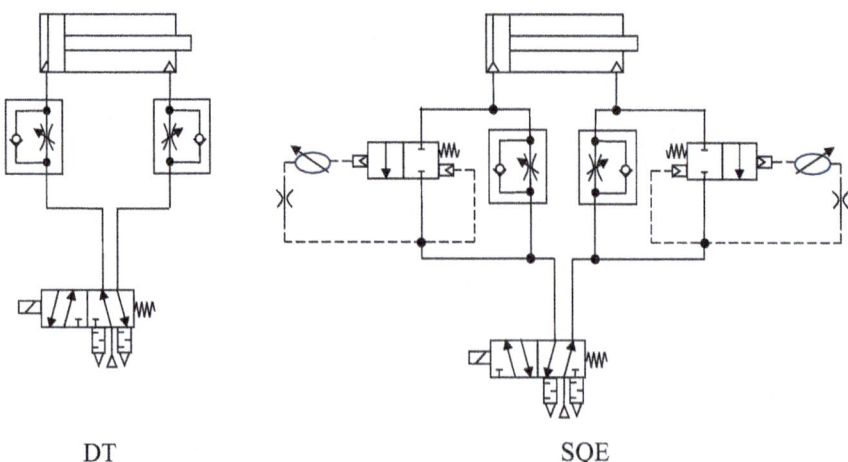

Fig. 2 Comparison of a downstream throttled cylinder drive with the new system with a switchable quick exhaust

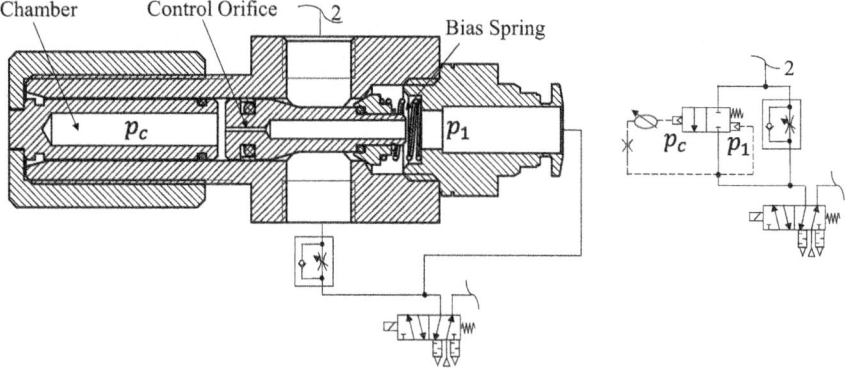

Fig. 3 Potential design of the switchable quick exhaust valve

size of the chamber volume and the size of the control orifice. The switching of the 5/2 control valve results in a sudden pressure drop in the pneumatic lines being exhausted. Due to the time delay circuit, a dynamic pressure imbalance is created, which opens the valve against the bias spring. After some time, the pressure p_c in the valve internal chamber reduces, as air flows through the control orifice to the exhausted line, and the bias spring closes the valve again. Thus, a time depended quick exhaust is achieved, which acting time window can be adjusted depending on the time constant of the time delay circuit.

4 Simulation Model

A lumped parameter simulation in Simcenter Amesim is conducted to investigate the system dynamics, taking into account factors such as friction and non-linear thermodynamic effects. Figure 4, the simulation model for the switchable quick exhaust system is presented. A comparable model is built-up for downstream throttling. The line model comprises the tube dead volume evenly split with an orifice connecting both halves. This orifice represents the pressure drops in the line and is parameterized using ISO 6358-3 [17]. The resistance of the throttle check valve is characterized through fixed orifices for each flow direction individually. The downstream resistance of the throttle check valve is utilized to regulate the speed of the drive, while its upstream resistance is parameterized to account for the flow losses of the check valves.

The thermodynamic model of the cylinder considers heat transfer through its walls. The friction force was implemented using a Stribeck curve, with parameters chosen based on the values presented by Raisch [18]. The drive model also incorporates the modeling of the air end-cushion, which was properly adjusted for each test case. Additionally, the air cushion was utilized to its maximum loading capacity. A properly adjusted end-cushion should result in a quick and smooth deceleration

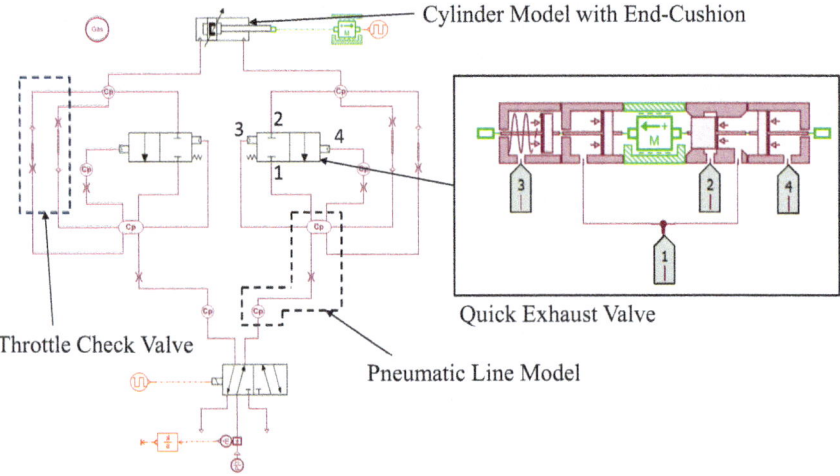

Fig. 4 Simulation model in Amesim

without oscillations. If the kinetic energy while entering the end-cushion is too high, such optimal set point cannot be achieved. The maximum loading capacity of the drive is therefore to be defined as the maximal kinetic energy that still allows for an optimal adjustment of the end-cushion. Due to the end-cushion being the limiting factor for the maximal cylinder speed in motion tasks, the minimal achievable cycle time was determined in simulation for each system, enabling an objective comparison. To achieve this, the downstream throttle, time constant of the switchable exhaust valve, and end cushion setting were iterated until the best performance was achieved.

Figure 4 also depicts the modeling of the switchable quick exhaust valve. No leakage was simulated, as a poppet valve design was chosen. However, the static pressure forces in the valve are balanced, resulting from the valve having equal acting areas for the pressures on both sides (see Fig. 3). The valve dynamics and friction are modeled, and the time delay circuit is represented externally as a dead volume and control orifice.

Table 1 provides an overview of the simulation model parameters. Two distinct cylinder lengths were tested. Since the objective of the system is to reduce the size of the cylinder drive while maintaining similar cycle times, the cylinder actuated with the novel quick exhaust valves was chosen to be one size smaller than the reference downstream-throttled cylinder. On the other hand, the mass being moved is chosen to be constant for all test cases.

The parameters used to characterize the valve are summarized in Table 2. The control orifice was implemented as a variable in this simulation study and used to adjust the time constant of the time delay circuit, while the chamber volume was held constant.

Table 1 Simulation system parameter

			DT		SQE	
	Parameter	Unit	Case 1	Case 2	Case 3	Case 4
Cylinder	Piston-ø	mm	32	32	25	25
	Rod-ø	mm	12	12	10	10
	Stroke	mm	200	350	200	350
	Mass	kg	10	10	10	10
	Dead volumes	ml	6	6	3.5	3.5
Tubes	Length	m	1	1	1	1
	Inner-ø	mm	4	4	4	4

Table 2 Simulation novel valve parameter

Quick exhaust valve		
Parameter	Unit	Value
Spring constant	N/mm	1
Spring preload	mm	4
Spool-ø	mm	8
Stroke	mm	1
Stiction force	N	2.1
Coulomb force	N	2
Viscous coefficient	Ns/m	2
Dead volume delay-circuit	ml	1
Conductance delay-circuit	Nl/(s·bar)	0.004–0.01

5 Results and Discussion

Figure 5 depicts simulation results achieved with a cylinder length of 200 mm, where the system with the switchable quick exhaust (SQE) utilizes a cylinder one size smaller compared to the conventional downstream throttling. The pressure curves shown in the upper diagrams include not only the pressure in the cylinder chamber but also the pressure of the air cushion for the two systems. The point at which the air cushion becomes active corresponds to the peak in braking pressure at the end of the movement. Furthermore, the opening area of the quick exhaust valve is presented in the lower diagram. The valve functions as expected, allowing for quicker air exhaust at the beginning of the movement and closing later. However, a stick-slip effect can be seen in the stroke curve.

Lower extension and retraction times were achieved with the novel system, even with a smaller cylinder. The strategy used to adjust the time delay circuit, which characterizes the opening characteristic of the quick exhaust, was to maximize its opening time. However, the higher cylinder speed achieved due to the quick exhaust needs to be reduced before entering the end cushion to prevent overloading. Consequently,

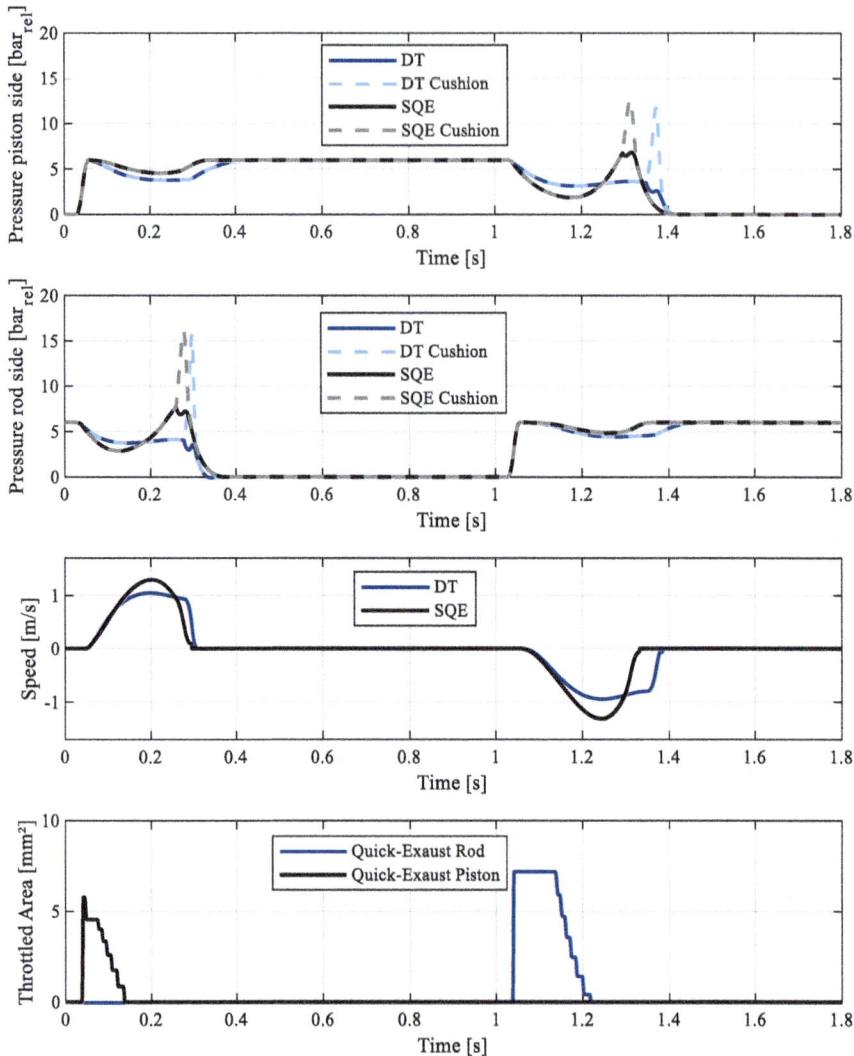

Fig. 5 Simulation results: stroke 200 mm, moving mass 10 kg

the downstream throttle in the novel system was adjusted to a higher resistance than in conventional downstream throttling. This adjustment leads to an increase in back pressure, braking the cylinder before it enters the end cushion. The higher back pressure and lower cylinder speed create favorable conditions for the proper functioning of the air cushion. On the extension stroke, the quick exhaust valve does not open fully, resulting in slower cylinder movement compared to the retraction stroke. A parameterization of the time delay circuit, aimed at fully opening the valve, however, led to the quick exhaust being active for too long. This situation doesn't allow sufficient time for the braking phase, overloading the end cushion.

The results with the longer cylinder are shown in Fig. 6. The novel system seems to work even better with longer cylinders due to the quick exhaust being active for a longer time. The pressure peak in this test case is even lower than with the conventional system, thus highlighting a lower demand on the end cushion due to pre-braking while the cylinder is still moving.

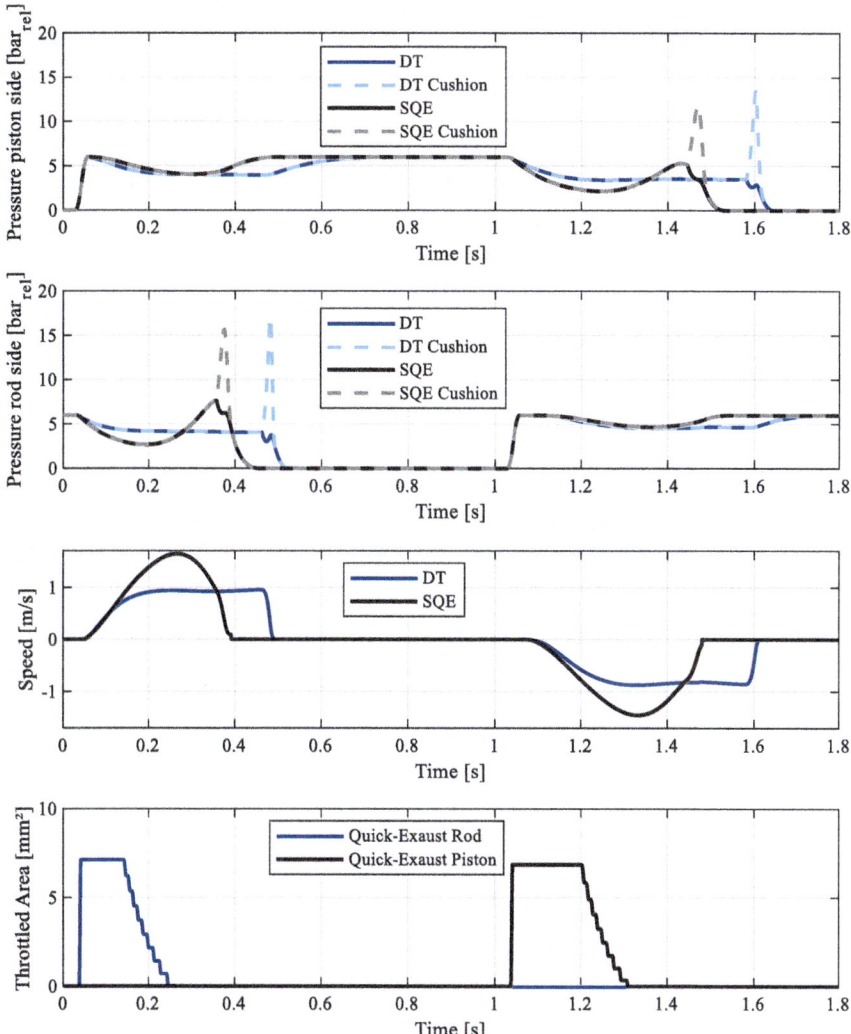

Fig. 6 Simulation results: stroke 350 mm, mass 10 kg

Table 3 Quantitative simulation results

		DT		SQE	
Result	Unit	Ø 32–200	Ø 32–350	Ø 25–200	Ø 25–350
Air consumption/cycle	g	2.616	4.403	1.66	2.78
Air savings	%	–	–	36,5%	36,8%
Extension time	ms	302	488	295	391
Retraction time	ms	382	610	337	480
Cycle time	ms	684	1098	632	871
Normalized air savings	%	–	–	42.75%	58.48%

The quantitative results for comparison are summarized in Table 3. Due to the smaller cylinder size, the novel system consumes about 35% less compressed air than the conventional downstream throttled system. Remarkably, both the retracting and extending times on all investigated test cases are lower with the quick exhaust despite utilizing a smaller cylinder. Thus, further reductions in air consumption are presumably possible, either by reducing the supply pressure or downsizing the cylinder diameter even further. Therefore, normalized air consumption, which is adjusted to changes in cycle time give a more accurate metric for objective evaluation. The normalized savings were calculated based on the proposed method by the authors in [14].

6 Impact of the Novel Valve on the Sizing Practice of Pneumatic Drives for Motion Tasks

While numerous energy-saving strategies achieve noteworthy reduction in compressed air consumption, most of them tend to raise installation and maintenance costs and/or diminish the dynamic performance of the system. Thus, only a limited number of this approaches result being economically viable and come to practical use. The integration of the new type of valve into all suitable drives of new systems to be built is certainly one way of reducing the energy consumption of the systems to a large extent. However, it is not yet clear what procurement costs and commissioning efforts this would entail. Thus, to date it is not yet clear whether the solution is better than the existing and commercially viable or not. For this reason, an alternative usage scenario is presented in the following.

The potential to enhance drive dynamics through the retrofit use of the novel quick exhaust valve makes a compelling case for a significant impact on the existing drive selection process. Current safety factors in drive sizing result in significant additional compressed air consumption during machine operation. These safety factors are introduced due to uncertainties in friction and process forces, which cannot be precisely specified in advance and are often omitted in many sizing tools. Additionally, pneumatic drives are only available in discrete diameter steps, necessitating oversizing when selecting the product [19]. Introducing the novel quick exhaust valve as a backup solution in case of an under-dimensioned conventional drive allows for a strong reduction in existing safety factors. From an economic perspective, this implies that smaller and less expensive drives could be generally employed. Only in cases of truly undersized drives, could the dynamic be retroactively improved while commissioning through the use of the novel valves. Consequently, significant reductions in air consumption become achievable without additional costs, as the expenses for the small number of retrofitted valves could be offset by the smaller sizing of multiple drives in the best-case scenario. This represents a key unique selling point of the proposed technology and, with correct functionality and moderate system costs, can be expected to lead to widespread use of the valves.

7 Conclusion and Outlook

In the scope of this paper, a novel approach to improve the dynamic behavior of downstream-throttled pneumatic drives has been presented. The system aims to achieve lower cycle times through a time-controlled quick exhaust valve, implemented as a pneumatic-mechanical solution. This is particularly relevant due to the relationship between cycle time and air consumption in the sizing of pneumatic drives. The lower cycle times achieved with the novel system can also be interpreted as a reduction in air consumption.

The simulation in this paper demonstrates that downsizing the cylinder drive by at least one size is feasible with the novel system. Furthermore, the results show a remarkable improvement in performance compared to the conventional system, even with the smaller cylinder size, resulting in shorter cycle times.

The main system characteristics contributing to the performance increase include:

- Higher cylinder speeds mid-stroke due to significantly lower downstream resistance while the quick exhaust is active.
- Braking of the cylinder before reaching the end cushion by timely closing the quick exhaust path, resulting in very favorable conditions for the end cushion to work properly (lower speed and significantly higher back pressure).

Despite these very promising simulative results of the proposed concept, there are still some challenges that need to be addressed in future work:

- Experimental investigations to validate the proper working of the novel system and simulated results are necessary. This should test the reliability of the valve and repeatability of the time control function under various conditions, such as moving masses and cylinder friction forces.

- The commissioning effort and adjustment difficulty of the novel system should be investigated, and ideally, an easy-to-implement strategy for optimally adjusting the system should be developed.
- The length of the pneumatic tubes is a highly variable parameter; therefore, the exhaust dependency on the air flow resistance of the pneumatic lines should be reduced. A possible optimization of the valve vents the exhaust air directly to the atmosphere, thus bypassing the pneumatic lines.

Acknowledgements The research and development project S-LEAP—Quick-Exhaust for increasing the performance and efficiency of exhaust throttled pneumatic drives—was funded by the European Union and the state of North Rhine-Westphalia. The Authors would like to express its sincere thanks to all project partners, the European Regional Development Fund and the state of NRW.

References

1. Raisch A, Hülsmann S, Sawodny O (2018) Saving energy by predictive supply air shutoff for pneumatic drives. In: European control conference (ECC). Limassol, Cyprus, pp. 965–970
2. Doll M, Neumann R, Sawodny O (2011) Energy efficient use of compressed air in pneumatic drive systems for motion tasks. In: International conference on fluid power and mechatronics (FPM). Beijing, China, pp. 340–345
3. Doll M, Sawodny O (2010) Energy optimal open loop control of standard pneumatic cylinders. In: 7th international fluid power conference (IFK). Aachen, Germany, pp. 256–270
4. Neumann R, Rager D, Doll M (2016) Verfahren zum Betreiben einer Ventileinrichtung, Ventileinrichtung und Datenträger mit einem Computerprogramm. Patent DE 10 2016 206 821 A1
5. Rager D, Doll M, Neumann R, Berner M (2018) New programmable valve terminal enables flexible and energy-efficient pneumatics for Industry 4.0. In: 11th international fluid power conference (IFK). Aachen, Germany, pp. 209–221
6. Festo Corporation, "Pneumatic Sizing". https://www.festo.com/eap/en-us_us/PneumaticSizing/. Accessed 05 Dec 2023
7. Emerson Corporation, "Numasing". https://www.asco.com/en-us/Pages/calculators-numasizing.aspx. Accessed 05 Dec 2023
8. Harris P, Nolan S, O'Donnell GE (2014) Energy optimisation of pneumatic actuator systems in manufacturing. J Clean Prod 72:35–45
9. Raisch A, Sawodny O (2019) Analysis and optimal sizing of pneumatic drive systems for handling tasks. Mechatronics 59:168–177
10. Doll M, Neumann R, Sawodny O (2015) Dimensioning of pneumatic cylinders for motion tasks. Int J Fluid Power 16(1):11–24
11. Doll M, Neumann R, Gauchel W (2024) Sizing of pneumatic drives under energy efficiendy aspects. In: 14th international fluid power conference (IFK). Dresden, Germany
12. Nazarov F, Weber J (200) Sensitivity Analysis and robust optimization of an integrated pneumatic end-position cushioning. In: 13th international fluid power conference (IFK). Aachen, Germany, pp. 974–988

13. Boyko V, Weber J (2024) Energy efficiency of pneumatic actuating systems with pressure-based air supply cut-off. Actuators 13(1):44
14. Reese C, Reinertz O, Schmitz K (2024) Feasibility study and experimental validation of a novel combined throttling approach. In: 14th international fluid power conference (IFK). Dresden, Germany
15. Festo Corporation, "Time delay valve", https://www.festo.com/media/pim/449/D15000100123449.PDF. Accessed 25 Jan 2023
16. Murrenhoff H (2014) Grundlagen der Fluidtechnik Teil 2: Pneumatik. Shaker Verlag, Aachen, Germany
17. International Standard (2014) ISO 6358-3: pneumatic fluid power—determination of flow-rate characteristics of components using compressible fluids—Part 3
18. Raisch A, Sawodny O (2020) Modeling and analysis of pneumatic cushioning systems under energy-saving measures. IEEE Trans Autom Sci Eng 17(3):1388–1398
19. Gauchel W (2006) Energiesparende Pneumatik: Konstruktive sowie schaltungs- und regelungstechnische Ansätze. In: O+P - Ölhydraulik und Pneumatik. Mainz, Germany, pp. 33–39

Open Access This chapter is licensed under the terms of the Creative Commons Attribution 4.0 International License (http://creativecommons.org/licenses/by/4.0/), which permits use, sharing, adaptation, distribution and reproduction in any medium or format, as long as you give appropriate credit to the original author(s) and the source, provide a link to the Creative Commons license and indicate if changes were made.

The images or other third party material in this chapter are included in the chapter's Creative Commons license, unless indicated otherwise in a credit line to the material. If material is not included in the chapter's Creative Commons license and your intended use is not permitted by statutory regulation or exceeds the permitted use, you will need to obtain permission directly from the copyright holder.

Subsystem-Based Learning Control of Hydraulically Driven Nonlinear Rotary Actuators with Unknown Input Backlash

Mahdi Hejrati and Jouni Mattila

1 Introduction

Having a remarkably larger power-to-weight ratio thanks to fluid power, hydraulically driven manipulators (HDMs) have been widely utilized for decades to drive working machines in various fields of construction, forestry, mining, and agriculture. Nowadays, the popularity of these machines in the industry is increasing; for instance, U.S. compact excavator sales in 2019 totaled about 46,000 units compared to 42,800 in 2018 [1]. Thanks to fluid power, we are heading toward a future in which these machines will become field-robotic systems, requiring minimal human supervision. Nonetheless, conventional hydraulic actuators employing cylinders restrict motion ranges below 180 degrees. In contrast, rotary hydraulic actuators (RHAs) with mechanisms such as rack and pinion, can extend the joint range up to 360 degrees. Despite this benefit, the existing backlash nonlinearity of the gears in RHAs compounds the inherent complexities of the HDMs, including structural complexity and governing fluid dynamics, amplifying system uncertainties. The mentioned issues make the stability analysis and control design of HDMs with RHAs significantly important and challenging.

Over the past decades, numerous studies have been conducted to control HDMs. It has been shown in [2] that nonlinear model-based control (NMBC) approaches are suitable for the control of HDMs. In [3], different NMBC methods have been designed and compared, such as sliding mode control (SMC), adaptive inverse dynamics controller (AIDC), and model-reference adaptive controllers with velocity measurement (MRACV). The experimental results showed that MRACV had better performance than others. In [4], a terminal SMC is employed to control the excavator. Moreover, in [5] an adaptive controller with gravity and friction identification is developed for n-link hydraulic manipulators. The virtual decomposition control (VDC)

M. Hejrati (✉) · J. Mattila
Department of Engineering and Natural Science, Tampere University, Tampere 7320, Finland
e-mail: mahdi.hejrati@tuni.fi

scheme proposed in [6], also, is widely utilized to address the control problem among HDMs [7–9]. Based on the comparison reported in [10], the VDC performed better than other approaches in real-world HDMs applications. In addition to its better performance, employing the VDC scheme as a baseline controller offers other benefits, as follows: (1) VDC decomposes the entire complex system into subsystems where the decentralized controller and stability analysis can be performed, and (2) the dynamics of each subsystem remains relatively simple which is invariant to the target system.

Although NMBC approaches consider the system model in control design, unknown uncertainties, such as unmodeled dynamics and input constraints, can degrade their performance and, in some cases, result in instability [11]. However, a limited amount of research has considered the mentioned issue in the field of HDMs. In [12], the backlash has been considered in the force control design of the electro-hydraulic load simulator. In [13], the backlash in an RHA with a rack and pinion mechanism is examined, and better performance in experiments is achieved by employing VDC. Further, in [11], the valve deadzone/backlash dynamics is considered in the adaptive backstepping controller design to improve the performance of the controller. In addition, the backlash of an RHA with a helical gear as a converter mechanism is tackled by utilizing VDC and adaptive inverse backlash controller in [14]. However, in the mentioned works, the model of the input constraint nonlinearity (backlash) is considered known, but the model can be much more complex or even unknown in real-world applications. Therefore, equipping the controller with a method that can tackle the unknown backlash nonlinearity will be significantly important in real-world applications. Due to their universality and perfect capabilities in function approximation, RBFNNs are widely utilized to estimate the model uncertainty of both electric [16] and hydraulic manipulators [15]. However, RBFNNs are commonly used to estimate the uncertainties in the coupled rigid body-actuator model based on the states of the coupled system. This approach, in the presence of some uncertainties, such as unknown backlash nonlinearity that exists at the actuator, performs poorly because it receives data from the coupled model, making it difficult to detect uncertainties specific to the actuator. Furthermore, as stated in [17], learning problems of complex systems should be solved using the decomposition and re-composition methods. The novel way of incorporating the RBFNNs into VDC would allow us to address the mentioned issue, resulting in a subsystem-based learning controller (SSL). By decomposing the entire model into subsystems, VDC enables the design of RBFNNs for each subsystem based on its states, whether it is required velocity and acceleration in rigid body subsystems or it is pressure and piston position in the actuator subsystem. Consequently, RBFNNs at the actuator subsystem level can detect backlash with greater accuracy, as they only receive data from the actuator. Such a novel incorporation, despite its mathematical challenges in stability analysis, is vital in the sense of performance improvement.

The rest of the paper is organized as follows. Section 2 expresses the fundamental mathematics of the VDC approach along with the essential lemmas and definitions utilized in this paper. Section 3 describes the modeling and control design procedure.

In Sect. 4, low-level voltage control is designed. Experimental results are provided in Sect. 5, and Sect. 6 concludes this study.

2 Mathematical Preliminaries

2.1 VDC Foundation

Consider $\{A\}$ and $\{B\}$ as frames that are attached to a rigid body. Then, the 6D linear/angular velocity vector ${}^A V \in \mathbb{R}^6$ and force/moment vector ${}^A F \in \mathbb{R}^6$ can be expressed as follows [6]:

$$ {}^A V = [{}^A v, {}^A \omega]^T, \quad {}^A F = [{}^A f, {}^A m]^T $$

where ${}^A v \in \mathbb{R}^3$ and ${}^A \omega \in \mathbb{R}^3$ are the linear and angular velocities of frame $\{A\}$, and ${}^A f \in \mathbb{R}^3$ and ${}^A m \in \mathbb{R}^3$ are the force and moment expressed in frame $\{A\}$, respectively. The transformation matrix that transforms force/moment vectors and velocity vectors between frames $\{A\}$ and $\{B\}$ is [6],

$$ {}^A U_B = \begin{bmatrix} {}^A R_B & \mathbf{0}_{3\times 3} \\ ({}^A r_{AB} \times) {}^A R_B & {}^A R_B \end{bmatrix} \tag{1} $$

where ${}^A R_B \in \mathbb{R}^{3\times 3}$ is a rotation matrix between frames $\{A\}$ and $\{B\}$, and $({}^A r_{AB} \times)$ is a skew-symmetric matrix operator defined in [6]. Based on (1), the force/moment and velocity vectors can be transformed between frames, as [6],

$$ {}^B V = {}^A U_B^T \, {}^A V, \quad {}^A F = {}^A U_B \, {}^B F. \tag{2} $$

The net force/moment vector of the rigid body in frame $\{A\}$, indicated as ${}^A F^* \in \mathbb{R}^6$, is expressed as follows:

$$ M_A \frac{d}{dt}({}^A V) + C_A({}^A \omega) {}^A V + G_A + {}^A \Delta_R = {}^A F^* \tag{3} $$

where ${}^A \Delta_R$ is model uncertainty, and $M_A \in \mathbb{R}^{6\times 6}$, $C_A({}^A \omega) \in \mathbb{R}^{6\times 6}$, and $G_A \in \mathbb{R}^6$ are expressed in [18].

Property 1 ([18]) *The equation (3) can be written in linear-in-parameter form as below*

$$ M_A \frac{d}{dt}({}^A V) + C_A({}^A \omega) {}^A V + G_A = \bar{Y}_A \phi_A \tag{4} $$

where $\bar{Y}_A \in \mathbb{R}^{6\times 10}$ is the compact regressor matrix and $\phi_A \in \mathbb{R}^{10}$ is the unique inertial parameter vector.

The design variable in the VDC approach is the required velocity, which encompasses the desired velocity trajectory along with one or two terms corresponding to control error. This control error can be a position error or force error for the motion and compliance control tasks, respectively [18, 19]. Since this study only examines the free motion task, the required joint velocity can be defined as

$$\dot{q}_r = \dot{q}_d + \lambda (q_d - q) \tag{5}$$

with $\dot{q}_d$ being desired joint velocity, q_d being desired joint trajectory, q being measured joint variable, and $\lambda > 0$ being a positive constant. Then, the linear/angular velocity vector, indicated as $^A V_r$, can be computed based on $\dot{q}_r$ by performing kinematic computation, thorough details of which will be provided later. The required force/moment vector, then, can be defined as

$$^A F_r^* = Y_A \hat{\phi}_A + {}^A F_c \tag{6}$$

where Y_A is in the sense of (4) by replacing $^A V$ with $^A V_r$, $\hat{\phi}_A$ is the estimation of ϕ_A, and $^A F_c$ is the regulating control term. The required force/moment vector in (6) demonstrates the amount of force/moment that must be applied to the system to achieve the control objectives.

Definition 1 ([6]) A virtual cutting point (VCP) is a directed separation interface that conceptually cuts through a rigid body. At the cutting point, the two parts resulting from the virtual cut maintain equal positions and orientations.

Definition 2 ([6]) Given the frame $\{A\}$, the virtual power flow (VPF) can be defined as

$$p_A = ({}^A V_r - {}^A V)^T ({}^A F_r - {}^A F).$$

Definition 3 ([6]) A single subsystem that is virtually decomposed from a complex system with affiliated functions $X(t)$ and $y(t)$ can be said to be virtually stable if and only if there exists a non-negative accompanying function $\nu(t)$ as

$$\nu(t) \geq \frac{1}{2} X(t)^T P X(t) \tag{7}$$

in a way that,

$$\dot{\nu}(t) \leq -y(t)^T Q y(t) + p_{\underline{A}} - p_{\overline{A}} \tag{8}$$

where P and Q are two block-diagonal positive-definite matrices, and $\underline{A}$ and $\overline{A}$ are two adjacent neighbors of A.

Theorem 1 ([6]) *Consider a complex system that is virtually decomposed into subsystems. If all the decomposed subsystems are virtually stable in the sense of definition 1, then the entire system is stable.*

2.2 Lemmas and Assumptions

Assumption 1 For unknown robot model uncertainties Δ_R and Δ_a corresponding to rigid body and actuator subsystems, respectively, we have

$$|D(t)| \leq \delta_1, \quad |\Delta_R| \leq \delta_2, \quad |\Delta_a| \leq \delta_3,$$

with $\delta_1, \delta_2, \delta_3 \geq 0$ being unknown constants.

Assumption 2 Only friction between the piston and cylinder is considered.

Lemma 1 ([20]) *RBFNNs can be utilized to estimate an unknown continuous function $Z(\chi) : \mathbb{R}^m \to \mathbb{R}$ with the approximation of*

$$Z(\chi) = \hat{W}^T \Psi(\chi) + \hat{\varepsilon}$$

where $\chi = [\chi_1, \chi_2, ..., \chi_m]^T \in \mathbb{R}^m$ is the input vector of the neural networks, $\hat{W}$ is the weight vector of the neural networks, $\Psi(\chi)$ is the basis function of the RBFNNs, and $\hat{\varepsilon}$ is the approximation error. The optimal weight vector W^* can be expressed by

$$W^* = \arg \min_{\hat{W} \in \Xi_N} \{ \sup_{\chi \in \Xi_T} |\hat{Z}(\chi|\hat{W}) - Z(\chi)| \}$$

where $\Xi_N = \{\hat{W} | \|\hat{W}\| \leq \kappa\}$ is a valid set of vectors with κ being a design value, Ξ_T is an allowable set of the state vectors, and $\hat{Z}(\chi|\hat{W}) = \hat{W}^T \Psi(\chi)$.

Lemma 2 ([18]) *For any inertial parameter vector ϕ_A, there is a one-to-one linear map $f : \mathbb{R}^{10} \to S(4)$ such that,*

$$f(\phi_A) = \mathcal{L}_A = \begin{bmatrix} 0.5tr(\bar{I}).\mathbf{1} - \bar{I} & h \\ h^T & m \end{bmatrix}$$

$$f^{-1}(\phi_A) = \phi_A(m, h, tr(\Sigma).\mathbf{1} - \Sigma)$$

where $\Sigma = 0.5tr(\bar{I}) - \bar{I}$, and m, h, and $\bar{I}$ are the mass, first mass moment, and rotational inertia matrix, respectively.

3 Modeling and Control of Rigid Body Subsystem

In this section, the equation of motion of the system is derived and the procedure of controller is expressed. Figure 1a demonstrates how the system is decomposed into objects in VDC context: object 1, which encompasses the base actuator and where the SSL controller is designed, and object 2, which is driven by object 1.

3.1 Kinematics and Dynamics Computations

Consider $^G V$ as the linear/angular velocity vector of the ground. The pillar velocity ^{P_1}V can be computed according to Fig. 1b:

$$^{P_1}V = {}^G U_{P_1}^T \, {}^G V + y_\tau \dot{\zeta}_1 \qquad (9)$$

where $^G U_{P_1}$ is in the sense of (1). In (9), the $\dot{\zeta}_1$ is the angular velocity of the pillar that can be computed by numerically differentiating the encoder angle data. This angular motion is generated by the piston's linear motion through the rack and pinion mechanism. To have a more accurate analysis, we need to consider the dynamics of the piston as well. According to Fig. 1, we have

$$^{P_{p2}}V = x_f \dot{x}_p \qquad (10)$$

with $\dot{x}_p$ being the linear velocity of the piston. The following relation between the angular velocity of the pillar and the linear velocity of the piston holds:

$$\dot{x}_p = r_p \dot{\zeta}_1 \qquad (11)$$

with r_p being the radius of the pinion in Fig. 1b. And $x_f = (1, 0, 0, 0, 0, 0)^T$, $y_\tau = (0, 0, 0, 0, 1, 0)^T$.

By considering ^{B_c}F as the force backpropagated from the rest of the manipulator, one can obtain the force/moment vector of the pillar as

$$^{P_1}F = {}^{P_1}F^* + {}^{P_1}U_{B_c} \, {}^{B_c}F \qquad (12)$$

which results in linear piston force in object 1 as below:

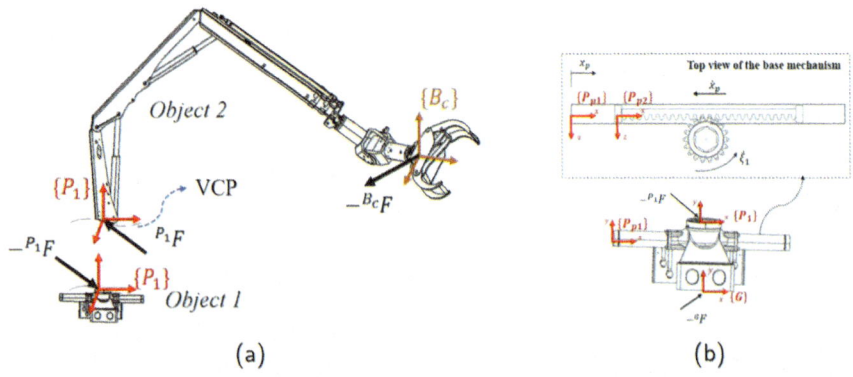

Fig. 1 **a** Heavy-duty hydraulic manipulator schematic decomposed into objects, **b** detail of the rotary hydraulic actuator with rack and pinion mechanism in object 1

$$f_c = \frac{1}{r_p} y_\tau^T {}^{P_1}F + x_f^T {}^{P_{p2}}F^*. \tag{13}$$

${}^{P_1}F^*$ and ${}^{P_{p2}}F^*$ are the net force/moment vector of the pillar and piston body, respectively, in the sense of (3). The actuator force in (13) has two terms: the first term shows the inertial force of the rigid bodies, and the second term demonstrates the inertial effect of the piston. Therefore, the (13) displays the unified force of the decomposed system, where each part is analyzed separately.

3.2 Subsystem-Based Learning Control Design

In this part, the SSL control design procedure is elaborated. First, the required joint velocity of the base rotation should be defined in the sense of (5) as

$$\dot{\zeta}_{1r} = \dot{\zeta}_{1d} + \lambda(\zeta_{1d} - \zeta_1), \tag{14}$$

where $\dot{\zeta}_{1d}$ is the desired angular velocity. Considering (9)–(11), the required linear/angular velocities can be computed as follows:

$$^{P_1}V_r = {}^G U_{P_1}^T {}^G V_r + y_\tau \dot{\zeta}_{1r} \tag{15}$$

$$^{P_{p2}}V_r = x_f \dot{x}_{pr} \tag{16}$$

The required joint velocity in (14) shows the required trajectory the base joint must follow to achieve the local control objective. Additionally, the required linear/angular velocity vector in (15) displays the velocity the rigid body should acquire to track the desired motion. It should also be mentioned that (14) and (16) are related through r_p.

The net required force/moment vector, which is the control signal for the rigid body subsystem, can be designed in the sense of (6) by taking advantage of (4):

$$^A F_r^* = Y_A \phi_A + K_A \left({}^A V_r - {}^A V\right) + {}^A \Delta_R \tag{17}$$

where ${}^A \Delta_R$ is the unknown model uncertainty and K_A is a positive-definite matrix. However, in real-world applications, the exact values of the inertial parameters ϕ_A and model uncertainty ${}^A \Delta_R$ are unknown, especially in industrial manipulators. Therefore, in this study, the adaptive control and RBFNN approaches are adopted to address the mentioned issues. According to Lemma 1, we can define

$$^A \Delta_R = {}^A W^T \Psi(\chi_A) + {}^A \varepsilon^* \tag{18}$$

where $^A W \in \mathbb{R}^{6 \times \bar{n}_A}$ is the RBFNNs weight, $\Psi(.)$ is Gaussian activation function, $\chi_A = [^A V^T, {}^A V_r^T, {}^A \dot{V}_r^T]^T \in \mathbb{R}^{18}$, $^A \varepsilon^* \in \mathbb{R}^6$ is the RBFNNs approximation error, and $\bar{n}_A$ is the number of neurons in rigid body subsystem $\{A\}$. By changing $\bar{n}_A$, one can set different numbers of nodes for different rigid body subsystems. Since the actual values for $^A W$ and $^A \varepsilon^*$ are unavailable, their estimation denoted $^A \hat{W}$ and $^A \hat{\varepsilon}$ are utilized in the control design, leading to the net required force/moment vector:

$$^A F_r^* = Y_A \hat{\phi}_A + K_A \left(^A V_r - {}^A V\right) + {}^A \hat{W}^T \Psi(\chi_A) + {}^A \hat{\varepsilon}. \tag{19}$$

Then, the required force/moment vectors can be computed by evoking force/moment vectors and using (19) as

$$^{P_1} F_r = {}^{P_1} F_r^* + {}^{P_1} U_{B_c} {}^{B_c} F_r \tag{20}$$

Finally, the required piston forces that must be applied at the actuator level in order to accomplish the control objective can be derived as

$$f_{cr} = \frac{1}{r_p} y_\tau^T \, {}^{P_1} F_r + x_f^T \, {}^{P_{p2}} F_r^* \tag{21}$$

where $^{P_1} F_r^*$ and $^{P_{p2}} F_r^*$ can be computed by replacing corresponding frames into (19). The first term in the required actuator force (21) compensates for the inertial effect of the rigid body while the second term establishes the joint tracking error. The unknown model uncertainty is considered in both subsystems through (19), improving the control performance.

Lemma 3 *The rigid body subsystem, shown in Fig. 1, having the dynamics of (3) in the presence of unknown model uncertainty under the SSL controller (19), with the following adaptation laws:*

$$^A \dot{\hat{\mathcal{L}}} = \frac{1}{\gamma} {}^A \hat{\mathcal{L}} \left(^A \mathcal{S}\right) {}^A \hat{\mathcal{L}} \tag{22}$$

$$^A \dot{\hat{W}} = {}^A \Gamma \left(\Psi(\chi_A) \left(^A V_r - {}^A V\right)^T\right) \tag{23}$$

$$^A \dot{\hat{\varepsilon}} = {}^A \pi \left(^A V_r - {}^A V\right), \tag{24}$$

is virtually stable, where $A \in \{P_1, P_{p2}\}$. *Moreover,* $^A \hat{\mathcal{L}}$ *is the estimation of* $^A \mathcal{L}$ *defined in Lemma (2) and* $^A \mathcal{S}$ *is a unique symmetric matrix defined in [16]. The* γ *and* $^A \pi$ *are positive constants, and* $^A \Gamma$ *is a positive-definite matrix.*

Proof 1 Define the non-negative accompanying function as

$$\nu_1 = \sum_A \frac{1}{2} \left({}^A V_r - {}^A V\right)^T M_A \left({}^A V_r - {}^A V\right) + \sum_A \gamma \mathcal{D}_F(\mathcal{L}_A \| \hat{\mathcal{L}}_A) \qquad (25)$$
$$+ \frac{1}{2} tr({}^A \tilde{W}^T {}^A \Gamma^{-1} {}^A \tilde{W}) + \frac{1}{2 {}^A \pi} {}^A \tilde{\varepsilon}^T {}^A \tilde{\varepsilon}.$$

By subtracting (3) from (19), and using (2), Definition 2, and velocity and force terms in Sects. 3.1 and 3.2 along with (22)-(24) and procedure expressed in [16], the time derivative of (25) can be written as

$$\dot{\nu}_1 = -\sum_A \left({}^A V_r - {}^A V\right)^T K_A \left({}^A V_r - {}^A V\right) + p_G - p_{Bc} + (\dot{x}_{pr} - \dot{x}_p)(f_{cr} - f_c), \qquad (26)$$

which is virtually stable in the sense of Definition 3.

4 Modeling and Control of Actuator Subsystem

In this section, the low-level voltage controller is designed to ensure that the required actuator force in (21) is generated by the actuator. Considering the friction model proposed in [7] as $f_f = Y_f \theta_f$, we can obtain

$$f_p = f_c + f_f. \qquad (27)$$

On the other hand, the piston force can be computed using the chamber pressure as

$$f_p = A_a p_a - A_b p_b \qquad (28)$$

with A_a and A_b being cross-sectional areas, and p_a and p_b being pressures in hydraulic actuator chamber. The cylinder chamber pressure can be generated by controlling the fluid flow rate entering the chamber. The equation of these fluid flow rates denoted as Q_a and Q_b can be written as

$$Q_a = c_{p1} \upsilon(p_s - p_a) u \mathcal{S}(u) + c_{n1} \upsilon(p_a - p_r) u \mathcal{S}(-u), \qquad (29)$$

$$Q_b = -c_{n2} \upsilon(p_b - p_r) u \mathcal{S}(u) - c_{p2} \upsilon(p_s - p_b) u \mathcal{S}(-u), \qquad (30)$$

where c_{p1}, c_{p2}, c_{n1} and c_{n2} are flow coefficients, p_s is the supply pressure, p_r is the return-line pressure, and $\mathcal{S}(u)$ and $\upsilon(\Delta p)$ are defined in [6]. By adopting continuity equations for hydraulic actuators, the pressure dynamics in the cylinder chamber can be expressed as

$$\dot{p}_a = \frac{\beta}{V_{0a} + A_a x} (Q_a - A_a \dot{x} - Q_l), \qquad (31)$$

$$\dot{p}_b = \frac{\beta}{V_{0b} + A_b(s - x)} (Q_b + A_b \dot{x} + Q_l), \tag{32}$$

where β denotes the oil bulk modulus, s is the maximum stroke of the piston, and the laminar leakage flow Q_l between the cylinder chambers can be modeled as

$$Q_l = c_l(p_a - p_b) \tag{33}$$

with c_l being the leakage coefficient. Taking the time derivative of the (28) and recalling (29)-(33), one can obtain

$$u_f = -Y_v \theta_v + \Delta_a \tag{34}$$

with Δ_a being the unknown model uncertainty in the actuator dynamics, and Y_v and θ_v defined in [8]. Moreover, Δ_a encompasses the compensation for the unknown backlash constraints. Then, for the given u_f in (34), a unique spool valve voltage signal u can be obtained as

$$u = \frac{u_f}{c_{p1} \frac{v(p_s - p_a)}{V_{0a}/A_a + x} + c_{n2} \frac{v(p_b - p_r)}{V_{0b}/A_b + (s - x))}} \mathcal{S}(u_f)$$
$$+ \frac{u_f}{c_{n1} \frac{v(p_a - p_r)}{V_{0a}/A_a + x} + c_{p2} \frac{v(p_s - p_b)}{V_{0b}/A_b + (s - x))}} \mathcal{S}(-u_f). \tag{35}$$

Considering (27), (28), and (34) along with Lemma 1, the voltage control law can be derived as

$$f_{pr} = f_{cr} + Y_f \hat{\theta}_f, \tag{36}$$

$$u_{fr} = Y_d \hat{\theta}_d + k_f(f_{pr} - f_p) + k_x(\dot{x}_{pr} - \dot{x}_p) + \hat{W}_a^T \Psi(\chi_a) + \hat{\varepsilon}_a \tag{37}$$

with, $\chi_a = [p_a, p_b, x_p, \dot{x}_p]^T$, and $\hat{W}_a \in \mathbb{R}^{\bar{n}_a \times 1}$ and $\hat{\varepsilon}_a \in \mathbb{R}$ are the estimation of W_a^* and ε_a^* in the sense of Lemma 1, respectively, to handle the unmodeled dynamics and unknown backlash in the hydraulic actuator. Y_d and θ_d are defined in [8]. It can be seen from (37) that the RBFNN at the actuator level is fed by pressure and piston motion data, increasing the accuracy of the unknown backlash estimation. Therefore, the unique spool valve voltage signal can be achieved by substituting (37) into (35), where we have

$$u_{fr} = -Y_v \hat{\theta}_v \tag{38}$$

with $\hat{(.)}$ being the estimation of $(.)$. The low-level control law designed in (37) not only compensates for unknown backlash uncertainties and unknown unmodeled dynamics at the actuator level, but it also makes the actuator produce the required piston force (21), resulting in precise trajectory tracking.

Lemma 4 *The actuator dynamics under required piston force (21) with a low-level voltage control signal (37) along with following adaptation functions:*

$$\dot{\hat{\theta}}_f = \delta_f \left(\frac{1}{k_{xi}} Y_f^T (\dot{x}_{pr} - \dot{x}_p) \right) \tag{39}$$

$$\dot{\hat{\theta}}_v = \delta_v \left(\frac{1}{k_x} Y_v^T (f_{pr} - f_p) \right) \tag{40}$$

$$\dot{\hat{\theta}}_d = \delta_d \left(\frac{1}{k_x} Y_d^T (f_{pr} - f_p) \right) \tag{41}$$

$$\dot{\hat{W}}_a = \delta_w \left(\frac{1}{k_x} (f_{pr} - f_p) \Psi(\chi_a) \right) \tag{42}$$

$$\dot{\hat{\varepsilon}}_a = \delta_\varepsilon \left(\frac{1}{k_x} (f_{pr} - f_p) \right) \tag{43}$$

is virtually stable.

Proof 2 Defining the accompanying function as

$$\begin{aligned} \nu_a = & (f_{pr} - f_p)^2 / (2\beta k_x) + \frac{1}{2\delta_d} \left(\tilde{\theta}_d^T \tilde{\theta}_d \right) + \frac{1}{2\delta_v} \left(\tilde{\theta}_v^T \tilde{\theta}_v \right) \\ & + \frac{1}{2\delta_f} \left(\tilde{\theta}_f^T \tilde{\theta}_f \right) + \frac{1}{2\delta_w} \left(\tilde{W}_a^T \tilde{W}_a \right) + \frac{1}{2\delta_\varepsilon} (\tilde{\varepsilon}_a)^2, \end{aligned} \tag{44}$$

and following the same procedure in Lemma 3, [8, 16] along with using (39)–(43), one can obtain

$$\dot{\nu}_a = -\frac{k_f}{k_x} (f_{pr} - f_p)^2 - (\dot{x}_{pr} - \dot{x}_p)(f_{cr} - f_c) \tag{45}$$

which is virtually stable in the sense of Definition 3.

Theorem 2 *If the accompanying function of the entire system is considered as $\nu(t) = \nu_a(t) + \nu_1(t)$, it follows*

$$\begin{aligned} \dot{\nu}(t) = & -\sum_A \left({}^A V_r - {}^A V \right)^T K_A \left({}^A V_r - {}^A V \right) - \frac{k_f}{k_x} (f_{pr} - f_p)^2 \\ & + p_G - p_{B_c}, \end{aligned} \tag{46}$$

showing that the entire system is asymptotically stable in the sense of Theorem 1.

Proof 3 Summing up (26), (45), and following the same procedure in [21], one can obtain (46).

5 Results

In this section, the experimental results are provided to evaluate the performance of the proposed controller. The commercial full-scale manipulator with a hydraulically actuated rotary joint is used for the performance evaluation of the presented robust controller. The following hardware components are utilized:

- Beckhoff and TwinCat 3 interface with a sample time of 1 ms
- Bosch Rexroth NG6 size servo solenoid valve with 12 l/min at $\Delta P = 3.5 MPa$ per notch
- Sick afS60 (18-bit) absolute encoders for joint angle measurements.
- Druck PTX1400 pressure transmitter (range 25 MPa) for pressure measurements.

The result of the proposed method is compared to the original VDC controller to better evaluate the control performance. Control gains are tuned to have the smallest tracking error and selected as $\gamma = 500$, $\lambda = 3$, $K_A = 50$, $k_{xp} = 0.02$, $k_{fp} = 1.2 \cdot 10^{-9}$, $^A\Gamma = 100 \cdot I$, $^A\pi = 10 \cdot I$, $\delta = 1.3 \cdot k_{xp} \cdot k_{fp}$. For the RBFNNs, the Gaussian activation function is used as $\Psi(\chi) = exp([-(\chi - c_j)^T(\chi - c_j)/(b_j^2)])$, with c_j and b_j denoting the center and width of the neural cell in **j**th unit. In this study the value for c_j is randomly selected in $[-1, 1]$ with $b_j = 0.5$ for actuator subsystem, and $b_j = 5$ for the rigid body. The aggregation of 50 nodes is utilized to estimate unknown uncertainties and unknown input constraints. Since the inputs of the RBFNN in the actuator subsystem were not in the same range, min-max method is utilized to normalize inputs.

Figure 2a demonstrates the trajectory tracking of the joint angle under the proposed controller and original VDC. As shown in Fig. 2b, the SSL controller tackled the unknown backlash and achieved a much lower steady-state error. The root-mean-square-error (RMSE) values of trajectory tracking with the VDC and SSL controller are 0.2 deg and 0.095 deg, respectively. Further, the steady-state error values with VDC and SSL controller are 0.22 deg and 0.018 deg, respectively. As the motion of the manipulator in the Z-direction of the Cartesian space is mostly generated by the base joint, we examined the impact of unknown backlash in Cartesian space, as well. Figure 3a, displays the path tracking of the manipulator end-effector in Z-direction at 4 meters' reach. As depicted in Fig. 3b, the SSL controller achieved much more accurate path tracking when compared to the original VDC. The RMSE values of path tracking with VDC and SSL controller are 1.5 cm and 0.6 cm, respectively, while the steady-state error is 1.67 cm and 0.17 cm, respectively. The provided experimental results demonstrate that the designed SSL controller perfectly handled the unknown backlash constraint, resulting in precise tracking performance of industrial heavy-duty hydraulic manipulator.

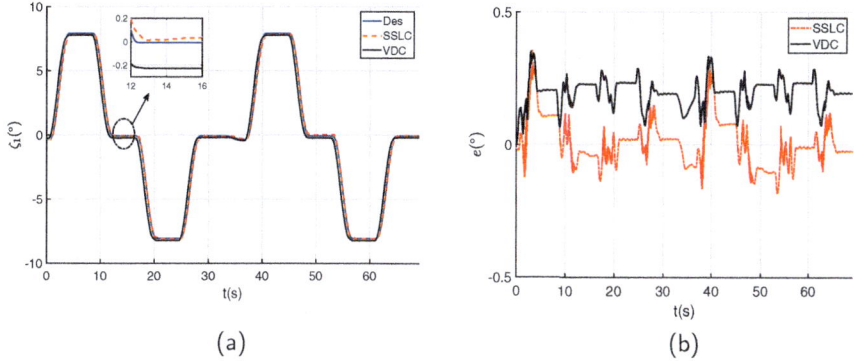

Fig. 2 Experimental results of joint tracking with ±8 deg range of motion, **a** trajectory tracking with proposed controller and VDC, **b** tracking error of controllers. (SSLC: subsystem-based learning control, Des: desired)

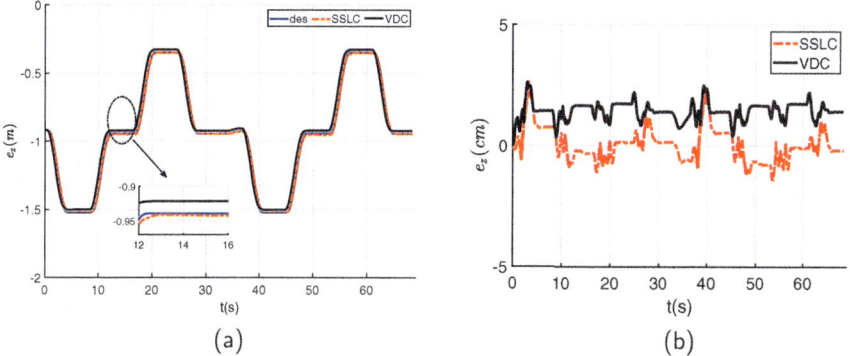

Fig. 3 Experimental results of path tracking in Z-direction, **a** path tracking with proposed controller and VDC, **b** tracking error of controllers

6 Conclusion

In this study, a subsystem-based learning controller was designed to control an HDM subjected to unknown input backlash. The VDC scheme was utilized as the baseline controller, which enabled us to decompose the system into the rigid body and actuator subsystems. Then, at each local subsystem, control design and stability analysis were performed. The RBFNNs were employed to estimate the uncertainties in the rigid body subsystem, while also tackling the unknown backlash in the actuator subsystem. As such, the novel way of exploiting RBFNNs presented herein allowed for an increase in control performance. Finally, the stability of the system under the designed controller was proved by means of virtual stability and VPFs. The provided experimental results perfectly demonstrated the effectiveness of the proposed method by reducing the steady-state tracking error from 0.22 degrees to 0.018 degrees by tackling the unknown backlash.

Acknowledgements This work is supported by Business Finland partnership project "Future all-electric rough terrain autonomous mobile manipulators" (Grant 2334/31/222).

References

1. Gribbins K (nd) The ultimate mini ex overview: a comprehensive analysis of the 2021 compact excavator market. https://compactequip.com/excavators/the-ultimate-mini-ex-overview-a-comprehensive-analysis-of-the-2021-compact-excavator-market/. Accessed 27 Sept 2023
2. Bonchis A, Corke PI, Rye DC (2002) Experimental evaluation of position control methods for hydraulic systems. IEEE Trans Control Syst Technol 10(6):876–882
3. Bech MM, Andersen TO, Pedersen HC, Schmidt L (2013) Experimental evaluation of control strategies for hydraulic servo robot. In: 2013 IEEE international conference on mechatronics and automation. IEEE, pp 342–347
4. Kim J, Jin M, Choi W, Lee J (2019) Discrete time delay control for hydraulic excavator motion control with terminal sliding mode control. Mechatronics 60:15–25
5. Liang X, Yao Z, Deng W, Yao J (2023) Adaptive control of n-link hydraulic manipulators with gravity and friction identification. Nonlinear Dyn 111(20):19093–19109
6. Zhu WH (2010) Virtual decomposition control: toward hyper degrees of freedom robots vol 60. Springer Science and Business Media
7. Zhu W, Piedboeuf J (2004) Adaptive output force tracking control of hydraulic cylinders with applications to robot manipulators. ASME J Dyn Sys Meas Control. June 2005; 127(2):206–217
8. Lampinen S, Koivumäki J, Mattila J (2019) Improved hydraulic cylinder model for the virtual decomposition control approach. In: 2019 IEEE international conference on cybernetics and intelligent systems (CIS) and IEEE Conference on Robotics, Automation and Mechatronics (RAM), pp 113–118. IEEE
9. Koivumäki J, Zhu WH, Mattila J (2019) Energy-efficient and high-precision control of hydraulic robots. Control Eng Pract 85:176–193
10. Mattila J, Koivumäki J, Caldwell DG, Semini C (2017) A survey on control of hydraulic robotic manipulators with projection to future trends. IEEE/ASME Trans Mechatron 22(2):669–680. https://doi.org/10.1109/TMECH.2017.2761462
11. Li L, Lin Z, Jiang Y, Yu C, Yao J (2021) Valve deadzone/backlash compensation for lifting motion control of hydraulic manipulators. Machines 9(3):57
12. Kang S, Nagamune R, Yan H (2020) Almost disturbance decoupling force control for the electro-hydraulic load simulator with mechanical backlash. Mech Syst Signal Proc 135:106400
13. Mustalahti P, Mattila J (2018) Nonlinear model-based controller design for a hydraulic rack and pinion gear actuator. In: Proceedings of the BATH/ASME 2018 symposium on fluid power and motion control. BATH/ASME 2018 Symposium on Fluid Power and Motion Control. Bath, UK, 12–14 Sept 2018. V001T01A020
14. Mustalahti P, Mattila J (2019) Nonlinear model-based control design for a hydraulically actuated spherical wrist. In: Proceedings of the ASME/BATH 2019 symposium on fluid power and motion control. ASME/BATH 2019 symposium on fluid power and motion control. Longboat Key, FL, USA
15. Yang, X., Deng, W. and Yao, J., 2022. Neural adaptive dynamic surface asymptotic tracking control of hydraulic manipulators with guaranteed transient performance. IEEE Transactions on Neural Networks and Learning Systems
16. Hejrati M, Mattila J (2023) Physical Human-Robot Interaction Control of an Upper Limb Exoskeleton With a Decentralized Neuroadaptive Control Scheme. IEEE Transactions on Control Systems Technology. https://doi.org/10.1109/TCST.2023.3338112

17. Sünderhauf N, Brock O, Scheirer W, Hadsell R, Fox D, Leitner J, Upcroft B, Abbeel P, Burgard W, Milford M, Corke P (2018) The limits and potentials of deep learning for robotics. The International journal of robotics research 37(4–5):405–420
18. Hejrati, M. and Mattila, J., 2022, November. Decentralized nonlinear control of redundant upper limb exoskeleton with natural adaptation law. In 2022 IEEE-RAS 21st International Conference on Humanoid Robots (Humanoids) (pp. 269-276). IEEE
19. Hejrati M, Mattila J (2023) Nonlinear Subsystem-Based Adaptive Impedance Control of Physical Human-Robot-Environment Interaction in Contact-Rich Tasks. In IEEE Robotics and Automation Letters 8(10):6083–6090. https://doi.org/10.1109/LRA.2023.3302616. Oct
20. Chen M, Ge SS, How B (2010) Robust adaptive neural network control for a class of uncertain MIMO nonlinear systems with input nonlinearities. IEEE Transactions on Neural Networks 21(5):796–812
21. Hejrati, M. and Mattila, J., 2023. Orchestrated Robust Controller for the Precision Control of Heavy-duty Hydraulic Manipulators. arXiv preprint arXiv:2312.06304

Open Access This chapter is licensed under the terms of the Creative Commons Attribution 4.0 International License (http://creativecommons.org/licenses/by/4.0/), which permits use, sharing, adaptation, distribution and reproduction in any medium or format, as long as you give appropriate credit to the original author(s) and the source, provide a link to the Creative Commons license and indicate if changes were made.

The images or other third party material in this chapter are included in the chapter's Creative Commons license, unless indicated otherwise in a credit line to the material. If material is not included in the chapter's Creative Commons license and your intended use is not permitted by statutory regulation or exceeds the permitted use, you will need to obtain permission directly from the copyright holder.

Evaluation of a Simple Method to Estimate the Shaft Torque in a Gerotor Pump

Giuseppe Totaro, Barbara Zardin, and Massimo Borghi

1 Introduction

The Gerotor pump is a positive displacement pump with characteristics of simplicity, compactness, and robustness. It finds use in many sectors such as aerospace (for lubrication, cooling, and as fuel pump), automotive (for engine or transmission lubrication circuits), and more standard hydraulic applications.

We have started working on this kind of pump realizing a lumped parameter fluid dynamic model with the integration of the calculation of the micro-motion of the external gear of the pump [1]. This model allowed the analysis of the pressure transient in the inter-teeth chambers, the instantaneous flow rate at the delivery and the volumetric losses. In this article, we want to describe the additional work developed to complete the model with the estimation of the torque losses adopting a simplified approach.

This type of pump has been studied in several publications over the years, exploiting the use of simulation models. A nice review can be found, for example, in [2]. Among the different modeling approaches [3–7], the lumped parameters model can provide good results with acceptable computational time [8, 9] and for this reason this last approach seems more suitable when simulation is integrated in the design process for a new prototype for example. The lumped parameter approach can be developed with different levels of details, considering or neglecting some aspects

---

G. Totaro (✉) · B. Zardin · M. Borghi
Engineering Department Enzo Ferrari DIEF, University of Modena and Reggio Emilia, Modena, Italy
e-mail: giuseppe.totaro@unimore.it

B. Zardin
e-mail: barbara.zardin@unimore.it

M. Borghi
e-mail: massimo.borghi@unimore.it

as thermal effects, cavitation, deformation, or micro-movements of the mechanical parts; some examples coming from the literature are discussed in [3].

The preceding literature introduced only simulation models that focused on the volumetric performance of the pump. Of course, there are examples in literature that also analyze the mechanical performance of the pump. Misty et al. [10] developed a 0D model capable of estimating both the fluid dynamic and mechanical performance of a Gerotor pump. They considered multiple contact points between the gears and used the elasto-hydrodynamic lubrication (EHL) theory to evaluate friction losses at the contact points between the gears. Harrison et al. developed a 1D Gerotor model to predict the volumetric efficiency and total efficiency of the pump [11]. Ivanovic et al. studied the influence of geometric and kinematic parameters of the Gerotor gears on the meshing instantaneous friction coefficient, using various empirical expressions found in the literature [12]. Inaguma [13] studied the impact of operating conditions, including pressures, speeds, and oil temperatures, on the friction torque characteristics of internal gear pumps for automobiles.

In conclusion, there are many works in literature discussing these pumps, but despite all the contributions and the apparent simplicity of the machine, there doesn't exist a unique simulation tool that can be used in the design process with enough confidence and that analyzes all the critical issues of the machine. Furthermore, there are no clear indications about the efficacy of simplified approaches applied to evaluate pump performance compared to more complex approaches, depending on the phenomena analyzed. This means that, to achieve a correct design of this kind of pump, the expertise of the designer and an extensive experimental activity are still necessary.

In this article, we propose a method to estimate the torque required for operating a Gerotor pump under steady-state conditions. In Sect. 2, we define the torque losses considered in the model and how these are estimated. To define these torque losses, only geometrical pump information and the working conditions (shaft speed and delivery/suction pressure levels) are needed. In Sect. 3, we apply the model to a specific Gerotor pump and analyze the contribution of different torque losses to the total torque, under various shaft speeds and pressure working conditions. In this section, we also analyze the influence of gaps heights on the torque losses. In Sect. 4, we compare the model-predicted torque values with the experimental data. In Sect. 5, we present our conclusions, delineating both the advantages and disadvantages associated with the adoption of our approach.

2 Pump Losses

The Gerotor pump consists of a few main elements: an inner gear, an outer gear, a port plate, and a housing (see Fig. 1) [14]. The pump inter-teeth chambers are defined between the two gears, and during their meshing, the volume of the chambers increases and decreases. These chambers are connected to the suction and delivery environments through appropriate ports realized on the port plate. The outer gears

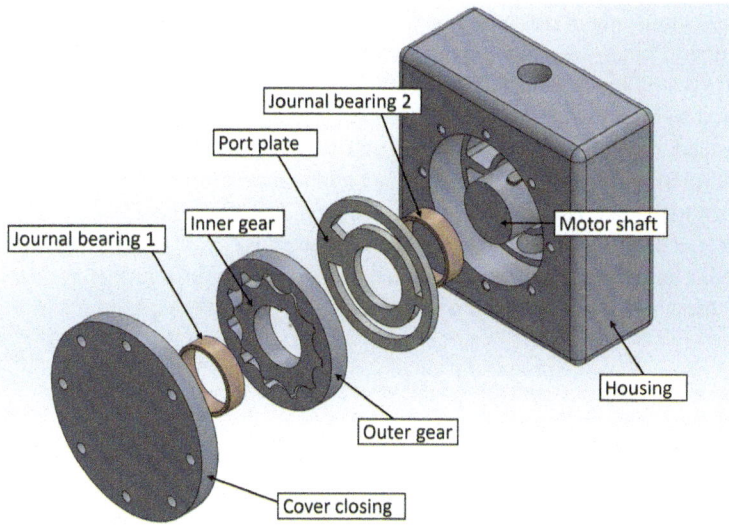

Fig. 1 Gerotor pump components

have a number of teeth denoted as Z, corresponding to the number of pump chambers. The inner gear has a number of teeth equal to Z−1. Therefore, the gear ratio between the two gears is

$$\tau_{ratio} = \frac{Z-1}{Z} = \frac{\omega_{outer}}{\omega_{inner}} \quad (1)$$

We estimate the torque needed to operate the pump (M_{shaft}) as the sum of the following contributions:

- M_{th}: Theoretical pump torque.
- $M_{outradial}$: Torque losses due to lubricated radial clearances between the outer gear and the pump's housing.
- $M_{outlateral}$: Torque losses due to lubricated lateral clearances between the outer gear and the port plate on one side and the cover closing on the other side. These viscous losses are determined considering always the situation of full film lubrication.
- $M_{inlateral}$: Torque losses due to lubricated lateral clearances between the inner gear and the port plate on one side and the cover closing on the other side. These viscous losses are determined considering always the situation of full film lubrication.
- M_{Jb}: Torque losses due to the journal bearings that support the pump shaft.

The calculated torque losses are referred to steady state working conditions. There are some contributions of losses that here are clearly neglected and this choice is clarified in the following.

Our previous model was based on the hypothesis of a single contact point between the inner and outer gears (as explained in [1]), hence between the other teeth there is always a small gap. Under this hypothesis, the meshing torque loss is very small compared to other torque loss contributions. The viscous torque losses at the gear tips, instead, depend on the relative angular velocity (both the external and internal gear are rotating), which is also influenced by the ratio of the number of teeth on the outer gear to the number of teeth on the inner gear. For the pump we analyzed, the two angular speeds are quite similar. Based on these considerations and the complexity involved in estimating the previously mentioned meshing losses, we have decided to neglect these two contributions in this work.

2.1 Theoretical Pump Torque

The theoretical pump torque is the torque needed to drive an ideal pump without losses, and it can be expressed as

$$M_{th} = \frac{V \Delta p}{2\pi} \qquad (2)$$

The term V is the pump displacement, while Δp is the pressure difference between the delivery environment and the suction environment.

The displacement is determined through geometric consideration and expressed as $V = (Z-1)(V_{max} - V_{min})$ [15] where V_{max} and V_{min} are maximum/minimum volume that a single chamber can achieve during the gears' rotations.

2.2 Outer Gear Radial Torque Losses

We modeled the coupling of the outer gear and the pump's housing as a hydrodynamic journal bearing, subjected to a constant load. The relationship between journal bearing load capacity (W_{out}) and journal eccentricity (ε_{out}), assuming the hypothesis of a short bearing and partial film assumption, is [16, 17]:

$$|W_{out}| = \frac{\mu \omega_{outer} R_{out} L_{gear}^3}{4 c_{out}^2} \frac{\varepsilon_{out}}{\left(1 - \varepsilon_{out}^2\right)^2} \left[\pi^2 \left(1 - \varepsilon_{out}^2\right) + 16 \varepsilon_{out}^2\right]^{1/2} \qquad (3)$$

Figure 2a shows the pump's chambers and the shape of the port plate. About half of the pump's chambers are subjected to delivery pressure, so we estimated the absolute mean value of the gear pressure load as:

$$|W_{out}| = |W_{in}| = \Delta p 2 R_W L_{gear} \qquad (4)$$

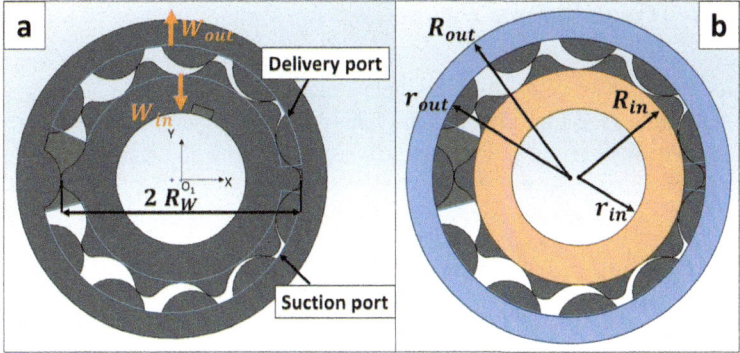

Fig. 2 (**a**) Pressure gears loads and delivery/suction ports shapes. (**b**) Shapes of lateral clearance inner/outer gears

Knowing the gear load, we can determine the relative journal eccentricity (ε_{out}) from Eq. 3. The journal eccentricity is used to estimate the torque viscous losses with the following formula [18]:

$$M_{outradial} = \frac{\mu \omega_{outer} R_{out}^3 L_{gear}}{c_{out}} \frac{2\pi}{\left(1 - \varepsilon_{out}^2\right)^{1/2}} \quad (5)$$

2.3 Outer Gear Lateral Torque Losses

To define these torque losses, we considered two lateral clearances: one between the outer face of the gear and the port plate and another between the outer face of the gear and the cover closing. The shapes of these two lubricated interfaces are equal because, in this pump, the cover reproduces the shape of the delivery/suction ports to favor the axial pressure balance [14]. We hypothesized that the heights of these clearances are equal, and we modeled the shape of the clearance as a simple ring (see Fig. 2b). By integrating the shear stress on the ring, we can obtain the torque losses [19]:

$$M_{outlateral} = 2\left(\frac{\mu \pi \omega_{outer}\left(R_{out}^4 - r_{out}^4\right)}{2h_l}\right) \quad (6)$$

2.4 Inner Gear Lateral Torque Losses

The considerations made for the lateral clearances of the outer gear are also applied to the inner gear (see Fig. 2b). For the lateral torque losses of the inner gear, we considered a different ring, and we can express the losses with the following formula:

$$M_{inlateral} = 2\left(\frac{\mu \pi \omega_{inner}\left(R_{in}^4 - r_{in}^4\right)}{2h_l}\right) \qquad (7)$$

2.5 Journal Bearings Torque Losses

The pump shaft is supported by two journal bearings, which were modeled again as short bearings with partial film assumption. The two journal bearings have different thicknesses, and their distances from the inner gear are not the same. We have considered the pressure load of the inner gear applied to the middle plane of the inner gear and the reactions of the journal bearings applied to the middle plane of the journal bearings (see Fig. 3a). By resolving the equilibrium for shaft translation along the Y-axis and rotation along the X-axis, we determine the reactions of the journal bearings (Fig. 3a):

$$\left|F_{jb1}\right| = |W_{in}|\frac{l_2}{l_1 + l_2} \qquad (8)$$

$$\left|F_{jb2}\right| = |W_{in}|\frac{l_1}{l_1 + l_2} \qquad (9)$$

We used the previous Eq. 3 to determine the eccentricity of the two journal bearings.

$$\left|F_{jb1}\right| = \frac{\mu \omega_{inner} r_{in} L_{jb1}^3}{4c_{jb1}^2} \frac{\varepsilon_{jb1}}{\left(1 - \varepsilon_{jb1}^2\right)^2}\left[\pi^2\left(1 - \varepsilon_{jb1}^2\right) + 16\varepsilon_{jb1}^2\right]^{1/2} \qquad (10)$$

$$\left|F_{jb2}\right| = \frac{\mu \omega_{inner} r_{in} L_{jb2}^3}{4c_{jb2}^2} \frac{\varepsilon_{jb2}}{\left(1 - \varepsilon_{jb2}^2\right)^2}\left[\pi^2\left(1 - \varepsilon_{jb2}^2\right) + 16\varepsilon_{jb2}^2\right]^{1/2} \qquad (11)$$

We used the previous Eq. 5, with the appropriate values, to determine the torque losses in the journal bearings.

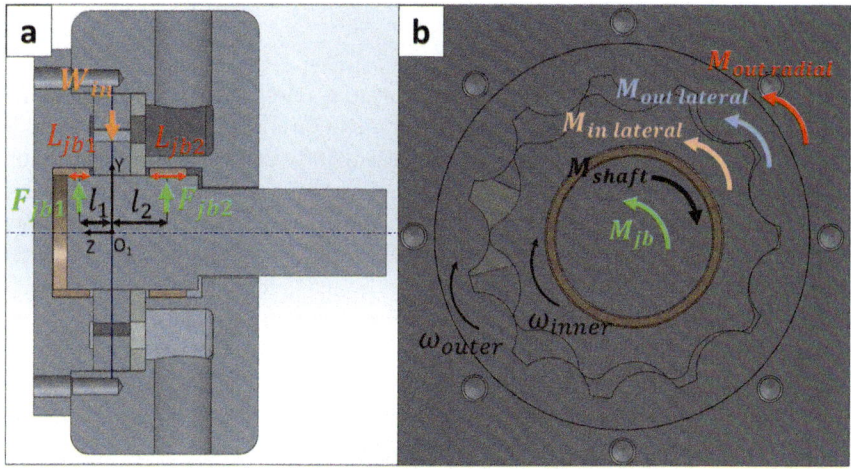

Fig. 3 (a) Journal bearings reaction. (b) All torques applied to the gears

$$M_{jb1} = \frac{\mu \omega_{inner} r_{in}^3 L_{jb1}}{c_{jb1}} \frac{2\pi}{\left(1 - \varepsilon_{jb1}^2\right)^{1/2}} \quad (12)$$

$$M_{jb2} = \frac{\mu \omega_{inner} r_{in}^3 L_{jb2}}{c_{jb2}} \frac{2\pi}{\left(1 - \varepsilon_{jb2}^2\right)^{1/2}} \quad (13)$$

The total torque losses are obtained as sum of the two contributions.

$$M_{jb} = M_{jb1} + M_{jb2} \quad (14)$$

We have determined the eccentricities of the pump shaft and outer gears with Eqs. 3, 10 and 11. However, as the two gears are meshing, this interaction potentially influences their micro-motions. Here, for the sake of simplicity, we have neglected this interaction.

2.6 Pump Shaft Torque

The pump shaft torque is obtained as the sum of the previous contributions. For the outer gear torques, we have taken into consideration the gear ratio between the two gears.

$$M_{shaft} = M_{th} + (M_{outradial} + M_{outlateral})\tau_{ratio} + M_{inradial} + M_{jb} \quad (15)$$

3 Model Application

In this section, we applied the model to analyze a Gerotor pump working at low delivery pressure values. The pump under consideration is designed to operate within a pressure range of 0–27.5 bar and a speed range of 500–3500 rpm. The fluid used for the simulation is ISO VG-46 at the temperature of 40 °C. The clearances utilized in the simulation were derived from the technical drawings of the pump's components, analyzing three distinct scenarios: one with maximum clearances (indicated in figures as Max), another with minimum clearances (indicated in figures as Min), and finally, a scenario with average clearances (indicated in figures as Avg).

Figure 4 shows the pump torque as a function of the pressure at different shaft speed values, whereas Fig. 6 shows the pump torque as a function of the shaft speed at different delivery pressure levels. The torque values have been normalized by dividing them by a reference torque value M_{ref}.

Figures 4 and 6 show the three clearances scenarios. The clearances values used in model impact in a significant way the calculated torque values, since the losses simulated in the model are exclusively of the viscous type. As we expected, the scenario with minimum clearances introduces more dissipations than the other two scenarios. Meanwhile, the torque values calculated in the scenario with average clearances fall between the torque values of the scenarios with minimum and maximum clearances.

Figure 4, the pump torque presents an increasing trend with pressure increasing at constant shaft speed. To explain the causes of this trend, we have analyzed the contribution of each individual torque loss.

Figure 5 shows the contribution of each individual torque as a function of pressure at a constant shaft speed of 2500 rpm in the average clearance scenario. From Fig. 5 and the torque loss equations defined in Sect. 1, the following observations emerge (see Fig. 5):

- The torque losses $M_{inlateral}$ and $M_{outlateral}$ are solely function of shaft speed, and their values remain constant with the pressure.
- The torque losses $M_{outradial}$ and M_{jb} are dependent on both pressure and shaft speed. Their values increase with an increase in pressure.
- The theoretical torque M_{th} is solely a function of pressure, and its values increase linearly with pressure

The trend of shaft torque shown in Fig. 4 is attributed to the contributions of M_{th}, $M_{outradial}$, and M_{jb}, while $M_{inlateral}$ and $M_{outlateral}$ remain constant with variations in pressure.

To explain the trend of the pump torque with increasing shaft speed and constant pressure (see Fig. 6), we refer to Fig. 7 that illustrates the individual torque contributions as a function of shaft speed, at a constant pressure of 25 bar in the average clearance scenario.

Based on the considerations made in Fig. 5 and observations from Fig. 7, the following conclusions emerge (see Fig. 7):

- The theoretical torque M_{th} is constant because the pressure is constant.

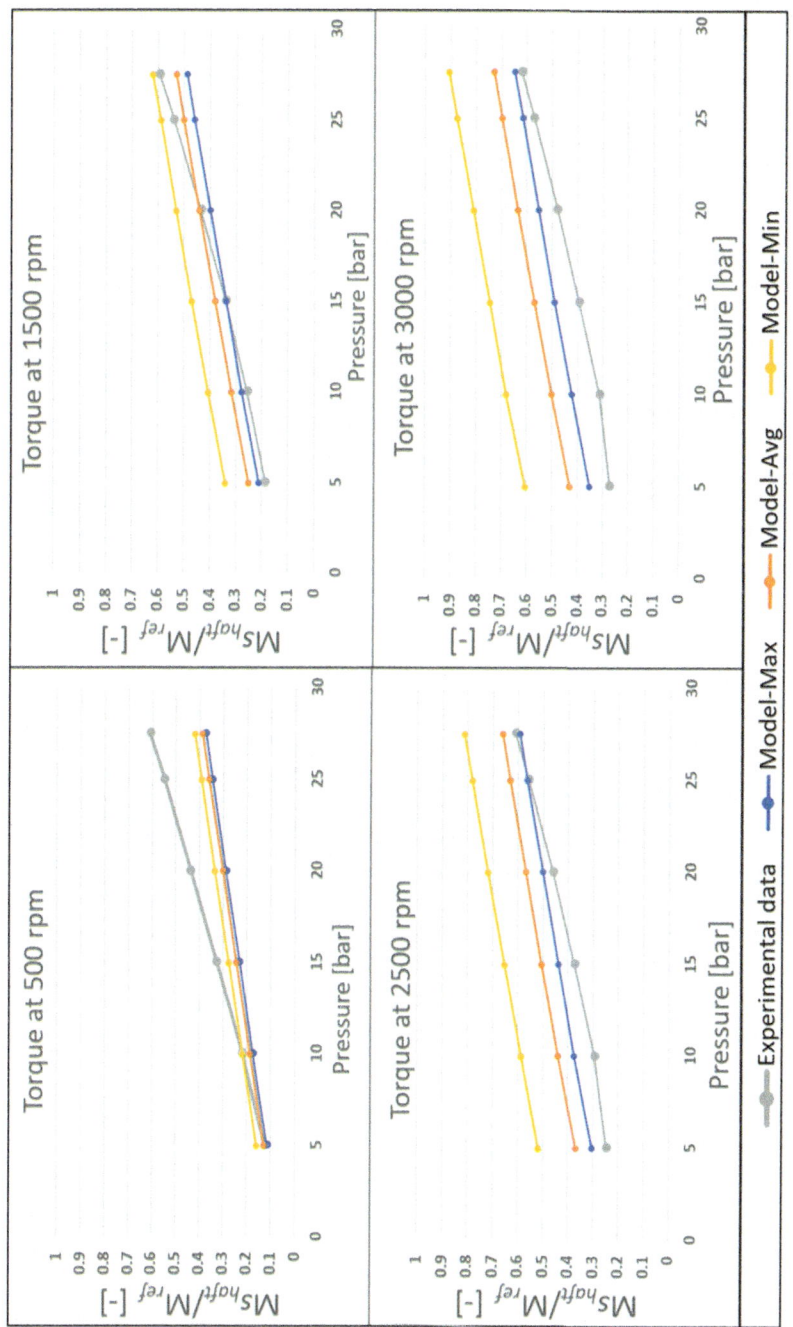

Fig. 4 M_{shaft} as a function of shaft speed at constant pressure value

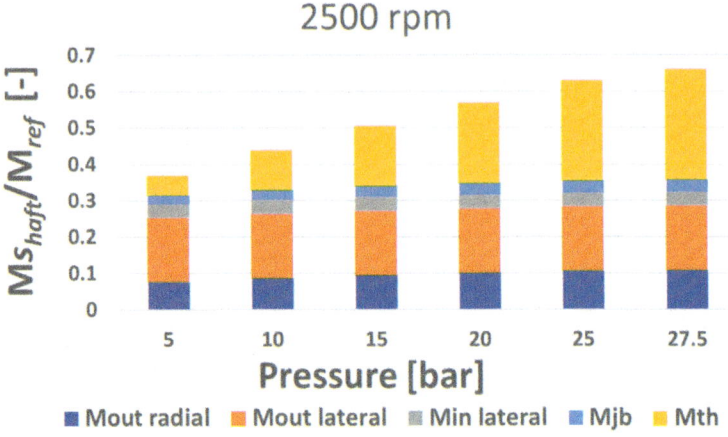

Fig. 5 Torque losses as a function of pressure at constant speed of 2500 rpm

- The torque losses $M_{inlateral}$ and $M_{outlateral}$ increase linearly with an increase in shaft speed.
- The torque losses $M_{outradial}$ and M_{jb} increase with an increase in shaft speed.

The trend of the shaft torque shown in Fig. 6, at each level of pressure considered, is solely attributed to the contribution of torque losses, since M_{th} is independent from the speed.

From the previous analysis, it emerged that the main torque losses are $M_{outlateral}$ and $M_{outradial}$ due to the difference in the arm of shear stress acting on the surface of the outer gear compared to that of the inner gear or pump shaft. One possible way to reduce these losses is to decrease the external radius of the outer gear (R_{out}). However, this action would affect the behavior of the lubrication clearance between the outer gear and the housing, as well as the mechanical resistance of the outer gear.

Figures 8 and 9 show the impact of the clearances on the individual losses. The influence of clearances is evident on the case of M_{jb}, where the range of variations in tolerances is greater than in other lubricated interfaces. In the model, the torque M_{th} is independent from the clearances.

The clearance value settings influence the model estimations significantly.

4 Comparison with Experimental Data

In this section, we compare the torque estimation model with experimental data using the same pump mentioned in Sect. 3. The experimental data are presented in Figs. 4 and 6. These have been provided by the company that provided the geometry information of the reference pump analyzed. The results obtained with the minimum clearance scenario deviates significantly from the experimental data,

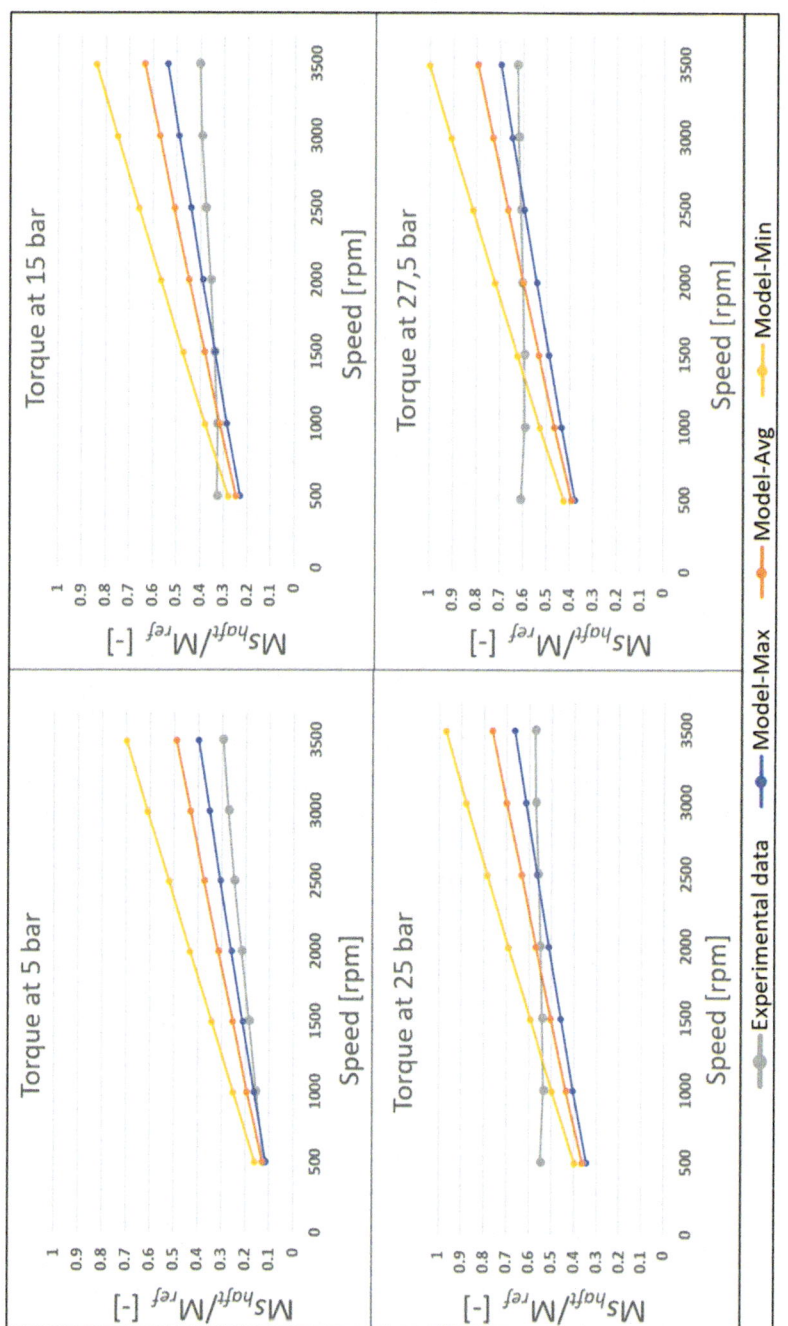

Fig. 6 M_{shaft} as a function of pressure at constant shaft speed value

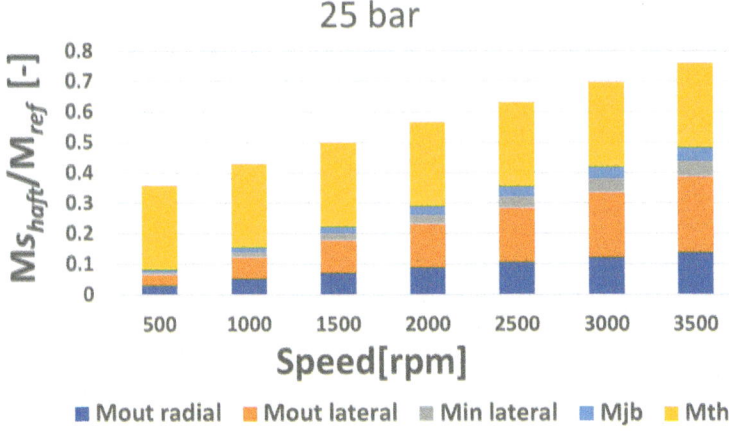

Fig. 7 Torque losses as a function of speed at constant pressure of 25 bar

while the torques obtained with the maximum clearance scenario match more closely the experimental data. For the discussion of the comparisons with experimental data, we are considering the torque losses obtained with the average clearances scenario.

Figure 4 observations:

- At a constant speed value, the experimental torque curve increases with pressure. The inclination of the experimental curve decreases as pressure increases. It's noteworthy that the inclination of the experimental curve at 500 rpm is greater than the curve at 3000 rpm.
- The model torque curves deviate from the experimental data at low speed (500 rpm) and high-pressure conditions, possibly due to a mixed lubrication regime. The presence of the mixed lubrication regime is also evident at 1500 rpm and high pressure value. This experimental behavior is also evidenced in [13]. This type of phenomenon is not considered in the model.
- At 1500 rpm, the model better follows the experimental curve. However, at high speed, the model overestimates the torque value, even though the inclination of the model curves is very similar to the experimental curves.

Observing Fig. 6 some considerations can be made:

- The experimental torque is influenced by speed variation at low pressure values, while at high pressure values, this effect is minimal. At 25–27.5 bar, the experimental torque curve is almost independent of speed. This type of behavior is not replicated by the model. These experimental results differ from what is also reported in [13], where the experimental torque losses increased linearly with the speed, at medium–high speed levels. Additional experimental tests on more samples of the pump would help us to understand better whether this was the behavior of a specific sample or whether it is confirmed.

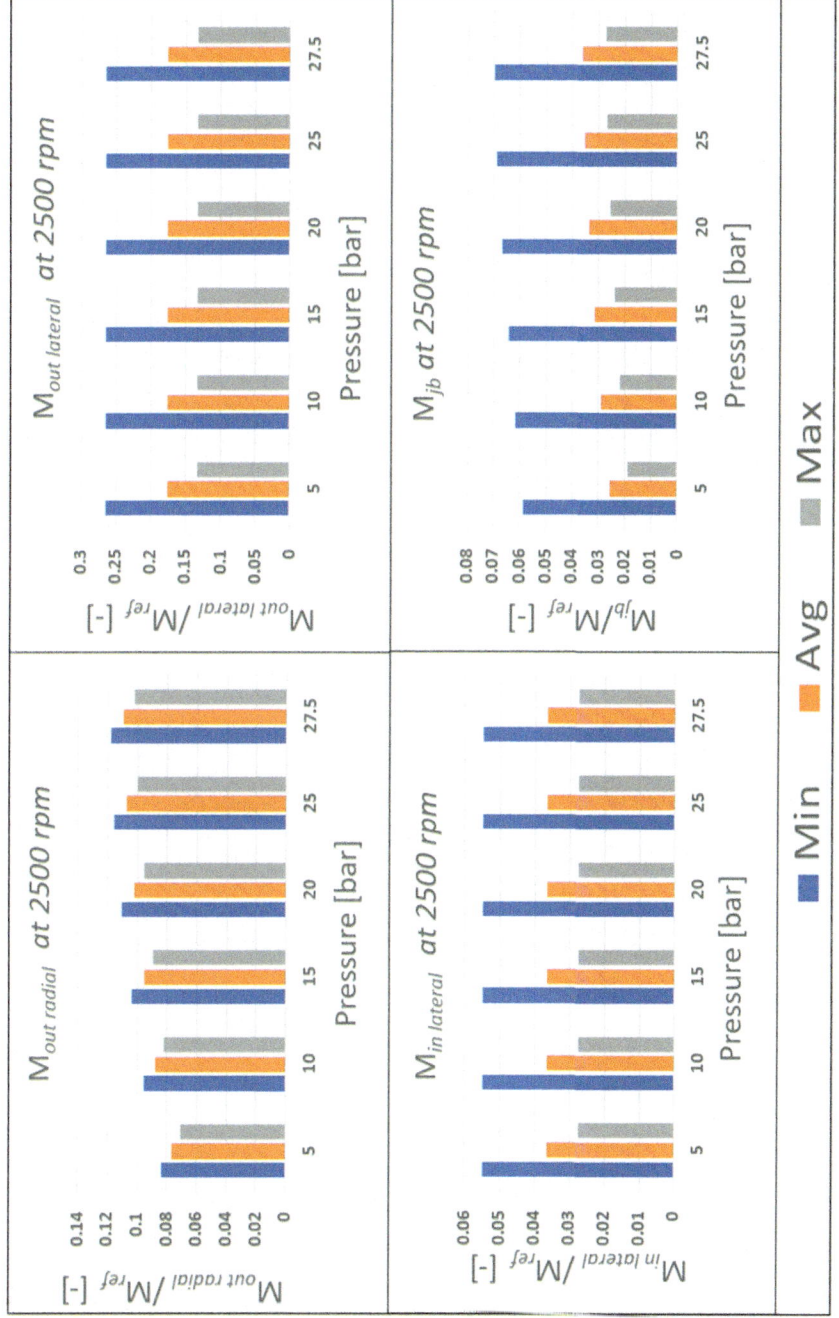

Fig. 8 Effect of clearances on torque losses at variable pressure and constant speed

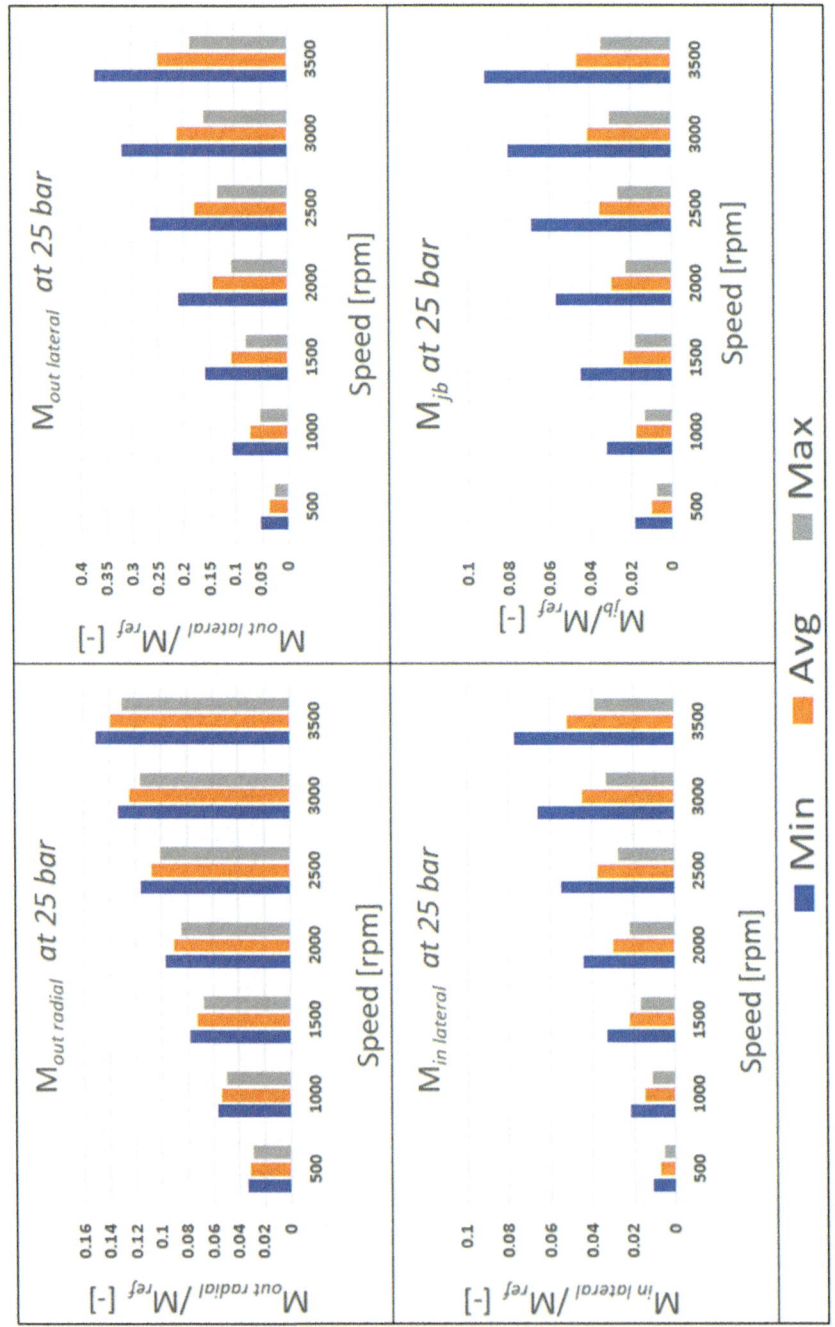

Fig. 9 Effect of clearances on torque losses at variable speed and constant pressure

- At high pressure and low speed, the model underestimates the torque, possibly due to a mixed lubrication regime occurring. Meanwhile, at high speed, the model overestimates the torque.

From this analysis, we conclude that the model is unsuitable for estimating torque at low speeds and high pressures. While it better follows the experimental data trend at medium and high speeds, but it tends to overestimate the torque. Although the current model does not allow for precise torque estimation, it still proves useful in identifying the main torque losses and understanding how variations in working conditions and geometrical modifications affect the pump's performance.

The deviations in the experimental data model can be attributed to the lack of knowledge regarding the real clearances of the tested machine. As showed in Figs. 4 and 6, these clearances have a significant impact on the model estimates. Additionally, the assumption of constant and symmetric lateral clearances contributes to these deviations. In reality, the lateral clearances are asymmetrical and vary depending on the operating conditions.

5 Conclusion

In this paper, we propose a simple model to estimate the torque required for the operation of a Gerotor pump. In Sect. 1, we describe the mathematical equations forming the basis of the model.

In Sect. 2, we apply the model to a Gerotor pump and analyze various torque loss contributions. The main torque losses are $M_{outradial}$ and $M_{outlateral}$, attributed to the major torque arm of the shear stress acting on the outer gear compared to the arm of the shear stress on the inner gear. Another aspect revealed in this section is the influence of clearance on the model's estimation capability.

In Sect. 3, through a comparison with experimental data, it becomes evident that the model underestimates the torque values at high pressure and low speed, likely due to a mixed lubrication regime, which is a phenomenon not considered in the model. Conversely, at higher speeds, the model overestimates the torque, and this overestimation tendency increases with speed.

In conclusion, the model is not suitable for precise torque estimation. However, it can be utilized to analyze the trend of pump torque at mild to high speeds, identify the main torque losses, and evaluate the effects of operating conditions and tolerances on them. Future work will involve applying the model to other samples of Gerotor pumps and conducting additional comparisons with experimental data to further understand the capabilities of the model. Moreover, the model will be completed with the addition of the losses at the gear tips and meshing losses.

Nomenclatures

$c_{jb1/2}$ Radial clearance journal bearing 1/2 [m]
R_{out} External radius of the lateral friction surface of outer gear [m]
c_{out} Radial clearance between outer gear and pump housing [m]
r_{out} Internal radius of the lateral friction surface of outer gear [m]
$F_{jb1/2}$ Force supported by journal bearing 1/2 [N]
W_{in} Load on inner gear due to fluid pressure [N]
h_l Height later gap [m]
W_{out} Load on outer gear due to fluid pressure [N]
$l_{1/2}$ Distance between gears middle plane and journal bearings 1/2 middle plane [m]
$\varepsilon_{jb1/2}$ Relative eccentricity journal bearing 1/2 [−]
L_{gear} Gear axial length [m]
ε_{out} Relative eccentricity outer gear [−]
R_W Distance between the center of the inner gear and the teeth tips of the inner gear [m]
μ Dynamic viscosity of oil [Pa s]
R_{in} External radius of the lateral friction surface of inner gear [m]
ω_{inner} Angular velocity inner gear [rad/s]
r_{in} Internal radius of the lateral friction surface of inner gear [m]
ω_{outer} Angular velocity outer gear [rad/s]

References

1. Totaro G, Zardin B, Borghi M, Scolari F (2023) Modelling of a Gerotor pump including the evaluation of the micro-movements of the external gear. J Phys Conf Ser 2648:012049. https://doi.org/10.1088/1742-6596/2648/1/012049
2. Rundo M (2017) Models for flow rate simulation in gear pumps: a review. Energies 10:1261. https://doi.org/10.3390/en10091261
3. Gamez-Montero PJ, Codina E, Castilla R (2019) A review of gerotor technology in hydraulic machines. Energies 12:2423. https://doi.org/10.3390/en12122423
4. Castilla R, Gamez-Montero PJ, Raush G, Codina E (2018) Three dimensional simulation of Gerotor with deforming mesh by using OpenFOAM
5. Altare G, Rundo M (2016) Computational fluid dynamics analysis of Gerotor lubricating pumps at high-speed: geometric features influencing the filling capability. J Fluids Eng 138:111101. https://doi.org/10.1115/1.4033675
6. Pellegri M, Vacca A, Frosina E, Buono D, Senatore A (2017) Numerical analysis and experimental validation of Gerotor pumps: a comparison between a lumped parameter and a computational fluid dynamics-based approach. Proc Inst Mech Eng C J Mech Eng Sci 231:4413–4430. https://doi.org/10.1177/0954406216666874
7. Pellegri M, Vacca A (2019) A simulation approach for the evaluation of power losses in the axial gap of Gerotor units. JFPS Int J Fluid Power Syst 11:55–62. https://doi.org/10.5739/jfpsij.11.55

8. Fabiani M, Mancò S, Nervegna N, Rundo M, Armenio G, Pachetti C, Trichilo R (1999) Modelling and simulation of Gerotor gearing in lubricating oil pumps. Presented at the international congress & exposition March 1. https://doi.org/10.4271/1999-01-0626
9. Pellegri M, Vacca A (2017) Numerical simulation of Gerotor pumps considering rotor micro-motions. Meccanica 52:1851–1870. https://doi.org/10.1007/s11012-016-0536-6
10. Mistry Z, Manne VHB, Vacca A, Dautry E, Petzold M (2020) A numerical model for the evaluation of gerotor torque considering multiple contact points and fluid-structure interactions. In: Volume 1—Symposium. Technische Universität Dresden, pp 409–418. https://doi.org/10.25368/2020.48
11. Harrison J, Aihara R, Eisele F (2016) Modeling Gerotor oil pumps in 1D to predict performance with known operating clearances. SAE Int J Engines 9:1839–1846. https://doi.org/10.4271/2016-01-1081
12. Ivanovic L, Mackic T, Stojanovic B (2016) Analysis of the instantaneous friction coefficient of the trochoidal gear pair
13. Inaguma Y (2011) Friction torque characteristics of an internal gear pump. Proc Inst Mech Eng C J Mech Eng Sci 225:1523–1534. https://doi.org/10.1177/0954406211399659
14. Gamez-Montero P, Castilla R, Codina E (2018) Methodology based on best practice rules to design a new-born trochoidal gear pump. Proc Inst Mech Eng C J Mech Eng Sci 232:1057–1068. https://doi.org/10.1177/0954406217697355
15. Rundo M, Nervegna N (2020) Passi nell'oleodinamica. EPICS
16. Ocvirk FW (1952) Short-bearing approximation for full journal bearings. NACA Tech Notes 2808:28
17. Ghosh MK, Majumdar BC, Sarangi M (2014) Fundamentals of fluid film lubrication. McGraw Hill LLC
18. Yukio H (2006) Hydrodynamic lubrication. Springer Tokyo, Tokyo
19. Hamrock BJ, Schmid BJ, Jacobson BO (2004) Fundamentals of fluid film lubrication. CRC Press

Open Access This chapter is licensed under the terms of the Creative Commons Attribution 4.0 International License (http://creativecommons.org/licenses/by/4.0/), which permits use, sharing, adaptation, distribution and reproduction in any medium or format, as long as you give appropriate credit to the original author(s) and the source, provide a link to the Creative Commons license and indicate if changes were made.

The images or other third party material in this chapter are included in the chapter's Creative Commons license, unless indicated otherwise in a credit line to the material. If material is not included in the chapter's Creative Commons license and your intended use is not permitted by statutory regulation or exceeds the permitted use, you will need to obtain permission directly from the copyright holder.

Optimization-Based Energy Efficient Power Transmission Design Methodology Applied to a Compact Excavator

Grégory Tardy, Éric Bideaux, Christophe Gostomski, and Armando Fonseca

Nomenclature

$\mathscr{P}$	Power [W]
P	Pressure $[Pa]$
Q	Flow $\left[m^3.s^{-1}\right]$
S_A, S_B	Sections $\left[m^2\right]$
D	Rotary actuators displacement $\left[m^3\right]$
F	Actuator load force $[N]$
v	Actuator linear velocity $\left[m.s^{-1}\right]$
T	Actuator load torque [N.m]
ω	Actuator rotational velocity $\left[rad.s^{-1}\right]$
ρ	Oil volumic-mass $\left[kg.m^{-3}\right]$
C_s	Flow coefficient $\left[kg^{0.5}.Pa^{-0.5}.m^{-0.5}.s^{-1}\right]$
n_{port_i}	Number of ports of an actuator i
x	Vector of the port pressures of the system actuators
u	Vector of the commands of the metering valves in the system

1 Introduction

In the current context of global warming and increasingly strict anti-pollution regulations, the heavy-duty mobile machine industry is heading toward carbon-free powertrains such as electric or fuel cell ones to power their machines. The low energy density of electric batteries and dihydrogen tanks makes it necessary to enhance the power transmission efficiency and define energy recuperation possibilities to increase autonomy. These machines widely use hydraulic transmission systems to convey the energy from the powertrain to the machine actuators: wheels, tracks, linear actuators, etc. The main advantages of hydraulic transmission are its power density, its natural shock absorption through the oil compression, and its electronic-free control making it a robust power transmission solution. However such systems have a low energy efficiency which would highly limit the potential of an electric powered machine with a low autonomy versus energy storage ratio.

Innovative hydraulic transmission systems for heavy-duty mobile machines can be found in the literature such as independent metering [1], digital hydraulics [2], pump control systems through displacement control [3] or speed control in electrohydraulic systems [4]. Also, non-hydraulic electromechanical systems are studied for such machines with characteristics close to electrohydraulic systems [5]. By increasing the number of power paths between machine actuators, the energy sources (pumps), and the storage units, new power transmission systems may benefit from better energy efficiency and enhanced energy recovery. It is especially the case for machines with multiple actuators like excavators, where kinetic and potential energy can be recovered or reused [6, 7].

At the design stage, the choice of one or another power transmission solution depends on multiple linked factors: the actuators load profile, control complexity, components cost, and available space in the machine. Individual metering and similar valve controlled systems offer interesting flow recovery capacities and remain quite inexpensive since the additional cost is limited to additional valves. However, these systems are complex to control because of the cross-influences of the valves on the flow metering. Electrohydraulic units are less complex to control but with one electric motor and a pump per actuator this solution becomes expensive and its implementation difficult due to the size of the components. Actually, there is a broad range of intermediate/alternative hydraulic power transmission systems between individual metering and the electrohydraulic solutions with different optimal performance, control complexity, and costs.

Among these solutions, the STEAM project introduced a dual pump individual metering concept [8] which provides more hydraulic power paths. Fassebender et al. [9] have proposed a downsized electrohydraulic system for multiple actuators with a shared hydraulic accumulator boost to meet the power pikes demand of the studied excavator. Rydberg et al. [10] concludes on the potential of systems with energy regeneration features like individual metering and accumulators for machines like excavators. These papers introduced new systems, their features, related hardware, and control schemes but no clear comparison was performed to rank these solutions.

In the Hybrid project however Linjama et al. [11] introduced an interesting approach which aims at comparing the optimal performance of multiple hydraulic power transmission systems with different valve layouts, one or more pumps and hydraulic accumulators over machine duty cycles. However, the proposed approach is based on rules considering the possible modes of operation for each architecture. This leads to a suboptimal result and limits the performance and energetic analysis.

In order to complete this approach enabling to draw out relevant hydraulic system benchmarks this paper introduces an approach, applicable to most of the conventional hydraulic architectures, which allows the calculation of the flow metering of the valves in the circuit that optimizes the energetic cost. Therefore it enables the analysis and comparison of the energetic performances and control complexity of different hydraulic topologies. The next section introduces the optimal flow metering problem and its combinatorial complexity. Then in Sect. 4 a graphical representation of hydraulic converters admissible operating conditions is introduced in order to limit the combinatorial complexity of the optimal flow metering problem. The last section illustrates this approach and compares the energetic performances of two variants of individual metering systems.

2 Optimal Flow Metering

A hydraulic power transmission system is a hydraulic assembly of passive (pipes) and active components (valves) distributing hydraulic energy between hydraulic converters: pumps, actuators, energy storage units like accumulators, etc. (Fig. 1). These

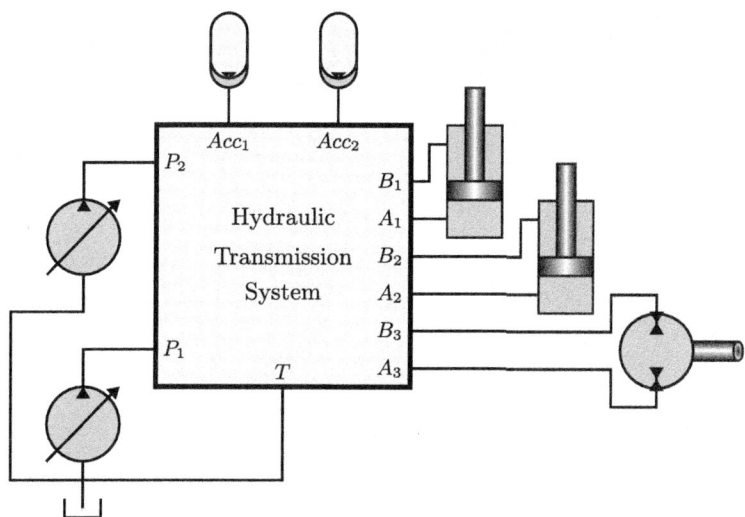

Fig. 1 Hydraulic transmission system

converters are connected to the hydraulic system through one or multiple ports, for instance, pumps, linear and rotary actuators have two ports (Fig. 2a, b) and hydraulic accumulators have one port. Depending on the system topology one or more hydraulic power paths are available to distribute the hydraulic energy between the converters so the actuators operate at the desired speed and effort. Each actuator operating point imposes constraints on the pressures and flows, as expressed in Eq. (1) where v_i (respectively ω_i) is the linear velocity of the actuator i (respectively the rotational velocity), F_i (respectively T_i) the force provided by the actuator (respectively the torque), Q_{A_i} and Q_{B_i} the port flows, S_{A_i} and S_{B_i} the piston and rod sections and D the displacement.

$$\boxed{\begin{array}{c|c} F_i = P_{A_i} \cdot S_{A_i} - P_{B_i} \cdot S_{B_i} & T_i = D_i \cdot (P_{A_i} - P_{B_i}) \\ v_i = \dfrac{Q_{A_i}}{S_{A_i}} = -\dfrac{Q_{B_i}}{S_{B_i}} & \omega_i = \dfrac{Q_{A_i}}{D_i} = -\dfrac{Q_{B_i}}{D_i} \end{array}} \quad (1)$$

To satisfy all actuator operating points the system's active components, i.e., pumps and valves, are controlled so the system flow metering meets the port flows. The valve flow metering is described in equation (2) with Q_{ij} the flow and $\Delta P_{ij} = P_i - P_j$ the pressure drop over the valve, S the valve opening section in the range $[0; S_{max}]$ set by the input u_{ij} in the range $[0; 1]$.

$$Q_{ij} = C_s \cdot S(u_{ij}) \cdot \sqrt{\dfrac{2}{\rho} \cdot |\Delta P_{ij}|} \cdot sgn(\Delta P_{ij}) \quad (2)$$

Defining an optimal system flow metering among the possible flow paths available depends on the chosen definition for the cost function. In a single pump powered system, for a given operating point, minimizing the energy consumption is equivalent to minimizing the power delivered by the pump $\mathscr{P}_P$. Considering systems with multiple pumps and energy storages there is a broad range of cost function definitions

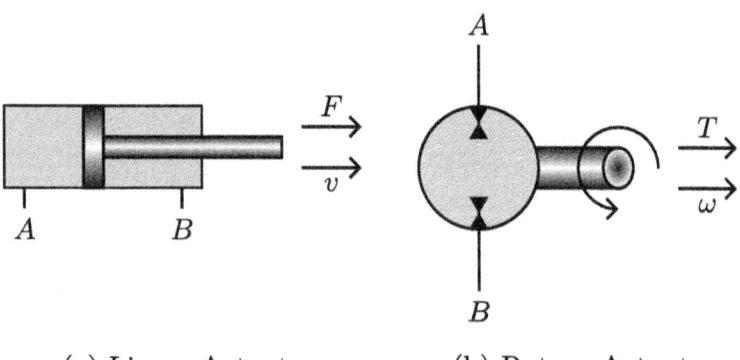

(a) Linear Actuator   (b) Rotary Actuator

Fig. 2 Actuators Schematic

depending on the desired system behavior (performance or energy savings oriented). Note that when energy storages are considered in a system, criteria based only on powers are no more adequate, the cost function has to evaluate the energy over a cycle.

A general definition of the optimal flow metering can be expressed as in (3) in which x is the vector of the pressures at the different nodes of the hydraulic circuit and u the active commands of the system (valves). At a given operating point the function f_C is the objective function to minimize, it is a function of the powers of the n_p pumps and n_{es} energy storages connected to the system which can be developed as a function of x and u. The functions f_h and f_g correspond to equality and inequality constraints applied to the system. Equality constraints are the model equations of the components expressed in equations (1) and (2), and the flow balance equation. In equation constraints are limitations applied to x and u, it imposes that all pressures and valve commands are in a given operating range.

$$\begin{aligned} \min\ & f_C\left(\mathscr{P}_1, \mathscr{P}_2, ..., \mathscr{P}_{n_p}, \mathscr{P}_1, \mathscr{P}_2, ..., \mathscr{P}_{n_{es}}\right) \\ such\ that\ & h = f_h(x, u) \\ & g \geq f_g(x, u) \end{aligned} \quad (3)$$

A fast computation of this problem would be promising to determine the optimal control of any hydraulic power transmission system however its nonlinear formulation makes this problem computationally complex. By setting a specific flow metering mode, each flow in the system will have its direction constrained, then translated in pressure constraints, i.e., $Q_{1\to 2} \geq 0 \Rightarrow P_1 \geq P_2$. If the pressure drops are neglected then the exhaustive search can be performed among linear problems resulting from the linear models of the actuators. This type of problem can be solved using efficient linear programming algorithms (such as simplex and interior-point). Moreover, depending on the hydraulic system topology the number of flow metering modes to consider can be high resulting in a combinatorial explosion of the problem dimension. However several of these flow metering modes can be identified as non-admissible. The next section introduces a graphical representation of hydraulic converters that is used to identify the admissible flow metering modes which are the only ones to be solved in a generic approach.

3 Graphical Representation of the Operating Point of Hydraulic Converters

The energetic balance of a n_{ports} hydraulic converter is the sum of all hydraulic powers at its ports which are the products of the pressures and flows at actuator ports, as given in (4), where the exponents C and S refers respectively to the converter and the hydraulic system.

$$\mathscr{P}_{net}^{C} = \sum_{i=1}^{N} \mathscr{P}_{i}^{C} = \sum_{i=1}^{N} P_{i} \cdot Q_{i}^{S \to C} \qquad (4)$$

The pressures and flows are set by a system of equations linking them to each actuator operating point, i.e., the mechanical effort and flow variables. These models always comprise an effort and n_{ports} flow equations (Eq. (1)). The effort equation expresses the relation between the converted (mechanical) effort and the required hydraulic pressures at each actuator port. The flow equations give the relations between the required flow (mechanical velocity) and the hydraulic flows at each port. From this, a general definition of the operating point of hydraulic converters can be given. These converters can have one, two, or multiple ports (Fig. 3 with e and f the converter effort and flow) and can be represented by a model including n_{ports} flow equations and an effort equation.

Since each converter effort is a combination of its ports pressure there are $n_{ports_i} - 1$ degrees of freedom in pressures available for a given operating point. In order to visualize this we propose a graphical representation of the converters based on pressure/power graphs (Fig. 5) in which the port pressures are attached to the y-axis and the power from the hydraulic system perspective is attached to the x-axis (positive powers represent system in-flows and reversely negative powers represent system out-flows). One graph can be defined for each operating quadrant of an actuator, quadrants 1 and 3 correspond to hydraulic to mechanical energy conversion, and reversely for quadrants 2 and 4.

For 2 ports actuator (a cylinder or a motor), this representation is given by Fig. 4. In this case, the straight lines on each side of the y-axis represent the A and B ports pressure combinations that satisfy a constant effort/net power. It is illustrated by the arrows under each graph representing the same net power for different A and B port pressures. These lines are always on each side of the y-axis for actuators with 2 ports

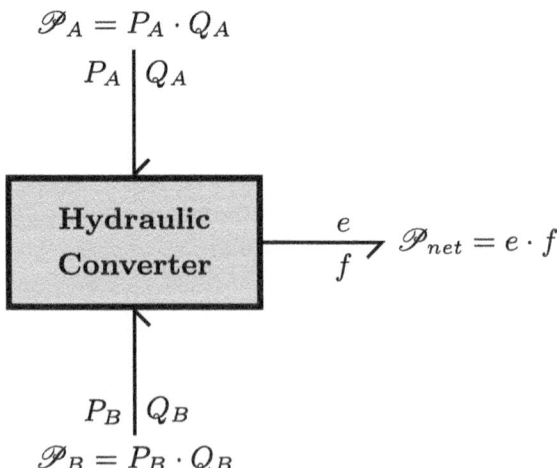

Fig. 3 General actuator

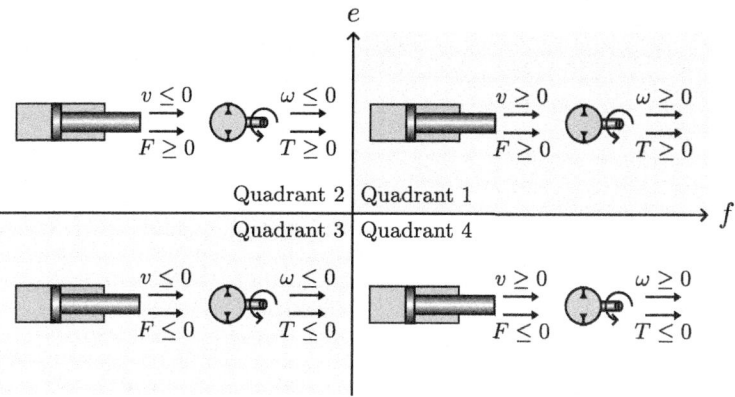

Fig. 4 Actuators quadrants

since the port flows have opposite signs. Each quadrant shows a minimum pressure set that enables the satisfaction of the required effort/net power. Note that the slopes on the A and B sides are generally different apart in the case of motors or symmetric cylinders. Then, in the case of an asymmetric actuator in quadrants 1 and 2, this representation highlights that there are two areas corresponding to opposite ports pressure inequalities, these areas are separated by a pressure threshold expressed in equation (5).

$$P_{th} = \frac{|F|}{S_A - S_B} \qquad (5)$$

This threshold is due to the linear actuator asymmetry; $S_A > S_B$, i.e., $|Q_A| > |Q_B|$, causes the slope on A side to be flatter than the slope on B side, the pressure at port B increases faster than the pressure at port A for the same required effort. Under this threshold, $P_A > P_B$, it enables the A side flow to be directed to the B side. If the operating point is above this threshold then the B side flow can be directed to the A side.

The main asset of this representation is to visualize the different pressure constraints according to the different operating quadrants. It highlights when flow regenerative metering modes can be activated in order to reduce the required pump flow. By identifying all the admissible flow metering modes for each quadrant, or identifying the non-admissible ones, it enables the cost of the brute-force computing to be reduced. An example of the optimal flow metering computation with reduced combinatorial complexity is given in the next section with two variants of individual metering systems in order to compare their optimal performance as a demonstration of the proposed methodology.

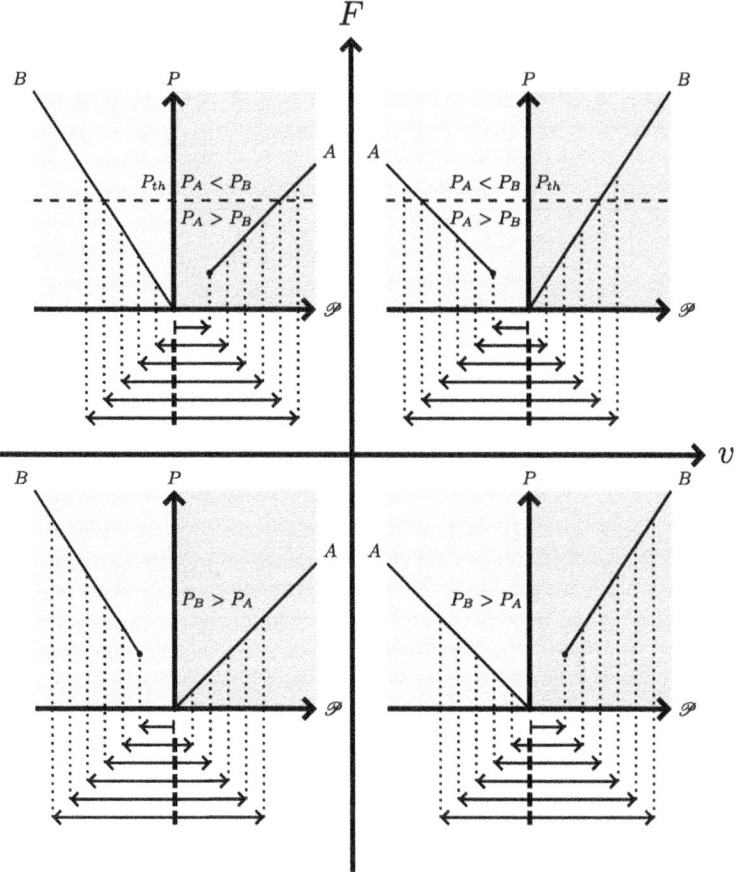

Fig. 5 Pressure/Power graph

4 Individual Metering Systems Comparison

In this section, two decentralized individual metering distribution systems are compared to a conventional centralized load-sensing system of a VOLVO ECR58 excavator (Fig. 6). These individual metering systems are referred to as the 4 valves system (respectively the 5 valves system) throughout the section. In the 5-valve system (Fig. 7) the A and B ports of each actuator are connected individually both to the pump and the tank thanks to 4 proportional 2/2 valves, and a last proportional 2/2 valve enables a direct connexion between port A and B (bypass valve). The 4 valves system is identical except it does not contain the bypass valve. The pumps of these systems are considered to be powered by an electric powertrain, so hydraulic energy may be recovered and stored in the excavator battery. The load cycle used for the comparison is a digging cycle, the acquired data are the pump power and the actua-

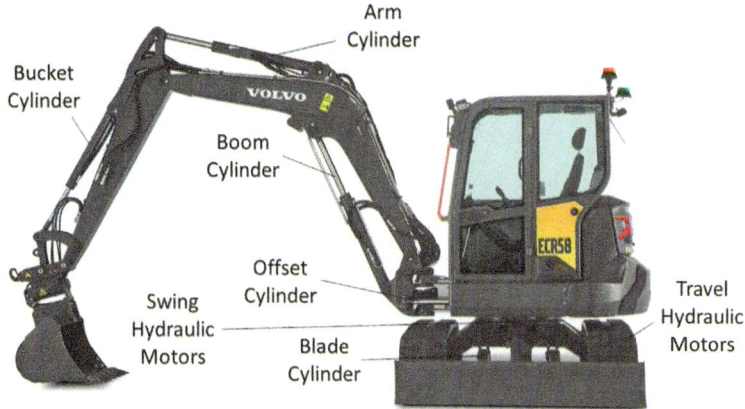

Fig. 6 Volvo ECR58 compact excavator

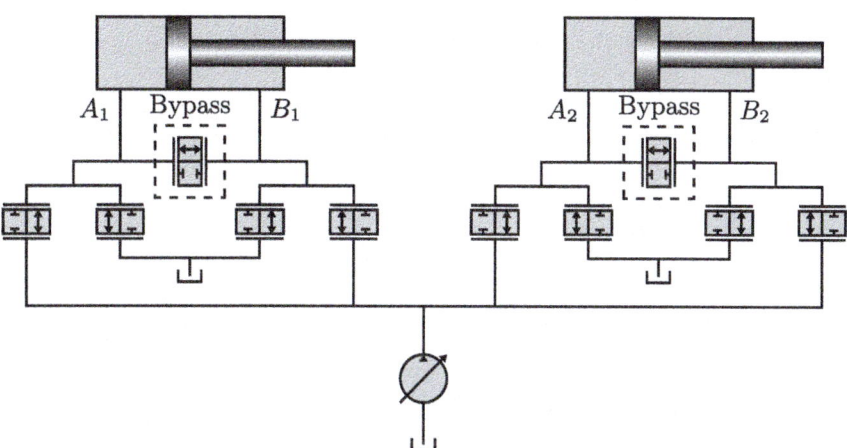

Fig. 7 Individual metering 5 valve system

tors' efforts and velocities on the current ECR58 excavator. The study is focused on the 4 actuators in motion during the cycle: bucket, arm, boom, and swing actuators.

There are four flow paths to be considered in both systems: $Q_{P \to A}$, $Q_{P \to B}$, $Q_{A \to T}$ and $Q_{B \to T}$, and an additional one in the 5 valves system: $Q_{A \to B}$. The tank pressure is null so all other pressures are greater or equal to the tank pressure; this assumption imposes that tank flows are always toward the tank, i.e., $Q_{A \to T} \geq 0$ and $Q_{B \to T} \geq 0$. The other flow signs are set according to the sign of the pressure drops (Table 1). Considering all the flow metering modes results in a significant number of cases to solve.

In order to greatly reduce the problem dimension all admissible solutions for each quadrant are identified using the Pressure/Power graphs. In the case of linear

Table 1 Actuator flow metering modes

System	Actuator flow metering mode	Pressure inequalities		$Q_{P \to A}$	$Q_{P \to B}$	$Q_{A \to B}$
5 Valves	1	$P_A \geq P_B \geq P_P$		–	–	+
	2	$P_A \geq P_P \geq P_B$		–	+	+
	3	$P_P \geq P_A \geq P_B$		+	+	+
	4	$P_B \geq P_A \geq P_P$		–	–	–
	5	$P_B \geq P_P \geq P_A$		+	–	–
	6	$P_P \geq P_B \geq P_A$		+	+	–
4 valves	1	$P_A \geq P_P$	$P_P \geq P_B$	–	+	
	2	$P_P \geq P_A$	$P_B \geq P_P$	+	–	
	3	$P_P \geq P_A$	$P_P \geq P_B$	+	+	

Table 2 Linear actuator 1st quadrant metering modes (5 valves system)

Actuator flow Metering modes	Admissible	Explanation
1	✗	$P_A > P_B$ so the A port flow request must be satisfied by the pump so only the 3rd flow metering mode is admissible
2	✗	
3	✓	
4	✗	$P_B > P_A$ so the B port flow can be regenerated to satisfy the A port flow request, however, due to the actuator asymmetry a fraction of the A port flow still must be satisfied by the pump making the 4th flow metering mode non-admissible
5	✓	
6	✓	

actuators operating in the first quadrant with the 5 valves system the graph analysis performed in Table 2 allows us to identify that only 3 flow metering modes are admissible. There is an additional constraint set by the 5th and 6th flow metering modes on P_P, because $P_B \geq P_A$ and $P_P \geq P_A$ in both cases then $P_P \geq P_{th}$ that is the actuator threshold pressure as expressed in Eq. 5.

By performing this analysis for the linear and rotary actuators for both 5- and 4-valves configurations all admissible flow metering modes can be identified (Tables 3, 4, 5 to 6). In these tables, the flow solutions are developed for each flow metering mode and expressed with absolute values with the right sign in order to make clear the flow direction. For rotary actuators, flow ports are not explicit due to the actuator symmetry $|Q| = |Q_A| = |Q_B|$.

These flow equations highlight the regenerative capacity of each flow metering mode, some of them enable flow regeneration between the actuator ports through the

Table 3 Linear actuator flow metering modes (5 Valves System)

Quadrant	Solution	Flow metering mode	$Q_{P \to A}$	$Q_{P \to B}$	$Q_{A \to T}$	$Q_{B \to T}$	$Q_{A \to B}$
1	S_1	3	$|Q_A|$	0	0	$|Q_B|$	0
	S_2	5	$|Q_A| - |Q_B|$	0	0	0	$-|Q_B|$
	S_3	6	$|Q_A| - |Q_B|$	0	0	0	$-|Q_B|$
2	S_4	1	$-|Q_A| + |Q_B|$	0	0	0	$|Q_B|$
	S_5	2	$-|Q_A| + |Q_B|$	0	0	0	$|Q_B|$
	S_6	3	0	0	$|Q_A| - |Q_B|$	0	$|Q_B|$
	S_7	6	0	$|Q_B|$	$|Q_A|$	0	0
3	S_8	6	0	$|Q_B|$	$|Q_A|$	0	0
4	S_9	5	$|Q_A| - |Q_B|$	0	0	0	$-|Q_B|$
	S_{10}	6	$|Q_A| - |Q_B|$	0	0	0	$-|Q_B|$

Table 4 Rotary Actuator flow metering modes (5 Valves System)

Quadrant	Solution	Flow metering mode	$Q_{P \to A}$	$Q_{P \to B}$	$Q_{A \to T}$	$Q_{B \to T}$	$Q_{A \to B}$
1	S_1	3	$|Q|$	0	0	$|Q|$	0
2	S_2	1	0	0	0	0	$|Q|$
	S_3	2	0	0	0	0	$|Q|$
	S_4	3	0	0	0	0	$|Q|$
3	S_5	6	0	$|Q|$	$|Q|$	0	0
4	S_6	4	0	0	0	0	$-|Q|$
	S_7	5	0	0	0	0	$-|Q|$
	S_8	6	0	0	0	0	$-|Q|$

bypass valve or the pump line, and some also allow to recovery of excess hydraulic energy at the pump. These flow metering modes match the ones described in the literature [12–14] and they allow to reduce the pump energy consumption and even to recover energy.

The optimal flow metering problem is then solved over the digging cycle for both studied setup configurations, that is with 4 or 5 valves. Power requirements of each actuator are shown in Fig. 8a, the experimental pump pressure of the ECRA8 load-sensing hydraulic system is given in Fig. 8b along with the optimized pump powers

Table 5 Linear Actuator flow metering modes (4 Valves System)

Quadrant	Solution	Flow metering mode	$Q_{P \to A}$	$Q_{P \to B}$	$Q_{A \to T}$	$Q_{B \to T}$
1	S_1	3	$|Q_A|$	0	0	$|Q_B|$
	S_2	2	$|Q_A|$	$-|Q_B|$	0	0
2	S_3	1	$-|Q_A|$	$|Q_B|$	0	0
	S_4	3	0	$|Q_B|$	$|Q_A|$	0
3	S_5	3	0	$|Q_B|$	$|Q_A|$	0
4	S_6	2	$|Q_A|$	$-|Q_B|$	0	0
	S_7	3	$|Q_A|$	0	0	$|Q_B|$

Table 6 Rotary Actuator flow metering modes (4 Valves System)

Quadrant	Solution	Flow metering mode	$Q_{P \to A}$	$Q_{P \to B}$	$Q_{A \to T}$	$Q_{B \to T}$
1	S_1	3	$|Q|$	0	0	$|Q|$
2	S_2	1	$-|Q|$	$|Q|$	0	0
	S_3	3	0	$|Q|$	$|Q|$	0
3	S_4	3	0	$|Q|$	$|Q|$	0
4	S_5	2	$|Q|$	$-|Q|$	0	0
	S_6	3	$|Q|$	0	0	$|Q|$

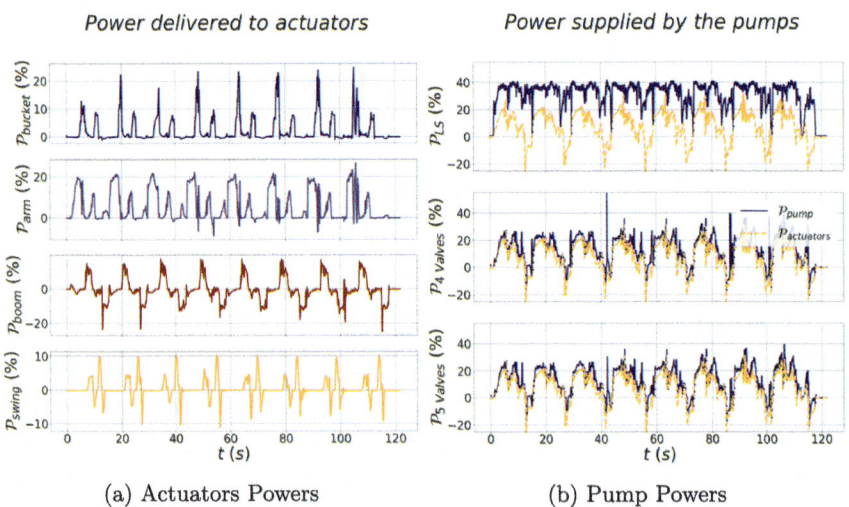

(a) Actuators Powers   (b) Pump Powers

Fig. 8 Digging cycle optimal powers results

Table 7 Digging cycle optimal energy results

System	LS	4 valves	5 valves
Energetic efficiency	25.9%	61.4%	63.4%
Mechanical energy recovery ratio	22.8%		
System Energy recovery fraction	0%	34.5%	34.6%

for 4 and 5 valves setups. Note that power is expressed as a % of the maximum pump power. The dotted line represents the sum of the power requirements of all the actuators. As shown in Table 7, both individual metering setups require globally less power than the current ECR58 load sensing system. However, it has to be noticed that both 4 and 5 valves' configurations have close energetic performances. The Load Sensing system's energetic efficiency over the cycle is 25.9 % whereas the 4-valve setup efficiency is 61.4 % and the 5-valve setup efficiency is 63.4%. Note that over the cycle, 22.8% of the actuators' mechanical energy is recoverable through the hydraulic circuit if the pump is reversible (pump/motor). However, these topologies do not allow all the potential recoverable energy to be recovered, both systems are able to recover nearly 35 % of this energy. Indeed additional recovery modes could have been taken into account by allowing flow from the tank (or a Low-Pressure Line) to actuators.

5 Conclusion

This paper introduced a graphical representation of the operating quadrants of hydraulic actuators. It enables the combinatorial complexity of the hydraulic optimal flow metering problem to be reduced. The proposed method was illustrated on two individual metering system variants and their energetic performances were compared to the conventional load sensing system of an ECR58 VOLVO excavator. Both 4 and 5 valves setups showed similar energetic performances and energy efficiencies of around 62%. In addition, both setups are able to recover at the main motor/pump nearly 35 % of the recoverable mechanical energy. However, these results correspond to ideal conditions since no pressure drops were taken into account in the valves or in the hydraulic circuit and the system dynamic is neglected. The main advantage of the proposed method is to determine at a low computational cost the best flow metering configuration and evaluate different hydraulic topologies very quickly. Due to the low computation cost, the method could efficiently be used in real-time control systems. In order to tackle the limitations of the proposed method, a new formulation taking into account the valve's pressure drops and the system dynamic is presently explored and will exploit efficient optimization algorithms such as quadratic and Linear programming.

References

1. Eriksson B, Palmberg JO (2010) Individual metering fluid power systems: Chal-lenges and opportunities. Autom Const 225(1). https://doi.org/10.1243/09596518JSCE1111
2. Linjama M (2011) Digital fluid power—state of the art. Tempere, Finland
3. Busquets E, Ivantysynova M (2014) The world's first displacement controlled ex-cavator prototype with pump switching—a study of the architecture and control. Matsue, Japan
4. Fassbender D, Zakharov V, Minav T (2021) Utilization of electric prime movers in hydraulic heavy-duty-mobile-machine implement systems. Autom Const 132. https://doi.org/10.1016/j.autcon.2021.103964
5. Hagen D, Padovani D, Choux M (2020) Guidelines to select between self-contained electro-hydraulic and electro-mechanical cylinders. Kristiansand, Norway
6. Jung T, Raduenz H, Krus P, De Negri VJ, Lee J (2022) Boom energy recupera-tion system and control strategy for hydraulic hybrid excavators. Autom Const 135. https://doi.org/10.1016/j.autcon.2021.104046
7. White NN, Howland J, Zhang H, Carl B (2022) Boom potential energy recovery of hydraulic excavator, pat 11,225,776
8. Vukovic M, Sgro S, Murrenhoff H (2013) "Steam—a mobile hydraulic system with engine integration, presented at the ASME-BATH 2013 symposium on fluid power and motion control. Sarasota, USA
9. Fassbender D, Fresia P, Rundo M, Altare G (2022) Downsizing the electric ma-chines of energy-efficient electrohydraulic drives for mobile hydraulics. J Phys: Conf Ser 2385 https://doi.org/10.1088/1742-6596/2385/1/012028
10. Rydberg KE (2005) Energy efficient hydraulic systems and regenerative capabilities. Linköping, Sweden
11. Linjama M, Huova M, Tammisto J et al (2019) Hydraulic hybrid working machines project—lessons learned. Tempere, Finland
12. Shenouda A (2006) Quasi-static hydraulic control systems and energy savings po-tential using independent metering four-valve assembly configuration, PhD dissertation, Woodruff School of Mechanical Engineering, Georgia Institute of Technology, Atlanta, USA
13. Eriksson B (2007) Control strategy for energy efficient fluid power actuators utiliz-ing individual metering. Linköping, Sweden
14. Ketonen M, Linjama M (2017) Simulation study of a digital hydraulic independent metering valve system on an excavator. Linköping, Sweden

Open Access This chapter is licensed under the terms of the Creative Commons Attribution 4.0 International License (http://creativecommons.org/licenses/by/4.0/), which permits use, sharing, adaptation, distribution and reproduction in any medium or format, as long as you give appropriate credit to the original author(s) and the source, provide a link to the Creative Commons license and indicate if changes were made.

The images or other third party material in this chapter are included in the chapter's Creative Commons license, unless indicated otherwise in a credit line to the material. If material is not included in the chapter's Creative Commons license and your intended use is not permitted by statutory regulation or exceeds the permitted use, you will need to obtain permission directly from the copyright holder.

Analysis of the Operating Point Method for Dimensioning of Pneumatic Drives Under Variable Loading Conditions

Vinícius Vigolo, Antonio Carlos Valdiero, and Victor Juliano De Negri

1 Introduction

The energy efficiency of compressed air systems has gained significant attention recently. Studies indicate that global energy efficiency in pneumatic systems can be as low as 2% of input electric energy, with losses occurring in the production, distribution, and usage of compressed air [1]. The energy consumption during the usage is characterized by the system design, being independent of its application [2]. This is explained by the fact that pneumatic drives are completely filled with compressed air to complete a given task, independently of the task itself.

This aspect has motivated several researchers to explore alternative design architectures and energy-saving strategies to improve the energy efficiency of pneumatic actuators. Examples of such strategies include recovering exhaust air for lower pressure applications [3, 4], shutting off the supply pressure when the task is complete [5–7], using different supply pressures for extending and retracting phases [8], and even heat recovery from the compressor unit [9], all of these examples resulting in remarkable improvements in energy efficiency (Fig. 1).

However, these strategies may introduce complexity and elevate the risk of production losses due to unplanned machine stops, thus reducing the attractiveness gained with higher energy efficiencies. Therefore, strategies requiring minimal intervention have also gained attention, including the reduction of dead volumes, the online regulation of supply pressure [10], and the development of specialized pneumatic devices for an user-friendly improvement operation of the drive [11, 12].

The optimization of drive dimensioning is also a topic of interest due to the common problem of oversized cylinders leading to excessive air consumption. To

---

V. Vigolo (✉) · A. C. Valdiero · V. J. De Negri
Federal University of Santa Catarina, Florianópolis, Brazil
e-mail: Vinicius.vigolo@laship.ufsc.br

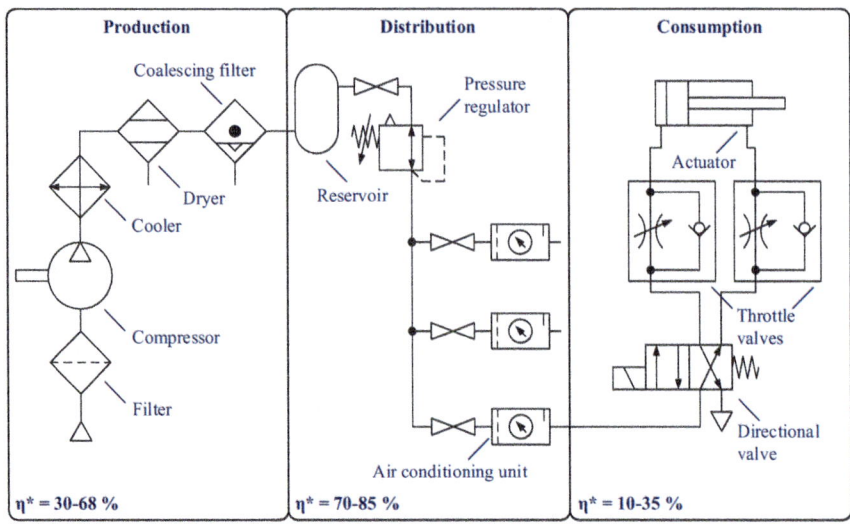

Fig. 1 Industrial pneumatic architecture. *Efficiency (η) values from [1]

address this, researchers have developed alternative procedures to improve the dimensioning process. Optimization algorithms, for instance, have been applied in [8, 13, 14], where the objective function was set to optimize the system parameters while minimizing air consumption and fulfilling the task requirements. However, the advanced mathematical models used in optimization solvers turn them into black-box tools, making the design choices not easily interpretable by the designers.

Analytical procedures have also been proposed. In [15], the authors developed a method based on the analysis of the eigenfrequency of the drive and the dynamic requirements of the task, resulting in a parameter called the pneumatic frequency ratio, used as reference to achieve a well-dimensioned cylinder in terms of dynamics. However, as shown in [16], this approach presents limitations for tasks that involve pressing forces or vertical displacements. An exergy-based approach is presented in [17], which considers the capacity of the consumed air to perform useful work. Nevertheless, in standard pneumatic drives, the expansion exergy cannot be transformed in work, resulting in an approach similar to a conventional force balance analysis [16].

In this context, a set of equations composing the operating point approach has been presented in the authors' previous work [18]. The goal is to size pneumatic drives for an optimum balance between energy efficiency and dynamic performance. In this paper, the application of the operating point method is assessed under the operation of non-constant load forces, where energy efficiency and the sensitivity of the cylinder are used to statistically define guidelines for an enhanced operation. Analytical expressions are presented to apply the operating point method for a wide range of working conditions, including different components of the load force. Simulations are carried out to assess the proposed dimensioning procedure, demonstrating its suitability for successfully sizing pneumatic drives for variable loading conditions, achieving a balance of energy efficiency, robustness, and dynamic performance.

2 Classification of Pneumatic Applications

Pneumatic systems have a wide range of applications with specific characteristics that must be considered during the dimensioning process. Based on the applications of pneumatic actuation systems described in [19–21] and taking into account the aspects outlined in [16] and [22], it is proposed the classification of pneumatic actuation systems into two categories: static and dynamic applications. Each category has distinct characteristics that define the requirements for cylinder dimensioning.

2.1 Static Applications

It covers applications where the force produced by the drive during the piston displacement is either small or nonexistent. In such cases, the maximum force that the cylinder is dimensioned for occurs when the piston is stationary. The main characteristic of a static application is the constant chamber volumes during the execution of the cylinder's primary task, leading to stable chamber pressures and, consequently, facilitating their determination.

A static application is not limited to scenarios without piston displacement, instead, it involves the requirement of maximum force at the moment of null or quasi-static displacement. Examples encompass tasks such as fixing parts, pressing, stamping, and forming. Load generation equipment, including tensile and compression testing machines, also falls into this category. Figure 2a presents a generic example of the main forces acting in a static application.

Since cylinder dimensioning in static applications often involves a simple force balance between the driving chamber force and the desired contact force, this paper is focused on the dimensioning of pneumatic drives for dynamic applications.

2.2 Dynamic Applications

This category includes the activities that require a significative force during the piston displacement. In these cases, the dimensioning of the cylinder area must consider variations in chamber pressures caused by changes in chamber volumes. This aspect is the main contribution of the operating point equations to be presented in Sect. 3.

Applications that fall within this category include the transport and manipulation of objects, actuation of articulated mechanisms, tasks of parts assembly, and various

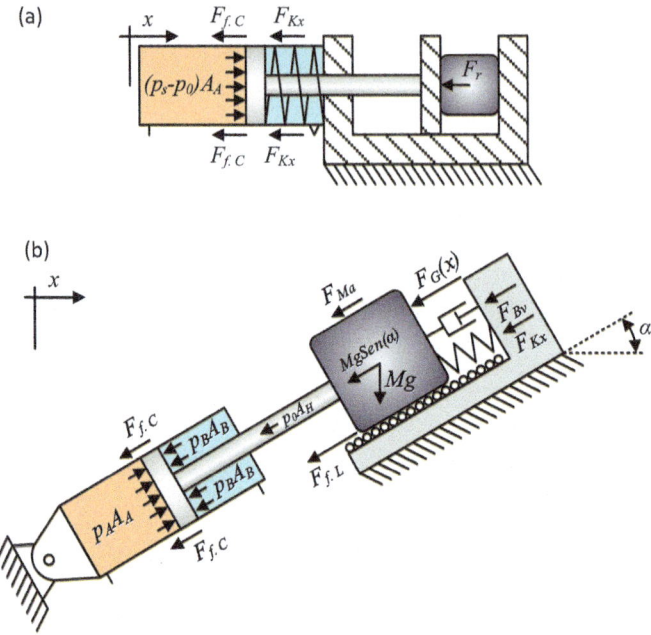

Fig. 2 Generic examples of pneumatic applications. (**a**) Static; (**b**) Dynamic

manufacturing processes such as folding, drilling, cutting, and machining, among others. Figure 2b shows the main forces acting in a dynamic application.

In Figure 2, p denotes pressure, and the subscripts $A, B, 0$, and S stand for chambers A, B, ambient, and supply pressures, respectively. A_A, A_B, and A_H are the chamber areas A, B, and the rod area, respectively. $F_{f.C}$ and $F_{f.L}$ are friction forces produced by the cylinder and the load, respectively. F_r, F_{Kx}, F_{Bv}, and F_{Ma} are the reaction, spring, viscous, and acceleration forces, respectively. $MgSen(\alpha)$. is the gravitational force, and $F_G(x)$ is a generic force representing force components that do not fall within remaining categories.

Applications that present both dynamic and static characteristics (e.g., moving and holding a workpiece at the stroke end) should be dimensioned using both the dynamic approach presented in this paper and a static approach. The most critical scenario must be considered when selecting the pneumatic drive.

3 The Operating Point Approach

Reducing the cylinder area is an effective alternative to improve the energy efficiency of pneumatic drives. However, undersizing pneumatic drives results in undesired effects on the system behavior, including reduced piston velocity, lower shock absorption resistance, and reduced capacity to withstand changes in the load force. Therefore, undersized pneumatic drives results in a poor dynamic performance and lower robustness, as shown in Fig. 3a. These results come from an experimentally validated simulation model [18], which is used to assess the impact of different cylinder diameters on air consumption, piston velocity, and sensitivity, considering the same task with a constant load force.

The sensitivity of the cylinder is a measure of the impact on the displacement time caused by a small increase on the load force. Therefore, small sensitivity values indicate a robust system. The results shown in Fig. 3a were obtained through simulations performed with the same load force for the different cylinder areas. The sensitivity of each cylinder diameter was determined by measuring the impact on the displacement time caused by an increase of 20% of the load force.

As observed, the air consumption increases with the cylinder diameter, which is due to the larger volume to be filled with compressed air. Since the same task is being performed in all cases, the energy efficiency decreases as the cylinder diameter increases. In contrast, the sensitivity decreases with bigger cylinder diameters, improving its robustness. However, it eventually reaches a plateau where further increases in diameter have a small impact on the cylinder sensitivity.

It is also evident that the stroke end velocity is impacted by the cylinder area. For every load force, there is a specific cylinder area that results in maximum velocity, leading to the minimum displacement time. In terms of sensitivity, Fig. 3a shows that the maximum velocity occurs near the sensitivity plateau, and it also corresponds to good energy efficiency. This combination results in an optimum balance of dynamic performance, robustness, and energy efficiency.

Aiming to analytically determine the cylinder area for optimum operation (maximum velocity), the operating point has been developed as a dimensioning procedure. Due to its relevance to the current contribution, the equations and curves that compose the operating point approach for an extending movement will be briefly introduced in this section, with further expression derivations provided in the appendix section. However, the authors encourage readers to refer to the authors' previous publications [18, 23, 24] for a detailed derivation and the expressions for the retracting movement.

The operating point approach is based on the analysis of the governing equations of the system assuming a steady-state behavior, where the time changes of pressure and temperature are neglected. As detailed in the appendix section, this hypothesis leads to an analytical correlation of chambers A and B pressures, resulting in the operating (Op) curve of the system, expressed as

$$\frac{p_0}{p_B} = \bar{b} + \sqrt{\bar{b}^2 - 2\bar{b} + 1 + r_A^2 + r_A^2(2\bar{b} - 1)\left(\frac{p_A}{p_S}\right)^{-2} - 2\bar{b}r_A^2\left(\frac{p_A}{p_S}\right)^{-1}} \quad (1)$$

where $\bar{b}$ is the average pressure ratio for all the restrictors on the system and r_A is the ratio of piston area B to area A.

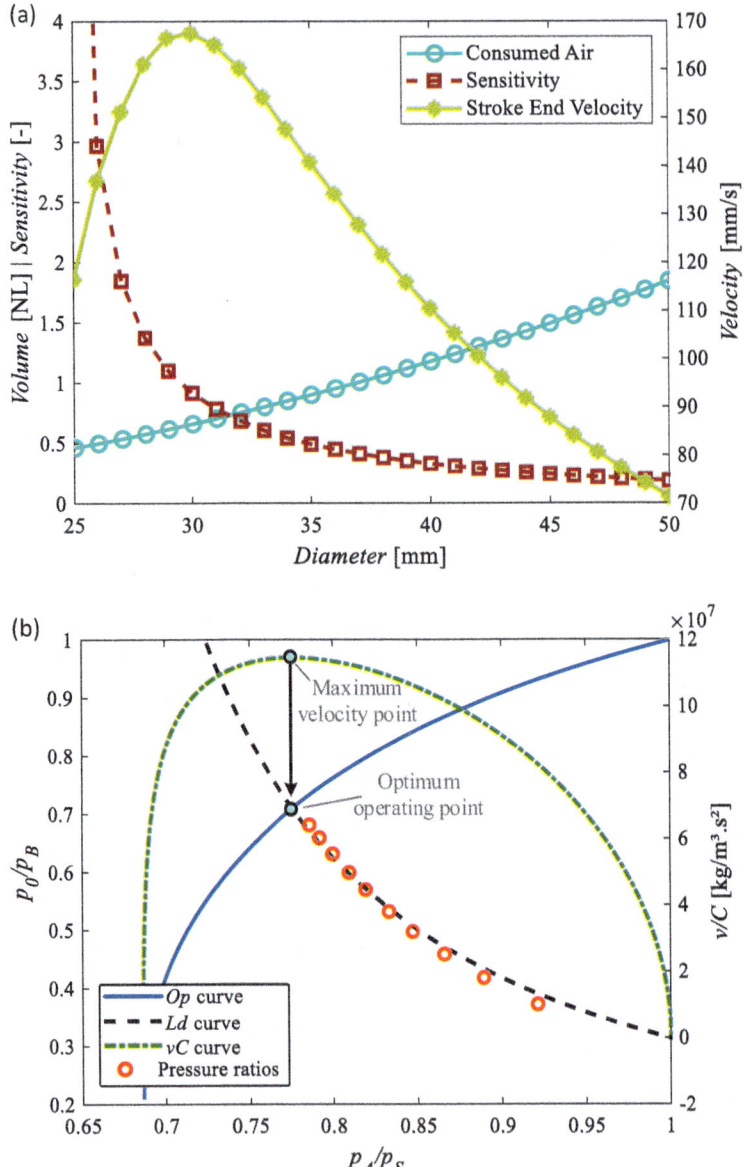

Fig. 3 (a) Sensitivity analysis of cylinder diameter; (b) Operating point chart

The *Op* curve (Fig. 3b) determines the potential combinations of chamber pressures during steady-state movement. Given that (1) is valid only for steady-state displacement and does not consider the load force, Newton's second law is expressed in the form of pressure ratios, leading to

$$\frac{p_A}{p_S} = \frac{F_L}{A_A p_S} + r_A \frac{p_0}{p_S}\left(\frac{p_0}{p_B}\right)^{-1} + \frac{p_0}{p_S}(1 - r_A) + \mu_d \qquad (2)$$

This equation describes the loading (*Ld*) curve of the system (Fig. 3b), where F_L represents the load and acceleration forces and μ_d is a dynamic friction coefficient [18].

The red circles that follow the trend of the *Ld* curve in Fig. 3b are effective pressure ratios obtained during the piston displacement. As the red circles approach the *Op* curve, the system approaches a steady-state behavior, and the intersection of the *Op* and *Ld* curves defines the operating point of the system.

Assuming a steady-state behavior, it is possible to analytically describe the extending piston velocity (v_{ess}) as function of pressure ratios, as detailed in the appendix section, resulting in

$$\frac{v_{ess}}{C} = \frac{p_S p_0}{\left(\frac{p_A}{p_S}\right) F_L} \sqrt{1 - \left(\frac{\left(\frac{p_A}{p_S}\right) - \bar{b}}{1 - \bar{b}}\right)^2} \left(\left(\frac{p_A}{p_S}\right) - r_A \frac{p_0}{p_S}\left(\frac{p_0}{p_B}\right)^{-1} - \frac{p_0}{p_S}(1 - r_A) - \mu_d\right) \qquad (3)$$

Since (1) defines $p_0/p_B = f(p_A/p_S)$, the piston velocity becomes a function of the driving chamber pressure ratio (p_A/p_S), and it is linearly dependent with the sonic conductance (C) of the valve. Therefore, the velocity is expressed as a ratio with the sonic conductance, leading to the *vC* curve in Fig. 3b.

Equation (3) demonstrates that at steady-state, the piston velocity increases linearly with the sonic conductance of the valve. Moreover, it also reveals that the velocity is dependent on the chamber pressures. For instance, during an extending movement, when chamber A pressure is close to the supply pressure ($p_A/p_S \approx 1$), the expected velocity of the system is lower (see Fig. 3b). Such cases occur when the cylinder is overloaded, requiring a maximum pressure difference across the cylinder chambers to perform the task.

Lowering the chamber A pressure during the movement (e.g., increasing the cylinder diameter) will increase the piston velocity, resulting in faster completion of the task. However, this behavior is not linear. The *vC* curve has a maximum point which occurs at a specific value of p_A/p_S. Therefore, excessively reducing the chamber A pressure (e.g., oversized cylinders) reduces the extending velocity, negatively impacting the dynamic performance of the system. This behavior is mathematically described by (3) and is demonstrated through simulations in Fig. 3.

In this way, the optimal operating condition is defined by a set of two pressure ratios: the first is the argument that maximizes the vC curve (p_A/p_S) for an extending movement), and the second corresponds to the chamber B pressure ratio (p_0/p_B), which is defined by (1). This optimal operating point, in turn, is applied in (2), which is valid for dynamic behavior and can be used to determine the optimum cylinder area.

4 Variable Loading Conditions

In most cases, the load force applied to the cylinder will present some degree of variability throughout the piston stroke. This variability may be caused by factors such as friction forces on cylinder guides or loads, the presence of springs and dampers, non-constant acceleration during displacement, or the actuation of mechanisms that lead to Cartesian force decomposition.

The amplitude of variability depends on each application. Tasks involving the displacement of objects are generally assumed to have constant load forces, although the acceleration force and friction between the load and the moving surface may impact the load force to some extent. In general, linear bearings have no significant viscous force and the spring of single-acting cylinders hardly accounts for more than 10% of the maximum cylinder force. Applications involving the actuation of mechanical arms, mechanisms, or large masses are more likely to have a strong dependence of the load force with the cylinder stroke.

In the context of cylinder dimensioning, addressing the variability of the load force raises the question of which force to consider during the dimensioning process. A conventional engineering approach involves analyzing the most critical scenario, often the condition of maximum load force. However, as detailed in Sect. 3, the operating point approach focuses on a balance between energy efficiency and robustness when a cylinder moves with a constant load force throughout its entire stroke. Consequently, assuming the maximum load force for applications where this condition occurs only in a small portion of the cylinder stroke, such as the compression of a spring, might be excessively conservative.

Since it is unfeasible to consider all potential pneumatic applications to analyze the load force behavior throughout the cylinder stroke, a Monte Carlo analysis was performed in this study. To generalize the results, random load force profiles (F_{pr}) were generated according to (4). Figure 4 presents a few examples of force profiles used during the analysis.

$$F_{pr} = A_1 \sin(2\pi f_1(x+\varphi)) + A_2 \sin(2\pi f_2(x+\varphi)) \\ + A_3 \sin(2\pi f_3(x+\varphi)) + A_4 x \qquad (4)$$

where A, f, and φ represent uniformly distributed random values for amplitudes, frequencies, and offset, respectively. x is the piston position.

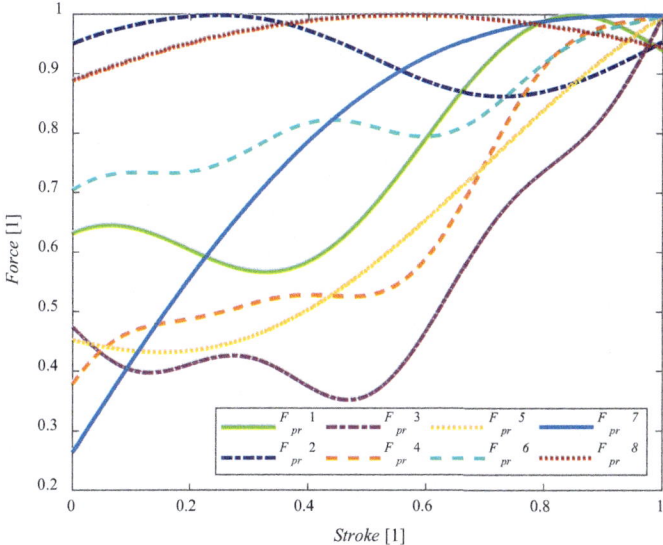

Fig. 4 Normalized force profiles

The load profiles were discretized along the piston stroke, and three different percentiles were established to analyze the impact on energy efficiency and sensitivity of the cylinder when the highest values of the load force are neglected during the dimensioning process. The adopted percentiles are

- F_L@100th percentile: The dimensioning load force is equal to the maximum load force;
- F_L@90th percentile: The dimensioning load force is equal to or greater than 90% of the discretized load force values;
- F_L@80th percentile: The dimensioning load force is equal to or greater than 80% of the discretized load force values.

A set of 1,000 distinct design requirements was randomly generated using a uniform distribution within the range presented in Table 1, which covers most of the commonly used pneumatic applications. For each design requirement, a random load force profile from (4) was assigned and multiplied by the maximum load force.

Table 1 Range of design requirements

Parameter	Range
Supply pressure (p_S)	[6–9] bar_{abs}
Maximum load force (F_{L_max})	[30–500] N
Displacement time (t_d)	[0.3–1.5] s
Stroke (L)	[0.05–0.5] m
Hose length (L_H)	[0.2–0.7] m

The operating point approach was used to select a commercially available cylinder diameter for each design requirement. The supply pressure was adjusted for the cases where the desired cylinder diameter did not match with a commercially available option. Taking into account the different percentiles, three different configurations of cylinder diameter and supply pressure were obtained for each design requirement.

Simulations were performed for the three configurations obtained in each design requirement using an experimentally validated dynamic model [18] developed in MATLAB/Simulink. The results for energy efficiency and sensitivity are shown in Fig. 5 in the form of Probability Density Functions (PDFs), derived from the means ($\bar{x}$) and standard deviations (s) of the obtained results. The energy efficiency in each simulation corresponds to the ratio of the performed mechanical work to the total supplied exergy on chamber A, as detailed in [18].

As expected, the mean energy efficiency increases as the adopted percentiles reduce, which is due to the lower load force considered for the design process, leading to a reduction in either the cylinder diameter or the supply pressure. The standard deviation for the energy efficiency results did not present a significative change between the analyzed percentiles. On the other hand, the standard deviation of the sensitivity showed a significative increase for smaller percentiles. This implies that the sensitivity uncertainty of systems dimensioned with lower percentiles is higher, which is the side effect of having a higher energy efficiency.

Based on the results shown in Fig. 5, it can be concluded that load forces percentiles lower than the 90th threshold excessively increase the uncertainty of the drive's sensitivity, not compensating the gain in energy efficiency. Therefore, the 90th percentile is recommended for the determination of the dimensioning load force of non-constant load force applications.

5 Analytical Dimensioning Procedure for Dynamic Applications

For the dimensioning of pneumatic drives for dynamic applications, the operating point equations and the observations presented in Sect. 4 will be used. The dimensioning procedure begins with an analysis of Newton's second law applied to the generic application shown in Fig. 2b.

$$p_A A_A - p_B A_B - p_0 A_H - F_{f \cdot C} - Mg Sen(\alpha) - F_{f.L} - F_{Bv} - F_{Kx} - F_G(x) = M \frac{d^2 x}{dt^2} \tag{5}$$

The goal of (5) is to consider all possible load components that might be present in an industrial application. Assuming $F_G(x) = 0$, a displacement with constant acceleration and considering the 90th percentile to determine the load force, the following expression are derived to account for acceleration, viscous, and spring

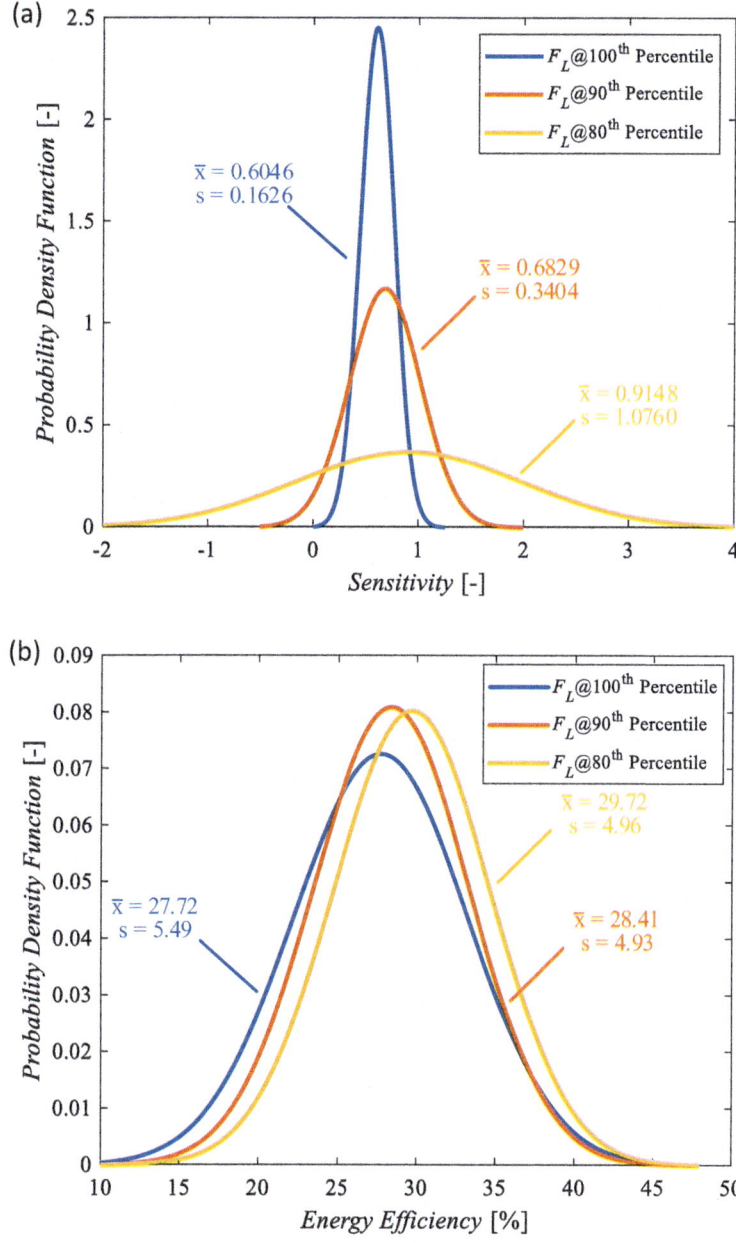

Fig. 5 Probability density functions. (**a**) Energy efficiency; (**b**) Sensitivity

forces, respectively:

$$M\frac{d^2x}{dt^2} = \frac{2ML}{t_d^2}, \tag{6}$$

$$F_{Bv} = \frac{2BL\sqrt{0.9}}{t_d}, \text{ and} \tag{7}$$

$$F_{Kx} = K(x_{pre} + 0.9L), \tag{8}$$

where t_d and L are the design requirements of displacement time and cylinder stroke, respectively. B, K, and x_{pre} corresponds to the viscous coefficient, spring constant, and spring preload, respectively, which are characteristics of applications with an eventual existence of a damper and/or a spring.

The friction forces produced by the cylinder ($F_{f.C}$) and the load ($F_{f.L}$) can be approximated by

$$F_{f.C} = p_S A_A \mu_d, \text{ and} \tag{9}$$

$$F_{f.L} = Mg\mathrm{Cos}(\alpha)\mu_{sli} \tag{10}$$

where the friction coefficient of the cylinder (μ_d) represents a percentage of the maximum force of the cylinder, with typical values ranging from 0.05 to 0.15 [18]. μ_{sli} is the friction coefficient of the surface where the object is sliding, and can be found in references such as [25] and [26].

Therefore, the total load force acting on the piston in a dynamic application is

$$F_L = Mg(\mathrm{Sen}(\alpha) + \mathrm{Cos}(\alpha)\mu_{sli}) + \frac{2BL\sqrt{0.9}}{t_d} + K(x_{pre} + 0.9L) + \frac{2ML}{t_d^2} \tag{11}$$

By considering (9) and (11), it is possible to rewrite (5), with $F_G(x) = 0$, in the form of pressure ratios, leading to an expression that can be used to determine the cylinder area:

$$A_A = \frac{F_L}{p_S\left(\left(\frac{p_A}{p_S}\right)_{opt} - r_A\left(\frac{p_0}{p_S}\right)\left(\frac{p_0}{p_B}\right)_{opt}^{-1} - \frac{p_0}{p_S}(1 - r_A) - \mu_d\right)} \tag{12}$$

where the pressure ratios $(p_A/p_S)_{opt}$ and $(p_0/p_B)_{opt}$ are defined by the operating point approach, corresponding to the optimum operating point defined by the maximum point of the vC curve in Fig. 3b.

However, due to the complexity of (3), it is not possible to use its derivative to analytically determine the $(p_A/p_S)_{opt}$. Therefore, it requires numerical calculations to iteratively define the pressure ratio that leads to the maximum vC value. Nevertheless, the characteristics of (3) allow for the determination of an approximate function.

In (3), its output is linearly dependent on the total load force (F_L), consequently, its value does not affect the $(p_A/p_S)_{opt}$ and it can be assumed equal to the unit. The average critical pressure ratio $\bar{b}$ is typically assumed as a constant parameter [16, 27]. Thus, a value of 0.3 can be considered for the dimensioning process.

The remaining parameters affecting $(p_A/p_S)_{opt}$ are the supply pressure (p_S), the area ratio (r_A), and the cylinder friction coefficient (μ_d). However, these values are often limited in pneumatic applications. Standardized cylinders have geometries defined by the ISO Standards 6432 [28] and 15,552 [29], thus defining the area ratio. The supply pressure in common industrial applications is constrained by commercially available pneumatic components, typically ranging between 3 and 13 bar_{abs}. The cylinder friction coefficient typically ranges from 0.05 to 0.15, but high-speed tasks results in higher friction forces. Therefore, a friction coefficient ranging from 0.05 to 0.50 should cover most practical applications.

Therefore, $(p_A/p_S)_{opt}$ can be numerically calculated using (3) for all possible combinations of discretized values of supply pressure, area ratio, and friction coefficients within common application ranges. The results of these calculations are compiled in Fig. 6.

To determine $(p_A/p_S)_{opt}$, one should select the supply pressure on the horizontal axis. For each supply pressure, a set of seven vertical bars is shown, with each bar corresponding to a value of area ratio, as indicated in the legend at the bottom left of the chart. Each bar is further divided into nine segments, each represented by a different color, indicating various friction coefficients as detailed in the legend. Once the relevant vertical bar and segment corresponding to the application is located, $(p_A/p_S)_{opt}$ can be read from the vertical axis.

Based on the results shown in Fig. 6, a non-linear regression can be performed, resulting in an approximated expression to calculate the $(p_A/p_S)_{opt}$, which is given by

$$\left(\frac{p_A}{p_S}\right)_{opt} = 0.571 + 0.2326 e^{\left(\frac{-0.2492 p_S}{1 \times 10^5}\right)} + 0.1678 r_A + 0.2483 \mu_d \qquad (13)$$

Comparing the results of (13) with the data shown in Fig. 6, 83% of the calculated values are within the ± 1% error margin. The maximum and minimum errors observed are 0.0208 and −0.0206, respectively, corresponding to maximum and minimum errors of 2.68% and −2.78% when compared with the numerical solution. This demonstrates that (13) is capable to of satisfactorily determining the $(p_A/p_S)_{opt}$ without the need to numerically solve (3).

Once $(p_A/p_S)_{opt}$ is determined, the ordinate value of the optimum operating point can be determined using the equation that describes the Op curve (1), that is

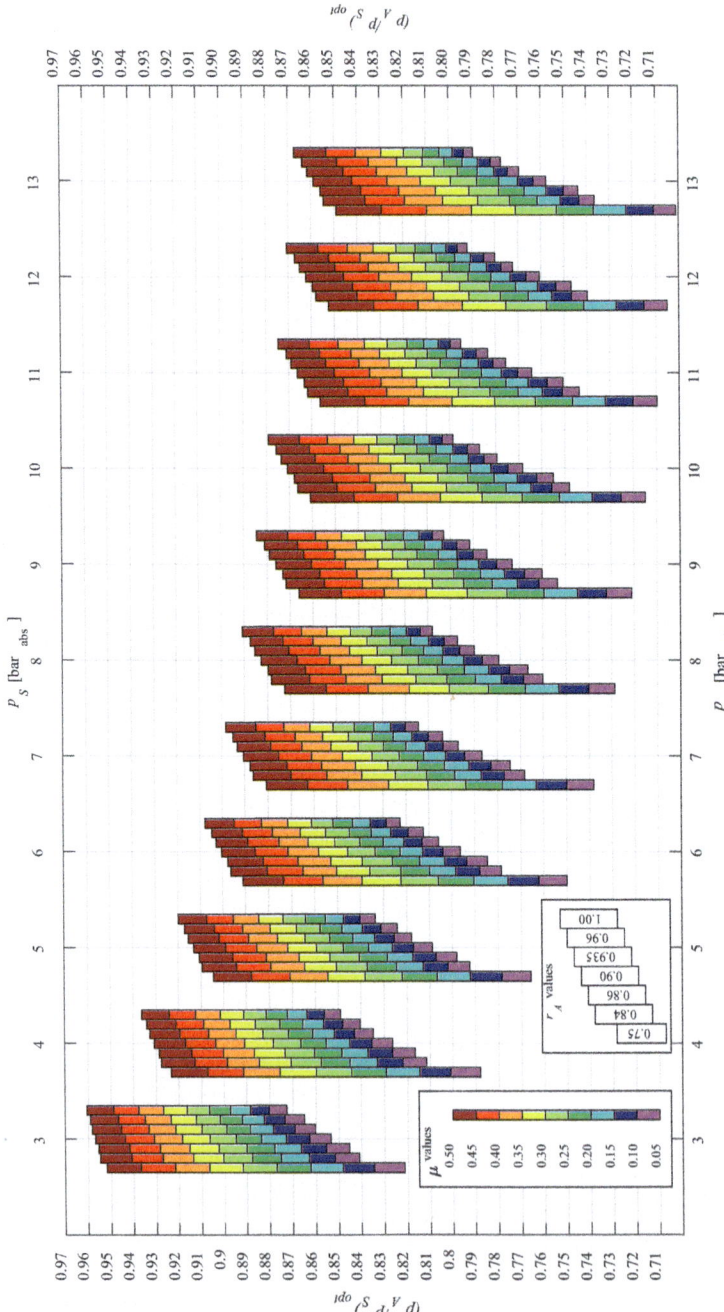

Fig. 6 $(p_A/p_S)_{opt}$ for an extending movement

$$\left(\frac{p_0}{p_B}\right)_{opt} = \bar{b} + \sqrt{\bar{b}^2 - 2\bar{b} + 1 + r_A^2 + r_A^2(2\bar{b} - 1)\left(\frac{p_A}{p_S}\right)_{opt}^{-2} - 2\bar{b}r_A^2\left(\frac{p_A}{p_S}\right)_{opt}^{-1}} \quad (14)$$

In summary, the cylinder area dimensioning process should be conducted using (12). The load force can be determined using (11), and the set of optimum pressure ratios will be provided by (13) and (14). If values for r_A and μ_d are not known in advance, initial estimates of 0.84 and 0.1, respectively, can be used. After selecting the cylinder, these values can be updated, and the calculation procedure can be reconducted to check for any potential need for cylinder re-dimensioning. It is also recommended to apply the proposed equations to recalculate the supply pressure in cases where the desired cylinder area does not match with commercially available options, ensuring that the system will operate in an optimum condition in terms of robustness and energy efficiency.

6 Assessment of the Proposed Approach

In this section, the results of simulations performed with various working conditions and different sizes of cylinder diameters are presented. These simulations show the impacts of oversized and undersized cylinders, as well as demonstrate the effectiveness of the proposed approach.

The objective is to replicate various scenarios that may occur during the operation of pneumatic drives, including different components and magnitudes of load forces, along with different design requirements of displacement time, cylinder stroke, and supply pressure. Table 2 presents the range of parameters considered for the performed simulations.

A total of 50 different working conditions were determined. Figure 7a presents a bar plot with the composition of the load force in terms of gravity, spring, viscous, and acceleration forces for each working condition. Simulations were conducted with four different commercially available cylinders in each working condition, beginning

Table 2 Design requirements parameter ranges

Parameter	Range
Load mass (M)	[0.5–665] kg
Spring constant (K)	[2–1,200] N/m
Damping coefficient (B)	[0–600] N.s/m
Orientation angle (α)	[0–90] °
Supply pressure (p_S)	[6–9]bar$_{abs}$
Displacement time (t_d)	[0.3–1.5] s
Stroke (L)	[0.05–0.5] m
Hose length (L_H)	[0.2–0.7] m

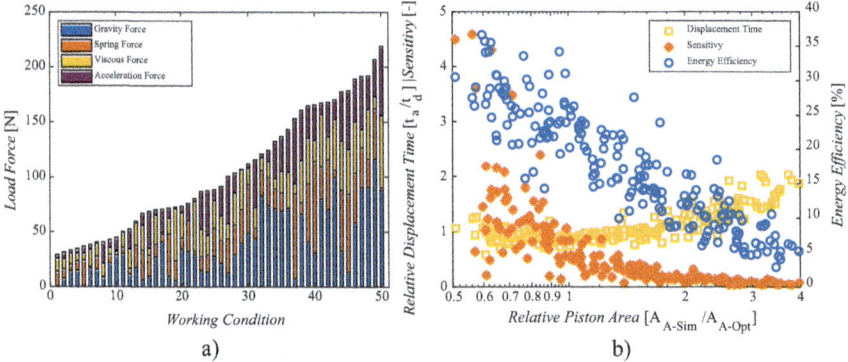

Fig. 7 Assessment of the proposed approach. (**a**) Characteristics of the working conditions; (**b**) Simulation results

with the smallest cylinder capable of completing the task and testing the subsequent standardized cylinder diameters.

The results shown in Fig. 7b are presented as a function of the relative piston area, that is, the ratio of the cylinder area used for the simulation to the optimum area, defined according to the approach shown in Sect. 5. Three characteristics were analyzed in the simulations: Energy efficiency, sensitivity, and the relative displacement time. The latter consists of the ratio of the actual displacement time (t_a) to the time specified on the design requirement (t_d).

Regarding energy efficiency, it is observed that it decreases as the relative piston area increases, with the highest efficiencies occurring with the smallest areas. In terms of sensitivity, simulations conducted with a relative area ratio smaller than 1 presented a significative variation in the drive's sensitivity. This indicates that under these circumstances, the behavior becomes unpredictable, possibly showing high changes on the displacement due to small changes in the load force. On the other hand, relative areas slightly larger than 1 reach the sensitivity plateau. Additionally, the relative displacement time shows a minimum value closer to the relative area equal to 1, with a larger variance for relative areas smaller than 1 and a clear uptrend for relative areas bigger than 1.

Therefore, it can be concluded that in the simulations where the cylinder area approaches the optimum area, the system presented the lowest displacement time, aligned with low sensitivity and good energy efficiency. Deviations from the optimum area resulted in drawbacks, including reduced energy efficiency, longer displacement times, or increased uncertainty in sensitivity. This highlights the effectiveness of the operating point approach for pneumatic cylinder dimensioning.

7 Conclusions

This paper presented a suitability analysis of the operating point approach as a dimensioning method for pneumatic applications operating with non-constant load forces. A classification of pneumatic applications between static and dynamic was proposed, and a statistical analysis was carried out to assess the impact of considering three distinct percentiles of the load force profiles as the dimensioning load forces, concluding that the 90th percentile aligns an enhanced energy efficiency while maintaining good robustness.

Analytical expressions were formulated to address different loading conditions considering the 90th percentile, and to overcome the need for numerical solvers for the operating point equations. This resulted in a set of equations easily applicable to size the actuation area of pneumatic drives for dynamic applications. Simulations performed with the derived equations evidenced the capability of the operating point method to align aspects of dynamic performance, robustness, and energy efficiency.

For the next research steps, the proposed dimensioning approach will be experimentally validated and compared with existent dimensioning methods, highlighting the advantages and drawbacks of each procedure. Moreover, guidelines for the determination of the cylinder friction coefficient will be developed, providing additional tools for an optimized selection of pneumatic drives.

Appendix: Derivation of the operating point equations

The operating point equations are derived from the governing equations of the system, namely the continuity Eq. (15), the ISO 6358 [30] mass flow rate Eq. (16), and the movement Eq. (17).

$$\frac{dp_i}{dt} = \frac{p_i}{T_i}\frac{dT_i}{dt} + \frac{1}{V_i}\left(q_{m_{in_i}}T_i R - q_{m_{out_i}}T_i R - p_i A_i \frac{dx}{dt}\right), \tag{15}$$

$$q_m = \begin{cases} p_1 C \rho_0 \sqrt{\frac{T_0}{T_1}} \sqrt{1 - \left(\frac{\frac{p_2}{p_1}-b}{1-b}\right)^2} & \text{for } \frac{p_2}{p_1} > b (Subsonic flow) \\ p_1 C \rho_0 \sqrt{\frac{T_0}{T_1}} & \text{for } \frac{p_2}{p_1} \leq b (Choked flow) \end{cases}, \text{ and} \tag{16}$$

$$p_A A_A - p_B A_B - p_0 A_H - F_{f.C} - MgSen(\alpha) - F_{f.L} - F_{Bv} - F_{Kx} - F_G(x) = M\frac{d^2 x}{dt^2} \tag{17}$$

In (15), the subscript i stands for either chamber A or chamber B, the subscripts *in* and *out* denote mass inflow our outflow of the chamber. T and R represent temperature and the ideal gas constant, respectively. In (16), ρ stands for air specific mass, and the subscripts 0, 1, and 2 represent ambient, upstream, and downstream conditions, respectively. The remaining variables and parameters have been previously defined on the paper.

Assuming a constant pressure and temperature, and considering that the piston velocity is the same for chambers A and B, (15) yields

$$v = \frac{q_{m_A}}{\rho_A A_A} = \frac{q_{m_B}}{\rho_B A_B} \tag{18}$$

Considering the ideal gas law, subsonic flow rate in each flow path, equal air temperature throughout the system, and assuming a symmetrical directional valve, it is possible to analytically correlate the chamber pressures in the form of pressure ratios by applying (16) in (15). This results on the first equation of the operating point approach, which defines the Op curve

$$\frac{p_0}{p_B} = \bar{b} + \sqrt{\bar{b}^2 - 2\bar{b} + 1 + r_A^2 + r_A^2(2\bar{b}-1)\left(\frac{p_A}{p_S}\right)^{-2} - 2\bar{b}r_A^2\left(\frac{p_A}{p_S}\right)^{-1}} \tag{19}$$

The second equation of the operating point approach is based on Newton's second law. It is assumed that all the loads acting against the movement (including inertial forces) are equal to F_L, the chamber areas are correlated by $r_A = A_B/A_A$, and the cylinder friction force is given by $F_{f.C} = \mu_d p_S A_A$. With these considerations, (17) is divided by $A_A p_S$, and its second term is multiplied by p_0/p_0. This leads to the equation that describes the Ld curve

$$\frac{p_A}{p_S} = \frac{F_L}{A_A p_S} + r_A \frac{p_0}{p_S}\left(\frac{p_0}{p_B}\right)^{-1} + \frac{p_0}{p_S}(1 - r_A) + \mu_d \tag{20}$$

The third equation of the operating point approach describes the steady state velocity of piston. The derivation is based on (18) and (16), with the mass flow rate to the chamber A of the cylinder and assuming $T_0 = T_1$, leading to

$$v_{ess} = \frac{p_S C p_0 \sqrt{1 - \left(\frac{\frac{p_A}{p_S} - b}{1 - b}\right)^2}}{\rho_A A_A} \tag{21}$$

The cylinder area (A_A) can be expressed as function of pressure ratios using (20) and the chamber B pressure ratio (p_0/p_B) is related to the chamber A pressure ratio (p_A/p_S) according to (19). Therefore, the ratio of steady state velocity (v_{ess}) by the sonic conductance of the directional valve (C) becomes a function of the chamber A pressure ratio, resulting on the vC curve used on the operating point approach.

$$\frac{v_{ess}}{C} = \frac{p_S p_0}{\left(\frac{p_A}{p_S}\right) F_L} \sqrt{1 - \left(\frac{\frac{p_A}{p_S} - \bar{b}}{1 - \bar{b}}\right)^2}$$

$$\left(\frac{p_A}{p_S} - \frac{p_0}{p_S}(1-r_A) - \mu_d - \frac{\frac{p_0}{p_S}}{\frac{\bar{b}}{r_A} + \sqrt{\frac{\bar{b}^2 - 2\bar{b}+1}{r_A^2} + 1 + \left(\frac{p_A}{p_S}\right)^{-1}\left(-2\bar{b} + \left(\frac{p_A}{p_S}\right)^{-1}(2\bar{b}-1)\right)}} \right) \tag{22}$$

References

1. Shi Y, Cai M, Xu W, Wang Y (2019) Methods to evaluate and measure power of pneumatic system and their applications. Chin J Mech Eng 32(1):1–11. https://doi.org/10.1186/s10033-019-0354-6
2. Energy efficiency in production in the drive and handling technology field. EnEffAH—Project Consortium. (2012). http://www.eneffah.de/EnEffAH_Broschuere_engl.pdf
3. Farias NM (2020) Reuse of compressed air in discrete pneumatic actuation systems using an intermediate reservoir (in Portuguese). M. S. Thesis, Federal University of Santa Catarina, Florianópolis, Brazil
4. Šešlija D, Šulc J, Reljić V (2013) Energy efficient pneumatic control scheme with recirculation of the used air. In: 2nd regional conference mechatronics in practice and education (MECHEDU2013). Subotica, Serbia
5. Raisch A, Hülsmann S, Sawodny O (2018) Saving energy by predictive supply air shutoff for pneumatic drives. In: 2018 European control conference. Limassol, Cyprus. https://doi.org/10.23919/ECC.2018.8550182
6. Raisch A, Sawodny O (2019) Modeling and analysis of pneumatic cushioning systems under energy-saving measures. IEEE Trans Autom Sci Eng 17(3):1388–1398. https://doi.org/10.1109/TASE.2019.2955806
7. Boyko V, Weber J (2024) Energy efficiency of pneumatic actuating systems with pressure-based air supply cut-off. Actuators 13(1):44. https://doi.org/10.3390/act13010044
8. Harris P, Nolan S, O'Donnell GE (2014) Energy optimisation of pneumatic actuator systems in manufacturing. J Clean Prod 72:35–45. https://doi.org/10.1016/j.jclepro.2014.03.011
9. Sarode RP, Vinchurkar SM (2023) An approach to recovering heat from the compressed air system based on waste heat recovery: a review. Energy Sources Part Recover Util Environ Eff 45(3):9465–9484
10. Vigolo V, Boyko V, Weber J, Valdiero AC, Negri VJD (2023) Online monitoring of pneumatic actuation system for energy efficiency and dynamic performance. In: ASME/BATH 2023 symposium on fluid power and motion control. Sarasota, USA. https://doi.org/10.1115/FPMC2023-111861
11. Festo (2017) Digital simplicity: new ways to increase productivity with "smart" systems. Esslingen, Germany: Festo AG & Co. KG
12. Mader (2017) Betriebsanleitung PCC blue sytem. Leinfelden-Echterdingen, Germany: Mader GmbH & Co. KG
13. Hepke J, Weber J (2013) Energy saving measures on pneumatic drive systems. In: The 13th Scandinavian international conference on fluid power (SICFP2013). Linköping, Sweden
14. Raisch A, Sawodny O (2019) Analysis and optimal sizing of pneumatic drive systems for handling tasks. Mechatronics 59:168–177. https://doi.org/10.1016/j.mechatronics.2019.04.003
15. Doll M, Neumann R, Sawodny O (2015) Dimensioning of pneumatic cylinders for motion tasks. Int J Fluid Power 16(1):11–24. https://doi.org/10.1080/14399776.2015.1012437
16. Boyko V, Hülsmann S, Weber J (2021) Comparative analysis of actuator dimensioning methods in pneumatics. In: BATH/ASME 2021 symposium on fluid power and motion control. Online conference. https://doi.org/10.1115/FPMC2021-68674

17. Rakova E, Weber J (2016) Exonomy analysis for the selection of the most cost-effective pneumatic drive solution. In: ASME 9th FPNI Ph.D symposium on fluid power. Florianópolis, Brazil. https://doi.org/10.1115/FPNI2016-1518
18. Vigolo V, De Negri VJ (2021) Sizing optimization of pneumatic actuation systems through operating point analysis. J Dyn Syst Meas Contr 143(5):051006. https://doi.org/10.1115/1.4049170
19. Bollmann A (1997) Fundamentals of pneutronic industrial automation—design of binary electropneumatic commands (in Portuguese). São Paulo, Brazil: ABHP
20. Hesse S (2001) 99 examples of pneumatic applications. Esslingen: Festo AG & Company
21. Prudente F (2000) Industrial pneumatic automation - theory and applications (in Portuguese). Grupo Gen-LTC, Rio de Janeiro
22. Fialho AB (2004) Pneumatic automation—design, dimensioning and analysis of circuits (in Portuguese), 2nd edn. São Paulo: Érica Ltda
23. Oliveira LG (2009) Determination of the operating points for pneumatic cylinder-valve set (in Portuguese). M.S. thesis, Federal University of Santa Catarina. Florianópolis, Brazil
24. Vigolo V (2018) Theoretical-experimental study to aid the sizing of pneumatic actuation systems (in Portuguese). M.S. thesis, Federal University of Santa Catarina, Florianópolis, Brazil
25. Avallone EA, Baumeister III T, Sadegh A (2007) Marks' standard handbook for mechanical engineers 11th edn. New York, USA: McGraw-Hill Education
26. Oberg E, Jones FD, Horton HL, Ryffel HH (2016) Machinery's handbook, 30th edn. Industrial Press, Norwal, USA
27. Beater P (2007) Pneumatic drives (system design, modelling and control). Springer, Berlin, Germany
28. ISO (2015) Pneumatic fluid power. Single rod cylinders, 1 000 kPa (10 bar) series, bores from 8mm to 25mm—Basic and mounting dimensions, ISO 6432
29. ISO (2004) Pneumatic fluid power. Cylinders with detachable mountings, 1 000 kPa (10 bar) series, bores from 32mm to 320mm—Basic, mounting and accessories dimensions, ISO 15552
30. ISO (2013) Pneumatic fluid power. Determination of flow rate characteristics of components using compressible fluids—Part 1: General rules and test methods for steady-state flow, ISO 6358-1

Open Access This chapter is licensed under the terms of the Creative Commons Attribution 4.0 International License (http://creativecommons.org/licenses/by/4.0/), which permits use, sharing, adaptation, distribution and reproduction in any medium or format, as long as you give appropriate credit to the original author(s) and the source, provide a link to the Creative Commons license and indicate if changes were made.

The images or other third party material in this chapter are included in the chapter's Creative Commons license, unless indicated otherwise in a credit line to the material. If material is not included in the chapter's Creative Commons license and your intended use is not permitted by statutory regulation or exceeds the permitted use, you will need to obtain permission directly from the copyright holder.

Experimental Analysis of Friction Forces of Hydraulic Rod Seals—Effect of Pressure, Sliding Speed, Sealing Type, and Different Rod Coatings

Kivi Knuuti and Olof Calonius

1 Introduction

Hydraulic cylinders are extremely widely used in mobile and industrial applications for creating linear movement. All these cylinders need tight sealing to keep the highly pressurized oil inside of the cylinders and to not let it leak past the piston. Tight sealing, however, always induces friction. Kühnlein et al. [1] have measured that the friction generated by the sealing can be as much as 4% of the nominal force of the cylinder, which means that frictional forces directly decrease its performance. Additionally, friction can cause stick–slip phenomenon and worsen positioning accuracy.

Friction is a difficult phenomenon to model and even after decades of research the current models for friction in lubricated contact typically either require a lot of experimental data from the system or do not provide very accurate results. Therefore, for creating better simulation models that can be used to predict sealing behavior and to develop better seals, more data and understanding of frictional behavior is needed.

Experimental measurements of friction of hydraulic seals have been conducted using whole differential cylinders. For example, Yanada et al. [2] and Tran et al. [3] used a whole hydraulic cylinder to measure and thereby verify their computational models for sealing friction. Kühnlein et al. [1, 4] have designed a system where constant pressure measurements are much easier to obtain. In more recent years, Pan et al. [5] have compared different sealing types in varying conditions. However, when investigating a whole cylinder, it is impossible to separate friction forces induced by piston and rod seals. Often frictional forces also have to be separated from the much higher pressure forces of the piston which causes uncertainties in the measurements.

K. Knuuti (✉) · O. Calonius
Aalto University, Espoo, Finland
e-mail: kivi.knuuti@aalto.fi

© The Author(s) 2025
L. Ericson and P. Krus (eds.), *Advancements in Fluid Power Technology: Sustainability, Electrification, and Digitalization*, Lecture Notes in Mechanical Engineering,
https://doi.org/10.1007/978-3-031-84505-5_6

Another approach is to measure friction using "non-cylinders". Angerhausen et al. [6] and Hess and Soom [7] have measured friction by pressing a strip of seal material against a rotating cylindrical surface. Muraki et al. [8] focused on stick–slip phenomenon of hydraulic seals by sliding a seal specimen on a plate.

The standard ISO 7986 [9] describes a test system for measuring hydraulic rod seal friction. Similar test rigs have been used, e.g., by Papatheodorou [10, 11] and Bhaumik et al. [12]. There are also papers studying rod seal friction with different test rigs. Bullock [13] has conducted wide-ranging measurements with constant speeds and sinusoidal movement in pressures under 80 bar. However, also his test system design requires computing out the pressure forces acting on the rod. Wang et al. [14] measured how rod surface roughness changes the frictional forces but only in pressures below 50 bar. Heipl and Murrenhoff [15] used a crank mechanism to achieve velocities up to almost 10 m/s. Crudu et al. [16] did simulations and measurements in pressures ranging from 50 to 200 bar with constant speeds ranging from 43 to 80 mm/s. Nikas et al. [17] have studied rectangular seals in very wide operating conditions: temperatures ranging from -54 to $+135$ °C and pressures ranging from 34 to 345 bar.

In this paper, friction forces of seals made of polyurethane and bronze-filled PTFE (polytetrafluoroethylene) are measured using three different rod coatings: standard hexavalent chromium, black nitride coating, and an alternative, more environmentally friendly, trivalent chromium coating. To the best of authors' knowledge, such comparison between mentioned rod coatings and sealing materials has not been conducted previously in the literature. The novelty of this research is the testing and comparison of these combinations with different relative velocities and at different pressures.

The measurements are conducted at constant pressures and with constant sliding speeds; dynamic changes in operating conditions are out of the scope of this article. The current test rig does not allow for measuring the direction dependency of the friction of the seals. Leakage and wear are also out of the scope of this article.

In chapter "Subsystem-Based Learning Control of Hydraulically Driven Nonlinear Rotary Actuators with Unknown Input Backlash" the test rig and the measurement plan are introduced. The results are shown in chapter "Evaluation of a Simple Method to Estimate the Shaft Torque in a Gerotor Pump" and discussed in chapter "Optimization-Based Energy Efficient Power Transmission Design Methodology Applied to a Compact Excavator". Chapter "Analysis of the Operating Point Method for Dimensioning of Pneumatic Drives Under Variable Loading Conditions" provides a conclusion.

2 Test Rig

The measurements in this paper are made using the test rig in Fig. 1. Section 2.1 is an overview of the physical construction of the test cylinder and the actuating cylinder. Section 2.2 goes over the seals and cylinder rods used for measurements and the measurement plan is introduced in Sect. 2.3.

2.1 Physical Construction

The test setup consists of a test cylinder and an actuating cylinder. A simplified schematic view is shown in Fig. 2 and a hydraulic schematic in Fig. 3. The actuating cylinder is custom-made to be low-friction to ensure smooth output. It does not have a piston seal and the rod seals are replaced with hydrostatic bearings. Low-friction PTFE seals act as scrapers to prevent excess leakage and air from getting into the cylinder. A proportional 4/4-type Bosch Rexroth 4WRPEH 6 C3B24L-2X/G24K0/A1M valve (Fig. 3, 2v5) is used for direction control.

The design of the test cylinder shares similarities with the one introduced in the standard ISO 7986 [9]. A section view of it is shown in Fig. 4. The design allows the

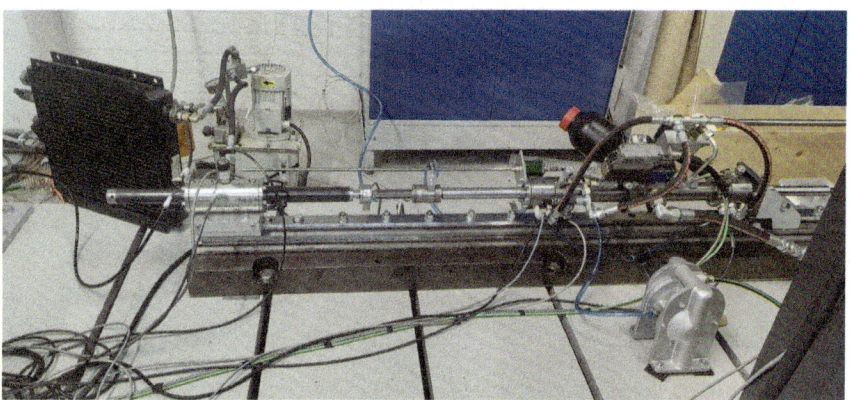

Fig. 1 Photo of the test rig

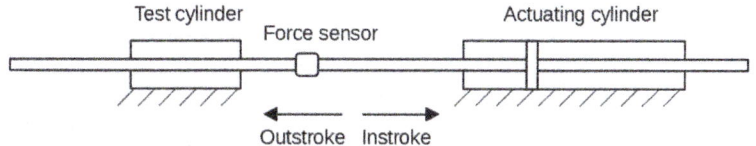

Fig. 2 Simplified schematic view of the test rig and the directions of instroke and outstroke

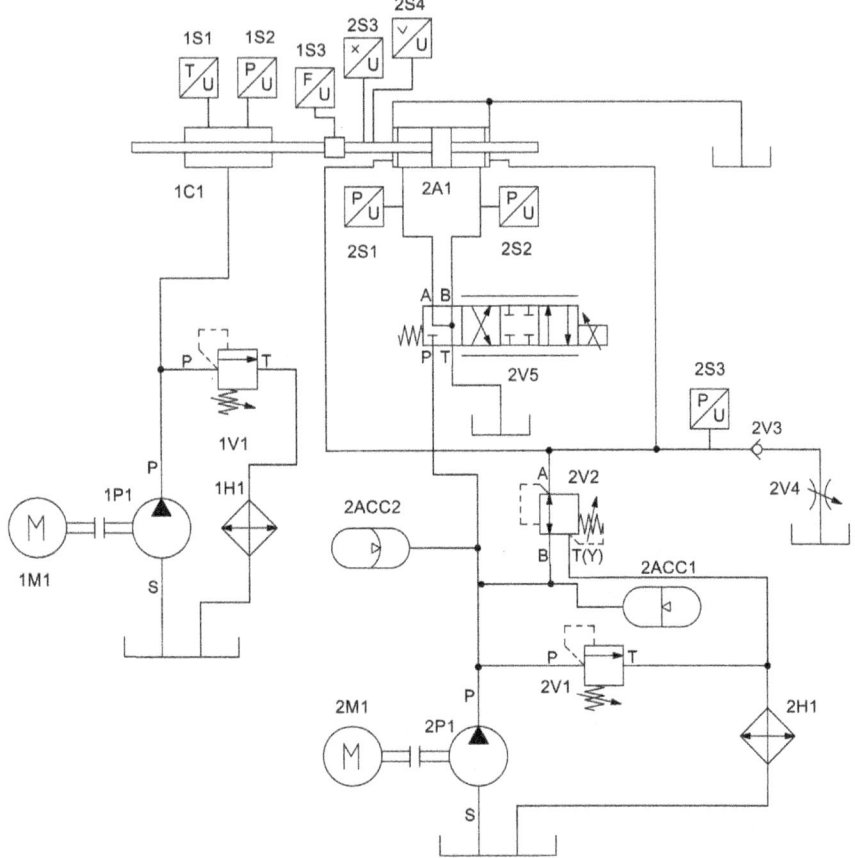

Fig. 3 Hydraulic circuit diagram of the test rig

seals to be kept in normal operating conditions (oil lubrication, pressure, temperature) while setting no limits for the stroke length. Additionally, there are no pressure forces acting on the rod in the axial direction. That means that the friction forces of the seals can be directly measured, and they do not have to be separated from pressure forces induced by a piston, for instance. The used force sensor is 5 kN HBM U2B (1S3).

Two opposing seals, one at each end of the test cylinder, means that the direction dependency of friction for the seals—which is to be expected [13]—cannot be measured using this setup and the friction value for a seal is a sum of its instroke and outstroke friction. In later chapters "instroke" and "outstroke" refer to the direction of movement of the actuating cylinder (Fig. 2).

The test seals and the guide rings are assembled in inserts machined to the manufacturer's specification. That makes it possible to use the same cylinder for testing different seals and allows for quick and easy exchange of seals.

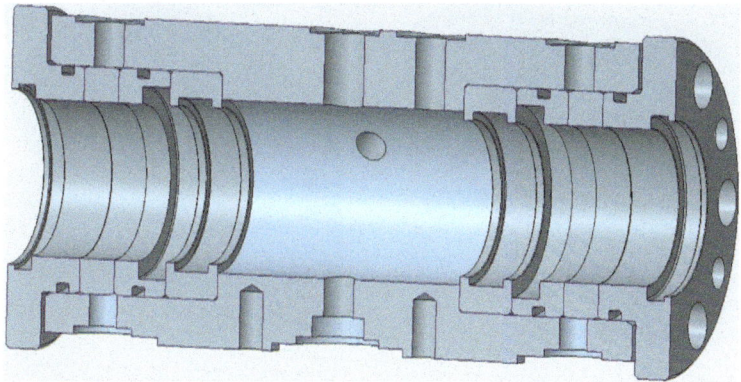

Fig. 4 Section view of the test cylinder

The test cylinder has drilled axial channels for temperature management and threaded holes for temperature and pressure sensors (Nokeval TRCP Pt100 (1S1), Hydac HDA 4746 (1S2), and Trafag 8251 (2S2, 2S3)).

Pressure inside of the test cylinder is generated by a hydraulic power pack (1M1 & 1P1, Fig. 2). The pressure is manually adjusted using a pressure relief valve 1V1. The viscosity of the oil used is ISO VG 32.

2.2 Test Seals and Rods

The measurements are made using two different seals: Merkel U-ring T20 made of polyurethane and Merkel Omegat OMS-MR made of PTFE-bronze compound. As is seen from the section views in Fig. 5, the cross-sectional profile of the seals is not similar. However, they represent a shape typical to their respective materials and therefore this study provides comparison data relevant in many applications.

The rods used are 40 mm in diameter and have a nominal length of 600 mm. Their specifications are in Table 1 and they represent what is currently commercially available. The device used for measuring the surface roughness values is Mahr MarSurf PS10.

2.3 Measurement Plan

In the standard ISO 7986, friction measurements are obtained in pressure levels of 63, 160, and 315 bar which are typical working pressures in applications [9]. However, the test cylinder used in this paper is not capable of reaching 315 bar. Therefore, 63

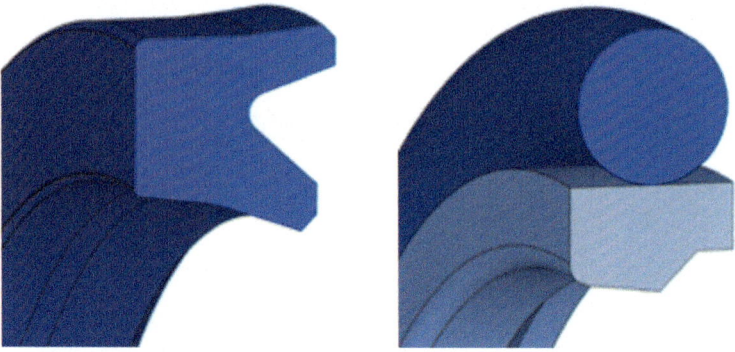

Fig. 5 Section views of the test seals. Merkel U-ring T20 (polyurethane) on the left and Merkel Omegat OMS-MR (PTFE-bronze + nitrile rubber O-ring) on the right

Table 1 Surface roughness properties of the test rods

Rod number	Rod 1		Rod 2		Rod 3	
Coating	Standard hexavalent chromium coating		Black nitride coating		Alternative coating	
Surface finish parameters before the tests	Parameter	Value (μm)	Parameter	Value (μm)	Parameter	Value (μm)
	Ra	0.140	Ra	0.411	Ra	0.251
	Rq	0.178	Rq	0.524	Rq	0.332
	Rz	1.037	Rz	3.778	Rz	2.704
	Rmax	1.350	Rmax	4.751	Rmax	3.802
	Rp	0.413	Rp	1.631	Rp	0.682
	Rt	1.350	Rt	5.839	Rt	3.993

and 160 bar are chosen from the standard and 110 bar, which is close to their average, is chosen as the third pressure point.

In the standard, data is collected in three constant sliding speeds: 50, 150, and 500 mm/s [9]. The hydraulic power pack available during the time of measurements was not capable of producing enough volume flow rate to reach 500 mm/s. Therefore, 500 mm/s was reduced to 450 mm/s. Additionally, to better capture the changes in friction as a function of velocity, an additional data point was added and data is collected in four constant velocities: 50, 150, 300, and 450 mm/s.

The speed profile for the constant speed measurements (Fig. 6) is made according to the standard. During one cycle, speed is constant for 80% of the time and for every direction change, a 10% ramp time is used. P-control of the rod position with velocity feedforward is used to implement the velocity command.

For each rod, one new set of both seals is used. Before taking the measurements, a 15-min-long run-in cycle is done. It is done at 110 bar pressure and 150 mm/s sliding speed, using the speed profile described above. After that, a measurement run lasting 150 s is done for each pressure and sliding speed combination.

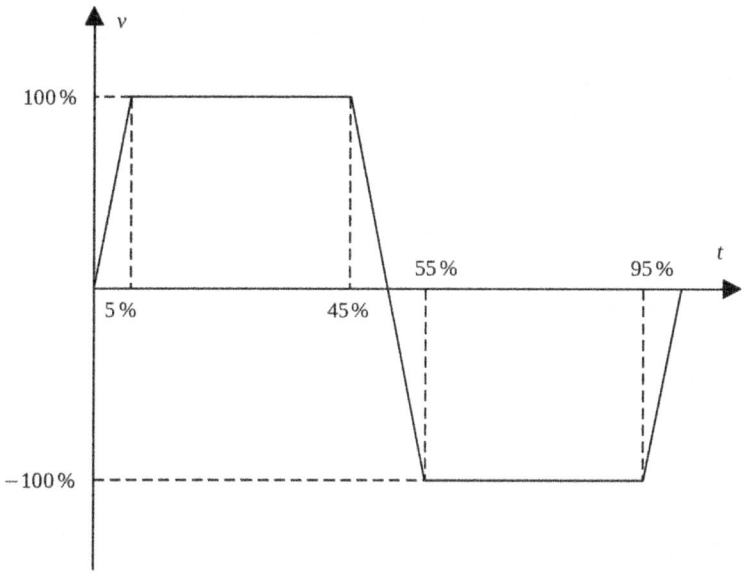

Fig. 6 Velocity command profile used in the tests. Positive velocity refers to outstroke motion as defined in Fig. 2

The temperature is held constant by controlling the temperature of water passing through the channels in the walls of the test cylinder. The controlling is done using Danfoss AVTA and AVTB thermostatic valves. The realized temperature during the tests was on average 42.4 °C.

Data acquisition is done using a National Instruments PCI 6259 card connected to a PC, where data is recorded to a MATLAB/Simulink environment at 100 Hz except for pressures and position, which are recorded at 10 Hz and 200 Hz, respectively. All signals are analog ± 10 V signals.

3 Results

A typical force–time graph of a measurement can be seen in Fig. 7. Especially with the polyurethane seal, some sticking happens in every direction change. Therefore, the friction changes into static friction for a brief moment and a peak in friction force is observed after each direction change. When the sliding speed settles to a constant value, friction stays rather constant. An arithmetic mean is taken from this rather constant section for four different cycles along the 150-s-long measurement. An example of such measurement interval is shown in the right graph in Fig. 7. An average of the four averages across the intervals is considered as the friction value for a specific measurement. The start and end points of each interval are chosen by hand.

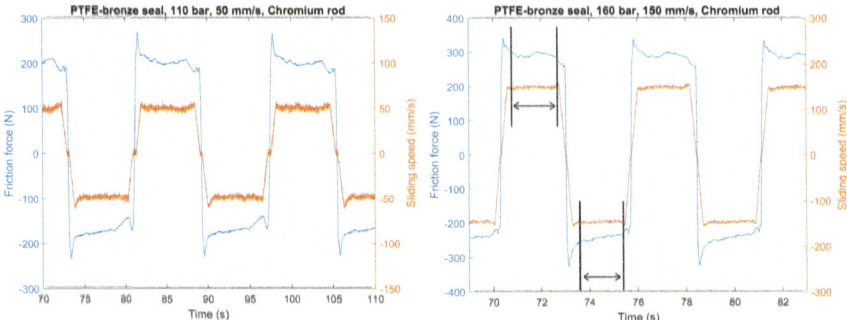

Fig. 7 Typical force–time graphs of measurements. One measurement interval for instroke and outstroke is shown in the right picture

The friction values are for an unknown reason not the same for outstroke and instroke. On average, the outstroke value of friction is 45 N higher than the instroke value. The discrepancy is of the same magnitude regardless of the friction force level but not constant enough to be treated as systematic error. Therefore, the average of absolute values of instroke and outstroke values is used.

The setup is carefully aligned using lasers but since the test rod is only supported by the soft seals and polymer wear bands, some misalignment can happen and cause the lack of symmetry of results.

3.1 Effect of Sealing Choice

Of the tested parameters, the sealing choice has the biggest effect on friction. The polyurethane seal always produces more friction force and on average the forces are 4.1 times higher. The greatest difference is 8.5 times higher force (50 mm/s, 63 bar, black nitride coating) and the smallest difference is 1.8 times higher force (450 mm/s, 160 bar, alternative coating).

No stick–slip behavior (as described, e.g., in [18]) was observed with the PTFE-bronze seals. However, with the polyurethane seal, on high pressures (110 bar and especially 160 bar) and low velocities (mainly on 50 mm/s but also on 150 mm/s with black nitride and alternative coating) different levels of stick–slip behavior ranging from minor to overwhelming was observed. Example force–time graphs are shown in Fig. 8. With the alternative coating rod, in 160 and 110 bar pressures and 50 mm/s sliding speed, no constant level of speed sliding was achieved, and these data points are omitted from calculations. Figure 8 the uneven movement also causes spikes to the force graph. Therefore, it is possible that values for high-pressure and low-speed tests are not fully representative.

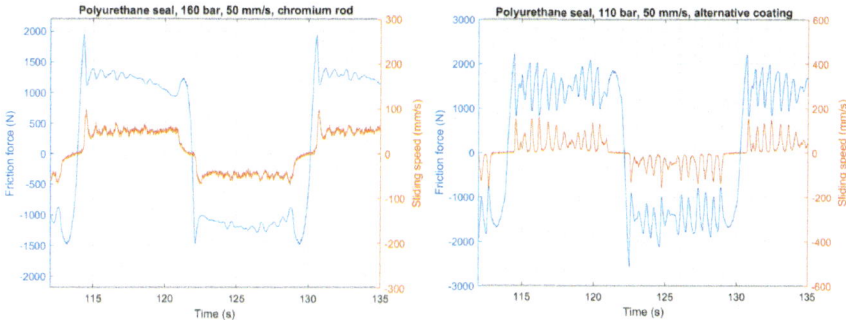

Fig. 8 Stick–slip observations. Minor on the left and major on the right

3.2 Effect of Choice of Rod Coating

The rod coating choice has a significant impact on the friction levels. In this chapter, comparisons between each coating are made for both sealing types.

Polyurethane sealing. For the polyurethane seal, the black nitride coating induces the most friction in all operating conditions, as seen in Fig. 9. On average, it generates 40.4% more friction than the standard chromium coating and 25.1% more than the alternative coating. The highest friction it generates is 1650 N at 50 mm/s sliding speed at 110 and 160 bar pressures and, at its lowest, it generates 836 N of force (450 mm/s, 63 bar).

The order between chromium and alternative coating changes depending on the sliding speed. On the slower speeds, 50 and 150 mm/s, chromium is the one with the least resistance to sliding with 49.2% less friction on average than the black nitride coating. On higher speeds, 300 and 450 mm/s, the alternative coating is the slipperiest one producing 39.8% less friction than the black nitride coating on average. The lowest friction measured with polyurethane seal was 535 N.

Interestingly, even though the black nitride coating generates the most friction, the alternative coating is more likely to induce stick–slip behavior. That could indicate that with this coating, ratio of the static and the dynamic friction is higher compared to the other tested coatings [8].

PTFE-bronze sealing. For the PTFE-bronze seal the chromium rod is the best choice in terms of friction in all measurement points. With it, the friction force values range from 120 to 315 N. In all but two measurement points, the alternative coating produces the most friction, on average 26.5% more than the chromium-coated rod. The maximum frictional force it creates is 380 N.

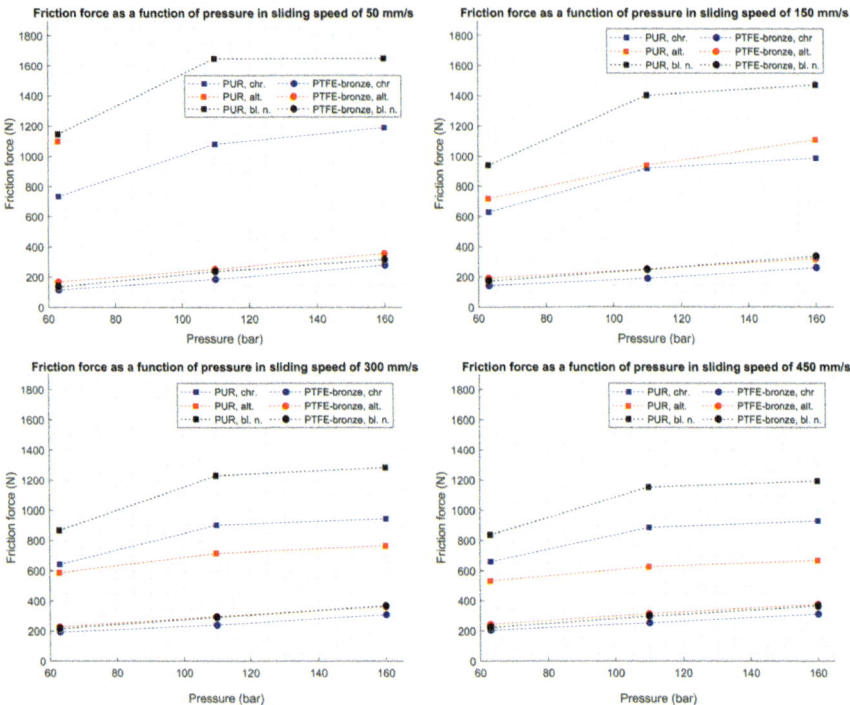

Fig. 9 Friction force as a function of pressure in different sliding speeds. Abbreviations: PUR = polyurethane sealing, "chr." = chromium coating, "alt." = alternative coating and "bl. n." = black nitride coating

3.3 Effect of Pressure

With all speeds, seals, and rod coatings, friction increases with increasing pressure. For the PTFE-bronze seal, the effect is more significant: when the pressure increases from 63 to 160 bar, friction increases 51.6–138.7% depending on sliding speed. The least increase is with 450 mm/s sliding speed, next least with 300 mm/s and the highest increase is with 50 mm/s.

For the polyurethane seal, increase in friction over the whole pressure range is 25.3–61.7% depending on sliding speed. The sliding speed has similar impact as with PTFE-bronze seals: the least increase is with 450 mm/s and the biggest is with 50 mm/s.

With the PTFE-bronze-seal, the change over the whole pressure range is rather linear: the change from 63 to 110 bar is of the same magnitude as from 110 to 160 bar (Fig. 9). With the polyurethane seal the increase of friction saturates: from 63 to 110 bar the change is on average 37.3% but from 110 to 160 bar only 6.5%. Papatheodorou and Igers have observed similar saturation behavior with polyurethane U-cup seals [11].

It is to be noted that the absolute values of the frictional forces and therefore the absolute change in them are higher for the polyurethane seal due to the higher starting values.

The increase in friction with increasing pressure is to be expected since the contact area between the seal and the rod increases with increasing pressure [11, 19]. Especially in lower sliding speeds, when the friction is in boundary or mixed lubrication regime [18], the bigger surface area means more touching surface asperities and therefore more friction. Additionally, higher pressures lead to lower film thickness with both PTFE [19] and polyurethane seals [16] which leads to more friction [16].

3.4 Effect of Sliding Speed

For the polyurethane seal, the Stribeck effect (as described, e.g., in [18]) can be observed in the force–velocity graphs (Fig. 10). The frictional force first decreases with increasing velocity. This is due to lubrication conditions improving in the contact between the seal and rod [18]. Based on theory, when velocity increases further, full lubrication regime is reached, and friction force should become dominated by the viscous friction of the oil film between the seal and the rod, and the friction should begin to increase linearly [20]. However, in these tests, no increase of friction is observed except for the 63 bar test with the chromium rod, suggesting that full lubrication regime is not reached with the sliding speeds used in this paper. Although, for example, Pan et al. [5] and Crudu et al. [16] have observed friction forces increasing with increasing velocity with polyurethane seals in constant velocities well under 100 mm/s.

For the PTFE-bronze seal, only in 160 bar pressure with chromium and alternative coating rod, the graph has (slight) characteristics of the Stribeck curve, i.e., first decreasing and then increasing friction for both instroke and outstroke. The decrease is small, 5.9% for the chromium rod (Fig. 10, bottom graph) and 9.8% for the alternative coating.

When increasing the sliding speed from 150 mm/s, friction forces increase with all rod coatings, which is to be expected when full lubrication regime is reached [20]. The friction increase, when sliding speed changes from 150 to 300 mm/s, is on average 19% and 3.9% when going from 300 to 450 mm/s.

3.5 Other Results

Leakage was not the object of this paper and thus was not measured, but with the black nitride coating and PTFE-bronze seal, a small amount (a drop for every two stokes) of leakage was observed. The leaking oil had very fine bronze-colored dust in it, presumably from the seal.

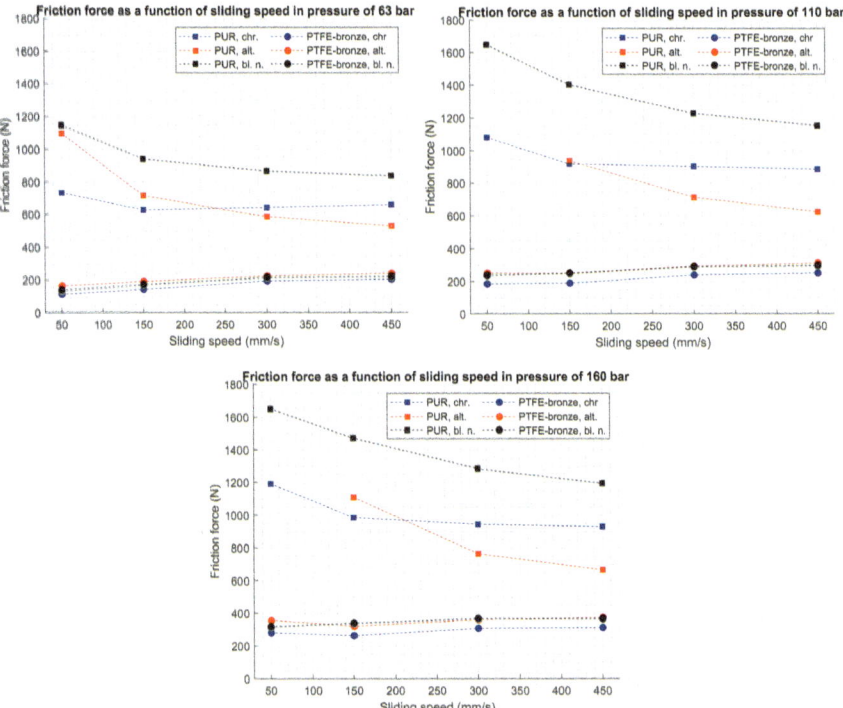

Fig. 10 Friction force as a function of sliding speed in different pressures. Abbreviations: PUR = polyurethane sealing, "chr." = chromium coating, "alt." = alternative coating and "bl. n." = black nitride coating

It was observed that especially during the runs with the polyurethane seal, the temperature in the test cylinder rose slightly (<1.5 °C). This is due to the friction heat accumulating on the contact interface [14].

4 Discussion

According to [19] and [11], PTFE-based seals can be considered as low-friction and these measurements support that claim. However, Papatheodorou and Igers [11] state that it holds true when the seals are new, but in worn-in condition and especially in high pressures and temperatures, the friction can increase a lot. In their measurements, Pan et al. [5] found polyurethane U-cup seals to generate less friction than PTFE seals. However, their measurements were done at much lower pressures (under 20 bar) which might affect the results. In tests conducted at very high sliding speeds (up to almost 10,000 mm/s), Heipl and Murrenhoff [15] found polyurethane seals to generate more friction compared to a PTFE seal.

The chromium rod is the only one filling the surface roughness requirements set by the seal manufacturer. For both seal types, Ra on the sliding surface should be 0.05–0.3 μm and Rmax should be less than 2.5 μm. The black nitride rod surpasses both of those numbers (Table 1). The Ra value of the alternative coating is within specifications, but the Rmax value is not. Increasing surface roughness causes greater frictional forces and leakage [14]. This means that some of the measured differences in friction can solely be from the greater surface roughness of the rods with black nitride and trivalent chromium coatings and not from the coating material itself. It should also be noted that the out-of-spec surface roughness can affect the lubrication conditions in the contact between the seal and the rod and cause the seal to work not as intended. Therefore, direct conclusions about comparing the frictional properties of these coating materials cannot be made based on this paper. However, these rods represent the current state of the art available commercially, making the comparison between the rods with as-purchased surface roughnesses justified.

The custom-made low-friction cylinder used in these tests proved to be quite problematic in regards to control. Possibly because the area of the piston is small compared to the diameter of the rod, or due to the lack of internal friction, stick–slip occurred easily and was hard to control. Different levels of stick–slip were observed with 50 and 150 mm/s runs at 110 and 160 bar pressures with polyurethane seals and that causes some uncertainties in the measured friction values. And some data points, such as 160 bar, 50 mm/s run with an alternative coating rod, could not be used.

The actuating cylinder can also be one of the reasons why the friction force and behavior of the system are different for the different stroke directions. One other possible reason could be the calibration of the force sensor. It was zeroed and calibrated only using tensile loads. However, since not only the friction force but also the behavior of the system is not symmetric, it is more likely that the reason has something to do with the construction of the system, for example, alignment.

Due to the lack of volume flow produced by the power pack, 500 mm/s sliding speed suggested by the standard had to be lowered to 450 mm/s. And still, with the most demanding run (450 mm/s, 160 bar, black nitride with polyurethane) the realized average sliding speed was 411 mm/s.

For the polyurethane seal, the speed value for the lowest point in the Stribeck curve, i.e., the point at which the decrease in friction stops, is quite high compared to literature. For example, in [5] the minimum point is roughly 15–30 mm/s depending on pressure for outstroke and under 20 mm/s for instroke with polyurethane U-cup seals. In [4], the minimum point is between 50 and 75 mm/s for compact seals (no material stated). In this paper, except for 63 bar tests with the chromium rod, no minimum was found, and friction forces were still decreasing at 450 mm/s. However, it is to be noted that the minimum can exist between 300 and 450 mm/s.

5 Conclusion

In this paper, friction forces of two different seal types were measured at three different pressure levels in four different constant sliding speeds using three different rod coatings. Especially the sealing material proved to have a significant effect on the friction. The friction force of a polyurethane seal was on average over 4 times higher than the friction force of a PTFE-bronze seal and therefore for an application where low friction is expected from the seal, a PTFE-bronze seal would be a better option.

Friction is seen to increase with increasing pressure with both seal types. With polyurethane, saturation behavior is observed, whereas with PTFE-bronze, the increase is more linear. Behavior as a function of sliding speed is not as uniform, however. With polyurethane seal, behavior is reminiscent of the Stribeck curve but with the PTFE-bronze seal friction increases with increasing velocity. That might, however, mean that the minimum of the Stribeck curve has been reached with lower sliding speeds and the whole measuring range is in full lubrication regime.

The surface roughness of the three tested rods differs so much that no clear conclusions can be made about comparing the frictional properties of the coating materials. However, the tested rods represent what is currently commercially available and this study shows that choice of rod coating has a significant effect on friction of the cylinder.

In further research, different rod coatings with similar surface roughness properties or seals with similar geometries could be compared. The comparisons could be expanded to dynamically changing velocities and breakout friction.

References

1. Kühnlein M et al (2011) Inner friction of large hydraulic differential cylinders. In: 12th Scandinavian international conference on fluid Power. Tampere, Finland, Tampere University of Technology
2. Yanada H, Sekikawa Y (2008) Modeling of dynanmic behaviors of friction. Mechatronics 18(7):330–339. ISSN: 0957-4158
3. Tran XB, Hafizah N, Yanada H (2012) Modeling of dynamic friction behaviors of hydraulic cylinders. Mechatronics 22(1):65–75. ISSN: 0957-4158
4. Kühnlein M et al Rapid parameterisation of a sealing friction model for hydraulic cylinders. In: 8th international fluid power conference. Dresden, Germany, Dresdner Verein zur Förderung der Fluidtechnik
5. Pan Q et al (2021) Experimental investigation of friction behaviors for double-acting hydraulic actuators with different reciprocating seals. Tribol Int 153:106506. ISSN: 0301-679X
6. Angerhausen J et al (2017) Influence of temperature and surface structure on the friction of dynamic hydraulic seals. In: The 10th JFPS international symposium on fluid power. Fukuoka, Japan, JFPS
7. Hess D, Soom A (1990) Friction at a lubricated line contact operating at oscillating sliding velocities. J Tribol 112(1):147–152. ISSN: 0742-4787
8. Muraki M, Kinbara E, Konishi T (2003) A laboratory simulation for stick-slip phenomena on the hydraulic cylinder of a construction machine. Tribol Int 36(10):739–744. ISSN: 0301-679X

9. ISO 7986: 1997(E) Standard test methods to assess the performance of seals used in oil hydraulic reciprocating applications. Geneva, Switzerland: International Organization for Standardization, 20 p
10. Papatheodorou T (2005) Influence of hard chrome plated rod surface treatments on sealing behavior of hydraulic rod seals. Seal Technol 2005(4):5–10
11. Papatheodorou T, Igers W (2011) Friction optimization on hydraulic piston rod seals. In: 52nd national conference on fluid power. Las Vegas, NV, USA. Milwaukee, WI, USA. National Fluid Power Assiociation, pp 833–840
12. Bhaumik S et al (2015) Investigation of friction in rectangular Nitrile-Butadiene Rubber (NBR) hydraulic rod seals for defence applications. J Mech Sci Technol 29(11):4793–4799. ISSN: 1738–494X
13. Bullock A (2010) Fundamental concepts associated with hydraulic seals for high bandwidth actuation. PhD Thesis. University of Bath, Department of Mechanical Engineering. Bath, Great Britain, 217 p
14. Wang B et al (2021) Experimental investigations on the effect of rod surface roughness on lubrication characteristics of a hydraulic O-ring seal. Tribol Int 156(106791):0301-679X
15. Heipl O, Murrenhoff H (2015) Friction of hydraulic rod seals at high velocities. Tribol Int 85:66–73. ISSN: 0301–679X
16. Crudu M et al A numerical and experimental friction analysis of reciprocating hydraulic 'U' rod seals. Proc Inst Mech Eng Part J J Eng Tribol 226(9):785–794. ISSN: 1350–6501
17. Nikas GK, Almond RV, Burridge G (2014) Experimental study of leakage and friction of rectangular, elastomeric hydraulic seals for reciprocating motion from −54 to + 135 °C and pressures from 3.4 to 34.5 MPa. Tribol Trans 57(5):846–865. ISSN: 1040–2004
18. Armstrong-Hélouvry B, Dupont P, De Wit CC (1994) A survey of models, analysis tools and compensation methods for the control of machines with friction. Automatica 30(7):1083–1138. ISSN: 0005–1098
19. Deaconescu A, Deaconescu T (2020) Tribological behavior of hydraulic cylinder coaxial sealing systems made from PTFE and PTFE compounds. Polymers 12(1). ISSN: 2073–4360
20. Márton L, Lantos B (2006) Identification and model-based compensation of striebeck friction. Acta Polytech Hung 3(3):45–58. ISSN: 1785–8860

Open Access This chapter is licensed under the terms of the Creative Commons Attribution 4.0 International License (http://creativecommons.org/licenses/by/4.0/), which permits use, sharing, adaptation, distribution and reproduction in any medium or format, as long as you give appropriate credit to the original author(s) and the source, provide a link to the Creative Commons license and indicate if changes were made.

The images or other third party material in this chapter are included in the chapter's Creative Commons license, unless indicated otherwise in a credit line to the material. If material is not included in the chapter's Creative Commons license and your intended use is not permitted by statutory regulation or exceeds the permitted use, you will need to obtain permission directly from the copyright holder.

Digital Twin-Based Classification of Hydraulic Excavator Duty Cycles in Road Construction

Johannes Sprink, Bernhard Sender, and Katharina Schmitz

1 Introduction

On modern construction sites, digitized machines have emerged as pivotal tools for enhancing productivity and efficiency. The integration of connectivity solutions and an array of sensors previously unavailable on mobile machines has unlocked a vast reservoir of data. While automation, condition monitoring, and fault detection are recognized applications of this data, there is a compelling opportunity to use process data to improve the productivity of machine use.

Hydraulic excavators, with their wide range of applications, are a particular focus of interest. Work processes are carried out manually, leading to a high dependence on the operator and a highly individualized approach. In road construction, understanding and classifying the duty cycles of hydraulic excavators can pave the way for significant advances in operational efficiency.

The aim of this paper is to introduce a methodology to classify duty cycles of excavators in road construction. For this purpose, a Digital Twin will be introduced and employed. The distinguished duty cycles can be used in different applications, one of which can be in construction management and the enhancement of existing publicly available geodata. In further work, this may be used to aggregate data from diverse sources to construct a Digital Twin of the environment, termed the Off-Highway Twin. This necessitates a comprehensive understanding of individual machines and their duty cycles before further derivations can be made. One other example of possible use for duty cycles within the digital twin could be the comparison of targets and actual volumes of materials handled on a construction site.

A concept for classifying duty cycles is proposed and its potential is shown through a comprehensive study. Also the concepts implications for automation, smart meter-

---

J. Sprink (✉) · B. Sender · K. Schmitz
RWTH Aachen University - Institute for Fluid Power Drives and Systems (ifas), Aachen, Germany
e-mail: johannes.sprink@ifas.rwth-aachen.de

ing capabilities, and overall impact on construction site operations are drawn out. The research question investigated in this paper is: How can the classification of work cycles positively interact with the creation of a digital twin?

This work is characterized by its use of real-world data, a deviation from the common use of laboratory experimental results in similar studies. It is characterized by its focus on hydraulic power transmissions, providing more in-depth analysis and practical insights. This focused approach distinguishes the research and makes it uniquely applicable to real-world working environments.

The paper begins with a discussion of the state of the research and what preliminary work has been done. Following this, the text describes the conducted case study and the methods used to collect data. Finally, the findings are discussed.

2 State of Research

Several existing studies merit discussion in the context of this paper. In this section, an overview of different approaches on classification of duty cycles and the use of digital twins will be given.

2.1 Standard Duty Cycles

Standard cycles have been published for various types of vehicles and machinery, including construction equipment, tractors, and automobiles. These cycles shall be applicable across different manufacturers ensuring comparability in performance evaluations. Standard cycles are used in particular with regard to energy efficiency. Especially for construction machinery, the definition of workloads is not universally applicable because of the many different cases of application the machines can be used in. For example an excavator could be used for grading the whole day, while on the next day it is only utilized for trench digging. These scenarios may also vary greatly from one company to the next. However, the machine must be suitable for all possible applications.

In his work *Fecke* describes various approaches for determining practical application scenarios. Special focus is placed on the evaluation and validation of mass data from fleet management systems from two manufacturers. The data set includes only the number of cycles and within the validation different load cycles have been investigated separately [1].

In [2], *Zarotti et al.* describe the energy use in a trench digging cycle. The comparison between a digging cycle and a simulated digging, where no earth is moved, is shown. It is outlined that the interactions between the machine and the surroundings may not be ignored for analysis, but also the adjustments of the machine itself have significant influence.

An architecture for the detection of cycles on a wheel loader and a tractor is presented by *Wünsche et al.* Within an Observer they use sensor data and Principal Component Analysis (PCA) to identify different work cycles. These are consecutive Y-cycles of a wheel loader and a tractor performing cycles of plowing, mowing, grubbering, and harrowing [3].

2.2 Machine Learning for Classification of Duty Cycles

In recent times, Artificial Intelligence has found applications in various domains. However, it is crucial to approach this topic with nuance. In our context, Machine Learning (ML) algorithms emerge as a viable option for pattern recognition. ML encompasses various classifiers, with Neural Networks being a noteworthy topic within this realm. It involves the use of algorithms that allow systems to learn patterns and make predictions based on data. While ML stands out as a robust approach, it is essential to explore alternatives. This might include rule-based systems, expert systems, and traditional programming methods, each with its own set of advantages and limitations compared to ML. Understanding these distinctions is pivotal in making informed decisions about the most suitable approach for a given application. In the following, different approaches in the context of classifying duty cycles on construction machines will be shown.

Keller shows how the cycles of a mini excavator's loading cycle can be identified. Different ML algorithms for classification like k-nearest neighbor and support vector machines are compared for this purpose. In a laboratory test, a simulated truck loading cycle was conducted and analyzed. Especially the command signals from the joysticks show a significant impact on the classification quality [4].

In [5] various ways of recognizing different cycles of an excavator and a wheel loader were investigated. In particular, the template-matching algorithm and the Hidden Markov algorithm were used for pattern recognition. Based on this, *Starke et al.* have shown that the performance of the model can be increased by using appropriate data preprocessing [6].

2.3 Digital Twins of Construction Machines

The concept of Digital Twins in the realm of construction machinery lacks a precise and universally agreed-upon definition, as it is employed with various objectives. One prominent application is Condition Monitoring for Predictive Maintenance. In this context, Digital Twins serve as dynamic virtual replicas of physical construction equipment, allowing real-time monitoring of their operational conditions. This multifaceted nature of the Digital Twin concept underscores its adaptability to diverse goals within the field, with a particular emphasis on enhancing predictive maintenance strategies through continuous monitoring and analysis.

In [7] a digital twin of a large mining excavator is created and utilized to calculate excavation trajectories. A framework for the creation of digital twins in large infrastructure projects is presented in [8]. *Rogage et al.* focus on monitoring earthwork operation to improve the productivity of earthwork equipment. They also present an extensive summary of related studies regarding digital twin technologies.

3 Case Study

As part of the presented research, a wheel excavator (shown in Fig. 1) on an active road construction site was equipped with sensors and measurement technology. The machine, a Caterpillar M317, is already equipped with many assistance systems, such as grade assistance and a payload display system. Due to a lack of access to the proprietary raw data of the installed sensors, the hydraulic drive train was equipped with own sensors, which are comparable to those on similar machines. The installed sensors include two pressure and temperature sensors at each cylinder and inertial measurement units (IMU) on the links of the excavator arm. More precise draw-wire sensors per cylinder were not installed due to the restrictions of mounting options and the rough conditions on the construction site. All sensors are connected via CAN bus to a device for recording and processing of the measurement data. For reference purposes, a visual camera was installed inside the cabin to periodically take pictures of the excavator's work space. In addition, a 5G router is installed, which allows remote access to the attached equipment and the measurement data in real time.

General machine data can also be retrieved via the machine's J1939 CAN interface. This interface provides the wheel-based speed of the vehicle, standardized information on fuel consumption, and similar data. Beyond that, a lot of in-depth information is usually only available on a proprietary basis. As much of the data on this bus is of undisclosed structure, it has not been included in the decoding.

Fig. 1 Excavator used for field study

3.1 Data Gathering

Over several weeks, numerous processes were recorded on the test vehicle described above during its productive use on a road construction site. This data is transmitted to a server in real time and can be stored for later analysis or used for immediate processing. Such processing includes the classification of specific work cycles, but this workflow also offers the possibility to abbreviate further information. For example, bucket loads can be calculated and environmental aspects such as soil type could be determined.

In preparation for the classification, the recorded data must be processed and additional characteristic values can be derived. Parts of the measurement values for an exemplary grading cycle are shown in Fig. 2. Within this time frame, a repetitive pattern can be spotted, which also has proven to be detected by ML algorithms in first trials. For further investigation, more features could be extracted from the raw data. Such features can be statistical values for specific time frames, but also data from follow-up calculations as for example shown in [9]. As the manual inspection of the data is a time-intensive work, an automated process helps saving valuable time.

Knowing the kinematics of the excavator arm and the angle information of the joints provided by the IMUs, the tool center point (TCP) can be determined. The position of the TCP can be used as a distinguishing feature, but also as a basis for manual analysis. A simplified visualization of the excavator's arm based on the positions is shown in Fig. 3. Even without a CAD model, an animated figure like this has proven to be a valuable resource for understanding the current state of the machine. It can even be used for manual labeling of duty cycles in the absence of

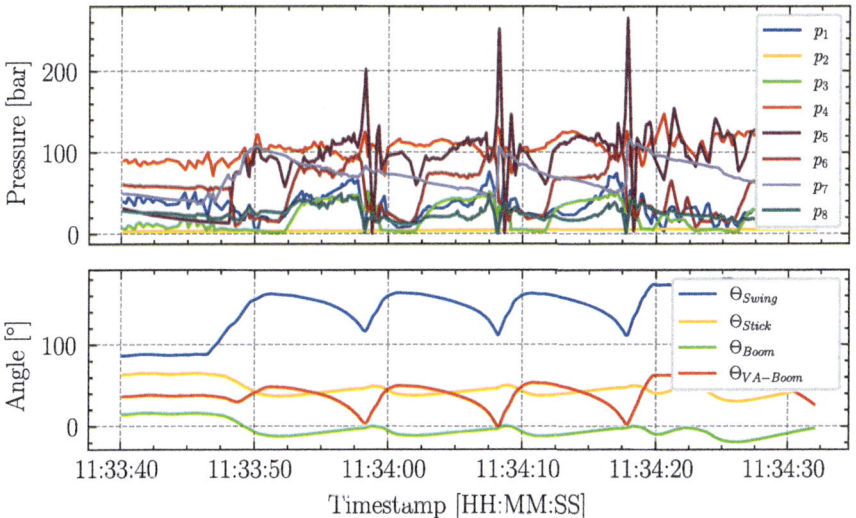

Fig. 2 Grading cycle: exemplary pressure and IMU data

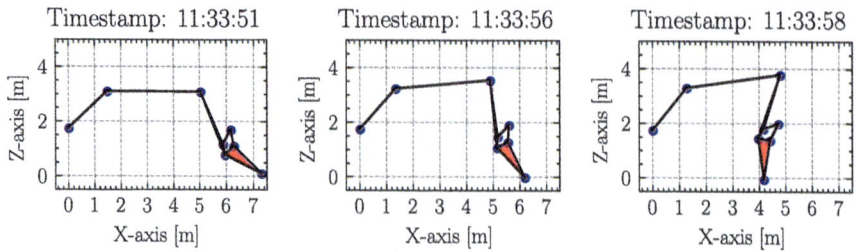

Fig. 3 Schematic visualization of the excavator arm with bucket at different times during a grading cycle

pictures, videos, or other data sources for reference. Graphs of raw data, as shown in Fig. 2, are interpretable only by experts with experience, so additional sources can be helpful in empowering more people for this activity.

3.2 Creation of Digital Twin

As part of creating the digital twin of a machine, the acquired data has to be stored and processed. In this study, some of the processing can already be performed on the measurement hardware on the machine before transmitting the raw data together with the derived features to an offsite server. This server acts as an data crate to archive data on the one hand but as an access point to the contained information on the other. The current state of the machine can be displayed on a dashboard in real time, but further analysis on the data can also be performed with a delay.

Such delayed actions may include daily fleet utilization logs, as well as diagnostic reports that can be used for predictive maintenance. Also, the classification of duty cycles does not necessarily need to be done in real time, but can be done at a later time. The classification of duty cycles is not the purpose itself but can be exploited for different purposes as stated in Sect. 1.

The digital twin of the machine presented here is therefore a collection of the respective measurement data and derived values. Interfaces for data transfer, renewed data access, data transfer, and triggering repeated processing are crucial here. In the future, this will be expanded to include a simulation model of the machine. A simulation can be used to implement virtual sensors, for example, to calculate measured values so that fewer real sensors are required or their accuracy can be improved. This significantly extends the functionality so that the concept described also lives up to the name digital twin.

The further research objective is to create a digital twin of the entire construction site. The digital representations of the individual construction machines are to be fed into this twin so that synergies can be used again and knowledge and benefits can be drawn from the overall data set. The integration of existing data sets is planned.

This means that geodatabases, for example, can be used as an additional data source on the one hand, but can also be enriched from the collected data on the other.

3.3 Data Classification

As shown in Sect. 2, different approaches for classification of duty cycles are available. These mostly rely on Machine Learning algorithms specialized on classification. For the use of supervised algorithms, the inspected data has to be labeled. Such labeling has been done for a part of the machine data. To do this, periods were manually labeled based on available camera images and significant data patterns. The manually classified tasks include

- Idle
- Travel
- Dig and Dump
- Grading
- Crane Operation.

Furthermore the change of attached tools or mixed cycles like a short travel phase during a Dig and Dump cycle has been observed and differentiated. Also undefined periods had to be included in the labels, because the work performed by the operators was not always fully comprehensible. Exemplary data of two days is shown in Fig. 4.

From this data, a huge drawback of the real-world data in comparison to the use of data generated in a laboratory environment becomes apparent, because if the data labels are already hard to determine, the automated classification will be even harder to achieve with a high level of accuracy.

For the input of ML algorithms, data sets of pressure, temperature, inclination sensor data, traction drive, GPS, and also derived characteristic values such as the TCP are used. The data is available with a frequency of 1 kHz. Such a high sampling rate has proven to be counterproductive for use as input for ML algorithms. Downscaling must therefore be carried out.

Another disadvantage of such heterogeneous data is the varying duration of the individual cycles. This is one of the reasons why it has not yet been possible to find a perfectly suitable algorithm for the classification of duty cycles in our scenario. However, the experience gained so far shows that promising candidates exist. The algorithms analyzed include different implementations of Random Forests and Neural Networks. For better performance, the use of an Autoencoder algorithm similar to the application shown in [10] is planned next. This type of algorithm can be used to reduce the high-dimensional input and provide a more concise input to the inspected algorithms.

With the knowledge of the specific task, certain statements about the environment can be made. For example after a grading task is finished, the ground in the machined area is flat at a specific level. As this is information that can otherwise be determined

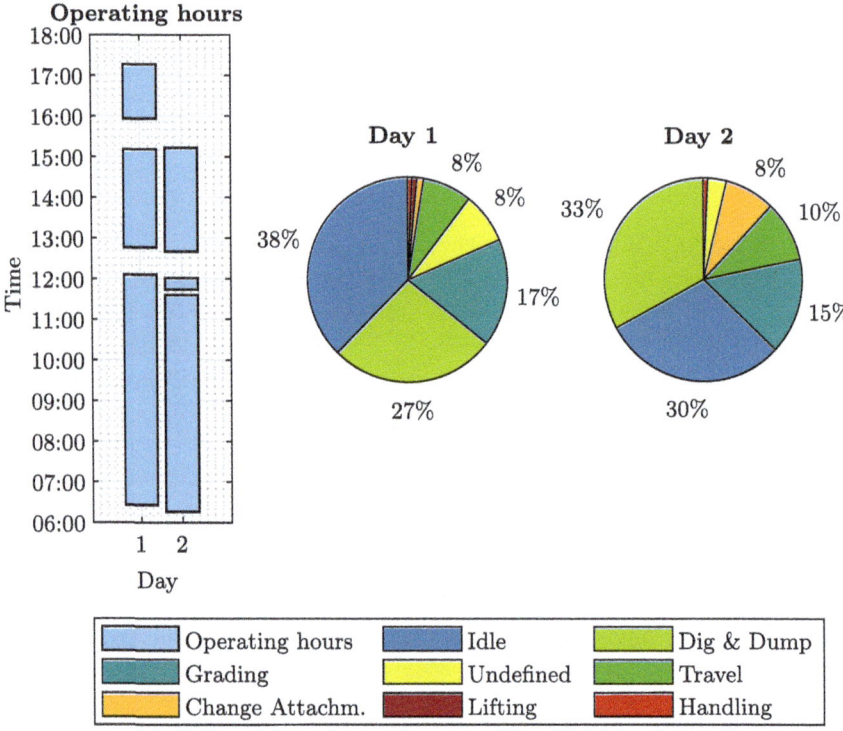

Fig. 4 Exemplary work cycles of two working days

by other means of measurement, there is a benefit in the processes on the construction site. Another example might be the accounting of materials which are moved by an excavator during a dig and dump cycle. Both scenarios are data points that shall be aggregated in the environment twin mentioned at the beginning and within this and be exposed and used by other applications.

The findings therefore show that the research question concerning the interaction between classification of work cycles and the creation of a digital twin can be answered positively as both benefit from each other. On the one hand, sensor data that is collected can be used for multiple purposes, not only the classification. On the other hand, the metadata that can be derived from the classification can improve and enlarge the application scenarios of the digital twin.

4 Conclusion and Outlook

A significant barrier to Off-Highway Twin technology is the prevalence of proprietary systems embedded in machines that inhibit interoperability and collaborative innovation. The inclination of manufacturers to promote closed ecosystems is understandable from an individual economic perspective. But sectors such as the IT industry and the implementation of ISOBUS in agricultural technology demonstrate the superior overall benefits of an open-source approach. As part of the MIC 4.0 initiative organized by the German Industry Association VDMA, several companies have pledged to work together to develop digital processes, communications, and data handling for construction machinery. However complete open-source frameworks foster a collaborative environment that encourages the sharing of knowledge and advances, ultimately resulting in more robust and adaptable solutions. As the Free Software Foundation aptly puts it, "think of *free* as in *free speech*, not as in *free beer*", emphasizing the freedom of ideas and collaboration that can coexist with a thriving economic model [11]. Our appeal to the industry is to make more information publicly accessible, encouraging the breakdown of proprietary barriers and nurturing a more inclusive and innovative ecosystem for the benefit of the entire industry in off-highway settings.

In the area of the Off-Highway Twin, our current progress has laid a foundation for further exploration and improvement. However, significant work remains to be done to unlock its full potential. Our immediate goals include in-depth analysis, enhancement of real-time capabilities, and refinement of the synergy between the Off-Highway Twin and simulation environments. Looking beyond the application of road construction, we aim to extend our framework to diverse scenarios, including fields such as underground construction, with a mid-term vision of integration into municipal operations. To achieve a comprehensive Off-Highway-Twin, we recognize the need to integrate different data sources to ensure a holistic representation of the physical environment. This roadmap outlines our commitment to advancing Off-Highway Twin technology by addressing these key areas, paving the way for a more versatile and effective tool in a wider range of commercial applications.

Funded by:

on the basis of a decision
by the German Bundestag

Acknowledgements This research is supported by the Federal Ministry for Digital and Transport of Germany within the research initiative mFUND (project number 19FS2036).

References

1. Fecke M (2018) Bewertung der Energieeffizienz von Baumaschinen mithilfe einer praxisnahen Lastzyklusentwicklung für einen In-Situ-Test. Dissertation, Wuppertal, Universität Wuppertal, 2018
2. Zarotti S, Paoluzzi R, Ganassi G, Terenzi F, Dardani P, Pietropaolo G (2009) Analysis of hydraulic excavator working cycle. In: Conference proceedings/11th European conference of ISTVS
3. Wünsche M, Mostaghim S, Schmeck H, Kautzmann T, Geimer M (2010) Organic computing in off-highway machines. In: Second international workshop on self-organizing architectures. Association for Computing Machinery (ACM), pp 51–58
4. Keller NJ, Ivantysynova M, Vacca A, Sun Y, Zhou Y, Lin G (2019) Classification of machine functions: a hydraulic excavator case study. In: Tampere University (ed) The sixteenth scandinavian international conference on fluid power (2019)
5. Technische Universität Dresden, Institut für Fluidtechnik, Professur für Baumaschinen: Methode zur online prozessmustererkennung für die ermittlung von kundenkollektiven an mobilen baumaschinen (processassist): Schlussbericht zu igf-vorhaben nr 18014 br : Berichtszeitraum: 01.01.2014 bis 31.12.2016. Frankfurt am Main (19 Apr 2017)
6. Starke M, Will F (2019) Automatic process pattern recognition for mobile machinery. In: Tampere University (ed) The sixteenth scandinavian international conference on fluid power
7. Fu T, Zhang T, Lv Y, Song X, Li G, Yue H (2023) Digital twin-based excavation trajectory generation of uncrewed excavators for autonomous mining. Autom Const 151:104–855
8. Rogage K, Mahamedi E, Brilakis I, Kassem M (2022) Beyond digital shadows: Digital twin used for monitoring earthwork operation in large infrastructure projects. AI Civil Eng 1(1):7
9. Makansi F, Schmitz K (2022) Data-driven condition monitoring of a hydraulic press using supervised learning and neural networks. Energies 15(17):6217
10. Brumand-Poor F, Makansi F, Schmitz K, Jiakun L (2022) Implementation of variational autoencoder for dimension reduction of a hydraulic system. In: [Global Fluid Power Symposium, GFPS, 2022-10-12 - 2022-10-14, Neapel, Germany]. https://publications.rwth-aachen.de/record/861032
11. Free Software Foundation: The free software definition. https://www.gnu.org/philosophy/free-sw.html.en

Open Access This chapter is licensed under the terms of the Creative Commons Attribution 4.0 International License (http://creativecommons.org/licenses/by/4.0/), which permits use, sharing, adaptation, distribution and reproduction in any medium or format, as long as you give appropriate credit to the original author(s) and the source, provide a link to the Creative Commons license and indicate if changes were made.

The images or other third party material in this chapter are included in the chapter's Creative Commons license, unless indicated otherwise in a credit line to the material. If material is not included in the chapter's Creative Commons license and your intended use is not permitted by statutory regulation or exceeds the permitted use, you will need to obtain permission directly from the copyright holder.

Design of a Cartesian Hybrid Force-Position Controller for a Hydraulic Manipulator

Lukas Bachmann, Paul Remde, and Jürgen Weber

1 Introduction

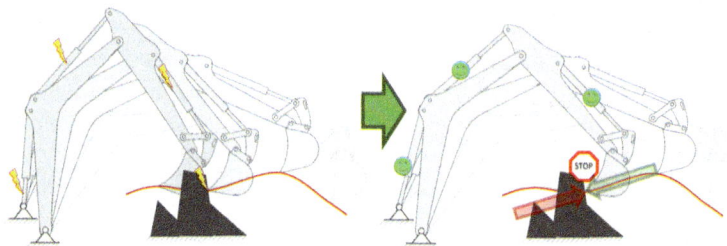

The automation of hydraulically actuated excavators for earthworks introduces multiple challenges in the fields of perception, navigation, guidance, and control. High process interaction forces, harsh and versatile environment conditions, cost pressure, and strongly nonlinear system behavior explain—among other reasons—why automation has not yet set its foot firmly in this domain.

Even though, a development toward automation is observable and some OEMs start incorporating assistance systems [3–5] for semi-autonomous functions, task planning, and on-site billing, those solutions are still costly, require many additional sensors and often still rely on the human operator.

Meanwhile, research projects at industry and universities focus on more complicated tasks, such as fully automated control of an excavator. This problem can be roughly divided into two aspects: The unconstrained movement through the air with or without load and the constrained movement through the ground while digging.

L. Bachmann · J. Weber (✉)
Institute of Mechatronic Engineering, Chair of Fluid-Mechatronic Systems, TUD Dresden University of Technology, Dresden, Germany
e-mail: fluidtronik@mailbox.tu-dresden.de

P. Remde
TUD Dresden University of Technology, Dresden, Germany

© The Author(s) 2025
L. Ericson and P. Krus (eds.), *Advancements in Fluid Power Technology: Sustainability, Electrification, and Digitalization*, Lecture Notes in Mechanical Engineering,
https://doi.org/10.1007/978-3-031-84505-5_8

In the first case, promising results can be achieved with position and velocity control as process interaction forces are minute. A team at ETH Zürich accomplished remarkable results in building a dry stone wall with a very sophisticated perception and guidance system. Visual localization, mapping, and motion planning made the process interaction observable and allowed for simple position control of the excavator tool [6]. But visual analysis of the task environment ranges from difficult to impossible when it comes to digging. During the excavation process variations in the soil characteristics or underground objects can lead to dangerously high interaction forces if pure position control is implemented. Depending on the situation, either sensitive underground infrastructure such as pipelines might be damaged or the machine might tip over.

A human operator reacts to such situations based on experience and reflexes, but a system with pure position control will not detect such situations without additional sensors. One approach to adapt the control strategy is by introducing force control. Process interaction forces are generally unknown and very difficult to estimate [7, 8] in advance, but a force controlled excavator can at least react to changing conditions and limit the resulting forces.

A solution without force control but rather with force monitoring is presented by Jud et al. [9]. They monitor the soil interaction forces during the position controlled digging phase and have developed an algorithm for autonomous excavation as well as an emergency routine in case underground objects are hit.

Combinations of force and position control can be distinguished in the fields of impedance control and hybrid force-position control.

Impedance Control is a concept where the controlled nonlinear manipulator is virtually transformed into a system with desired linear dynamics through a linearizing feedback. According to Hogan [10] neither force nor position is directly controlled but rather the acceleration requirements that are needed to achieve the desired system dynamic. Multiple publications have successfully implemented such controllers, like Ha et al. [11] and specifically Heinrichs et al. [12] obtained good results when adapting the impedance along the trajectory in order to reduce contact forces.

Very similar to the impedance control approach is the principle of hybrid force-position control. In fact, both of the two are equivalent in case of second-order impedance control and proportional force control [12, 13]. Apart from this specific case, in general, with a hybrid force-position control the task space of the TCP is divided into position controlled dimensions and force controlled dimensions. An early proposal for this controller exists by Raibert et al. [1] and since then numerous researchers have been working on this [14–18]. Both position and force controller work in parallel and a selection matrix decides which dimensions in task space are used to compute the required accelerations.

Generally, a computed torque controller (CTC) [2, p. 174 et sqq.], [17, p. 185 et sqq.] is implemented to achieve those accelerations of the TCP by corresponding torques in joint space. The CTC is also used to compensate for inertia, Coriolis force, and conservative forces such as gravitational force. However, with the considered system being actuated by prismatic, hydraulic cylinders, those torque requirements have to be converted into corresponding forces or cylinder chamber pressures, respectively. The difficulty lies in the fact, while in task space position and force control are strictly separated, in joint space force and velocity control are required for all actuators.

To the authors' knowledge, all other researchers who have worked on hybrid force-position or impedance control in hydraulically actuated systems used simple servo valves with common metering and simply added the force control and position control law for each valve [11, 16]. This approach is strongly influenced by electro-mechanical drives where the only available control input is the motor current. However, in hydraulics, systems are inherently suited for simultaneous velocity, i.e., flow, and pressure control. Furthermore, there is no necessity for additional force sensors if cylinder chamber pressures are measured.[1]

There exist various hydraulic system structures allowing for individual flow and pressure control. One of which is the principle of independent metering valve control (IMVC) [20, 21] which has been the focus of several scientific publications at the institute of fluid-mechatronic systems at TU Dresden [22–25]. Specifically, the realized valve control regime is a modification of the load pressure feedback strategy [23, 25] where the required torques from the CTC are used to calculate the required pressure trajectories for a nonlinear feed forward valve control based on inverse flow mapping.

2 System

Manipulator Kinematics
Kinematic Simplifications For simplicity, and in accordance with our lab demonstrator (See Figs. 1 and 13), the slewing gear is neglected. By introducing the task space coordinates x and y for the Cartesian position of the TCP in the global frame $\{0\}$ and φ_{TCP} as the cutting angle of the bucket, i.e., TCP, the manipulators Jacobian has full rank for three parallel revolute joints.

Figure 1 shows the general coordinates in joint space $\boldsymbol{\theta} = [\theta_1, \theta_2, \theta_3]^T$ and task space $\boldsymbol{x} = [x, y, \varphi_{TCP}]^T$ of the excavator arm. The z-axes of all coordinate frames $\{i\}$ with $i = 0, \ldots, 4$ points out of the picture plane, hence the angles θ_2, θ_3 and φ_{TCP} have negative sign in the given configuration.

Given the definition of the coordinate frames one can derive the Denavit-Hartenberg Parameters[26] in Table 1. Those allow for a very structured development of the systems' forward kinematics, i.e., the mapping from joint space variables to task space variables in the form of holonomic transformation matrices

---

[1] To be precise, pressure induced and actual cylinder force are not equal due to non-conservative forces such as friction, but friction models can be introduced to reduce this effect [19]. This publication, however, will not consider friction models in the controller design.

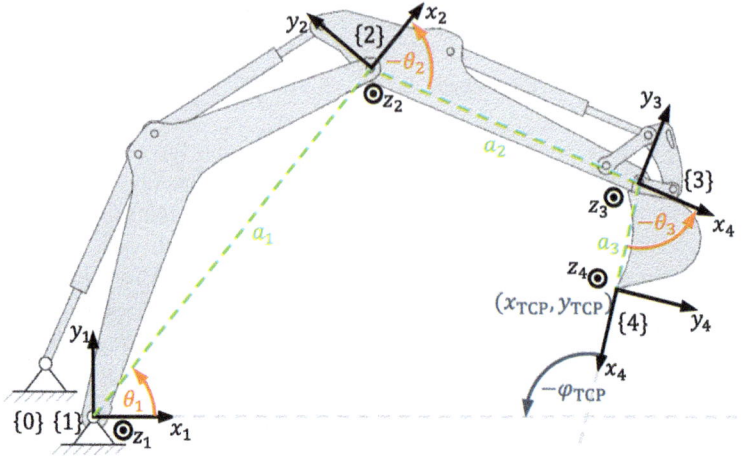

Fig. 1 Planar excavator model with generalized coordinates in JS and TCP coordinates in TS

Table 1 Denavit-Hartenberg Parameters of the planar excavator manipulator[26]

Joint [i]	θ_i [rad]	d_i [mm]	a_i [mm]	α_i [rad]
1	θ_1	0	2050	0
2	θ_2	0	1350	0
3	θ_3	0	502	0

$$_i^{i-1}T = \begin{bmatrix} \cos\theta_i & -\sin\theta_i \cos\alpha_i & \sin\theta_i \sin\alpha_i & a_i \cos\theta_i \\ \sin\theta_i & \cos\theta_i \cos\alpha_i & -\cos\theta_i \sin\alpha_i & a_i \cos\theta_i \\ 0 & \sin\alpha_i & \cos\alpha_i & d_i \\ 0 & 0 & 0 & 1 \end{bmatrix} \quad (1)$$

and

$$_N^0 T = \prod_{i=0}^{N} {}_i^{i-1}T. \quad (2)$$

Derivation leads to the differential mapping between joint space and task space and yields the analytic Jacobian $\boldsymbol{J}_A$ with

$$\dot{\boldsymbol{x}} = \boldsymbol{J}_A(\boldsymbol{\theta})\dot{\boldsymbol{\theta}} \quad \text{and} \quad \ddot{\boldsymbol{x}} = \dot{\boldsymbol{J}}_A(\boldsymbol{\theta}, \dot{\boldsymbol{\theta}})\dot{\boldsymbol{\theta}}. \quad (3)$$

The transposed Jacobian matrix also yields the relation between joint space torques and task space forces at the TCP by

$$\boldsymbol{\tau} = \boldsymbol{J}_A^T(\boldsymbol{\theta}) \cdot \boldsymbol{f}. \quad (4)$$

Manipulator Dynamics

Dynamic Simplifications The inertial properties of all body elements and cylinders were calculated based on CAD data. Cylinder friction is modeled as the sum of Coulomb and viscous friction with the individual parameters determined experimentally. This was done by moving the cylinders of the excavator in the lab at a constant speed and estimating the acting cylinder forces based on pressure measurements (See Table 2). Inertial, Coriolis, and conservative forces were subtracted based on the calculated mass and configuration properties based on Eq. 5. However, due to the fact, that the mass parameters were only estimated based on CAD data, deviations from the actual mass and friction parameters will occur. Even though, this proceeding will lead to a more realistic modeling of the dynamic system.

Further simplifications are pressure independent friction forces, frictionless joints—friction effects of the joints are included in the cylinder friction—and no bearing clearance. The general equation for a serial manipulator is given by

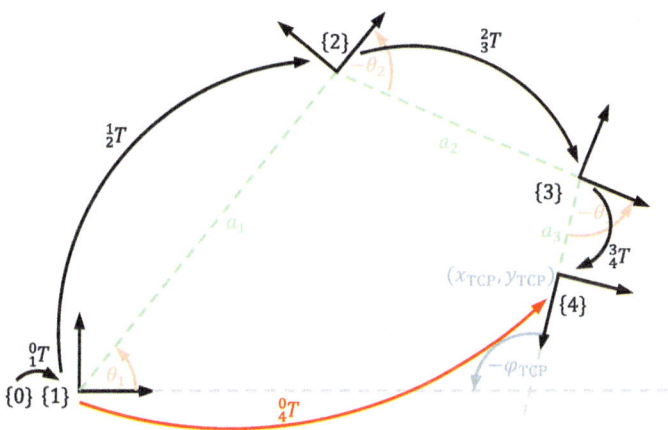

Fig. 2 Holonomic transformations in between joints and between {0} and {4} or {TCP}, respectively

Table 2 Mass and friction parameters

	Parameter		Body element		
			Boom	Stick	Bucket
CAD	m_{body}	/[kg]	100.95	62.98	51.72
	m_{cyl}	/[kg]	45.3	36.6	30
	m_{add} †	/[kg]			8.8/10.7
EXP	F_C	/[N]	250	500	250
	k_V	/[N s/m]	21000	13940	17000

† Additional kinematic elements, i.e., swing and coupler of four-link joint

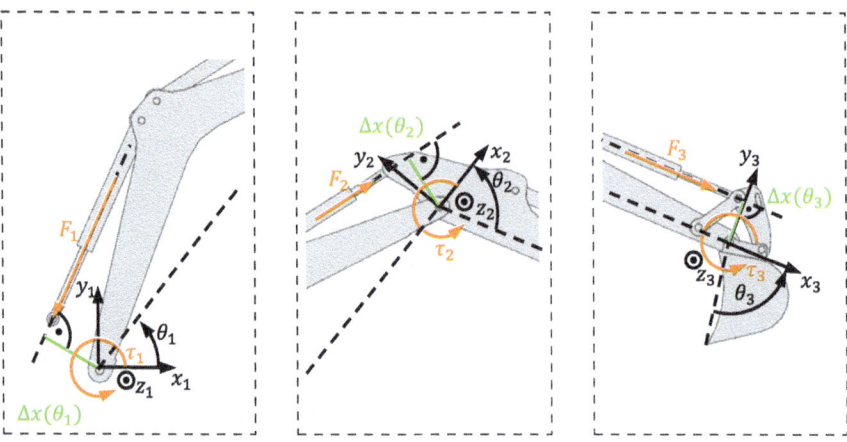

Fig. 3 Torque to cylinder force transformation

$$\underbrace{M(\theta) \cdot \ddot{\theta}}_{\text{inertia forces}} + \underbrace{C(\theta, \dot{\theta}) \cdot \dot{\theta}}_{\text{Coriolis forces}} + \underbrace{g(\theta)}_{\substack{\text{conservative} \\ \text{forces}}} + \underbrace{k_V \cdot \dot{\theta} + F_C \cdot \text{sign}(\dot{\theta}) = \tau - J_A^T \cdot f}_{\text{non-conservative forces}} \quad (5)$$

with τ being the torques required at the manipulator joints to compensate for external forces f and gravity g and to enable a particular movement specified by θ and its derivatives.

The joint torques τ are provided by linear acting hydraulic cylinders and the transformation between cylinder forces and joint torques is computed based on the angle dependent distance $\Delta x(\theta_i)$ by

$$f = \Delta x(\theta) \times \tau \quad \rightarrow \quad F_i = \tau_i / \Delta x(\theta_i) | i = 1, \ldots, 3 \quad (6)$$

and it is assumed that there are no transformation losses between the delivered cylinder force and received joint torque. For the sake of clarity and lack of space in this publication, the trigonometric computations of $\Delta x(\theta_i)$ are omitted (Fig. 3).

Hydraulic System

Hydraulic Simplifications The hydraulic system incorporates one common constant pressure source, temperature independent oil properties, and leakage free cylinders. Each cylinder is connected to a 4/3 switching valve for speed reversal and two 2/2 proportional seat valves for control with a linear spool position to flow and turbulent pressure to flow characteristic:

$$Q(y, \Delta p) = K_{Qy} \cdot y \cdot \text{sign}(\Delta p) \sqrt{2 \cdot |\Delta p| / \rho}. \quad (7)$$

The dynamic of the switching valve is ignored and the control valves have a second-order dynamic behavior with an opening time of 100 ms, no hysteresis, zero

Fig. 4 Scheme of the hydraulic system for one actuator

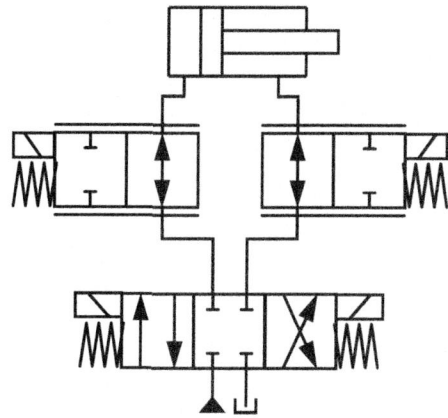

overlap, and no leakage. Pipes are assumed to be rigid and inflexible and pressure losses along pipes are neglected.

3 Controller Design

The controller design incorporates three controller elements. The *Hybrid Force-Position Controller (HFPC)* is the high-level controller which specifies the position and force control directions and control laws. A *Computed Torque Controller (CTC)* is used in order to calculate the cylinder forces for trajectory tracking with compensation of inertia, Coriolis, and gravitational forces. The given force and velocity requirements are the inputs for the *Independent Metering Valve Controller (IMVC)* which tracks the necessary chamber pressures and flows by controlling the inlet and outlet cylinder valves.

Hybrid Force-Position Controller

The main novelty of the implemented HFPC compared to the original version of Raibert et al. [1] is the adaption toward arbitrary, time-varying directions for the position and force controlled directions. One approach leading in this direction can already be observed in Lewis [17, p. 489, 497 et sqq.]. The goal is not to be limited in the original base frame coordinates for tangent space and normal space selection for force or position control. There are two possibilities to achieve this. One is to use a coordinate transformation, e.g., for tracking a circle to use polar coordinates and compute the Jacobian matrices based on those coordinates. The disadvantage is that this limits the movement on concentric circles and requires complicated reformulating for more complex trajectories.

The other, more general approach is to use holonomic transformations of the base coordinate frame so that the base frames orientation follows the tangent of the

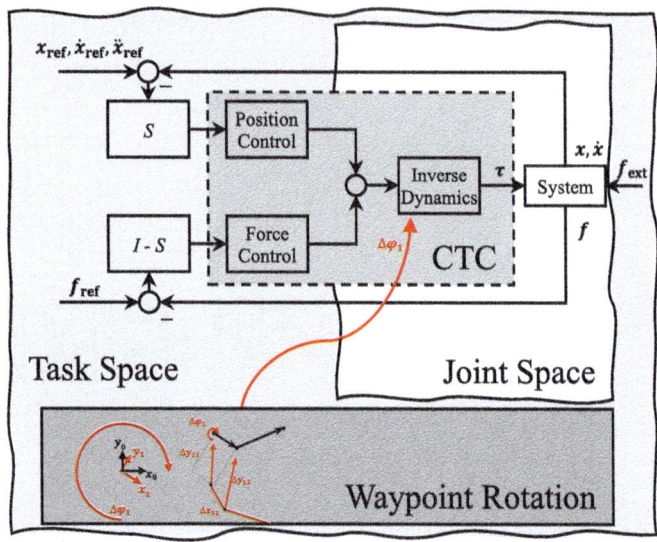

Fig. 5 Structure of the Hybrid Controller

trajectory and the problem of normal and tangent space selection is reduced to the standard case and can be performed by the selection matrix S (See Fig. 5).

In general, this would require recomputing the Jacobian matrices at every transformation of the base frame because of the introduction of an additional time-varying transformation into the chain of holonomic transformations in Fig. 2. With certain assumptions and requirements for the kinematic structure though, this can be avoided. If the base frame is located inside a joint or group of concurrent joints which are able to follow the rotation transformation of the base frame introduced by the trajectory tangent, then a simple offset of the joint angles defined by the frame rotation is sufficient to alter the Jacobian matrices.

Apart from the forward and inverse kinematics, the waypoints of the trajectory have to be transformed likewise by counter-rotating them in the base frame in order to relocate them correctly in the "un-rotated" global frame (See Fig. 6).

Computed Torque Controller

The CTC plays an important central role in the control structure as it is used to compute the required torques in joint space in order to complete the specified force and position control task in the task space. The superimposed HFPC interacts with the CTC by altering the Jacobian matrices depending on the aforementioned rotation of the base frame and by selecting the reference position and force trajectories.

In the current implementation investigated in this paper, we used a simple feedforward path with no control feedback for the force control and a PD-controller for position control. For the compensation of gravitational and inertial forces in the CTC the mass, Coriolis, and gravitation matrix have to be computed. Typically,

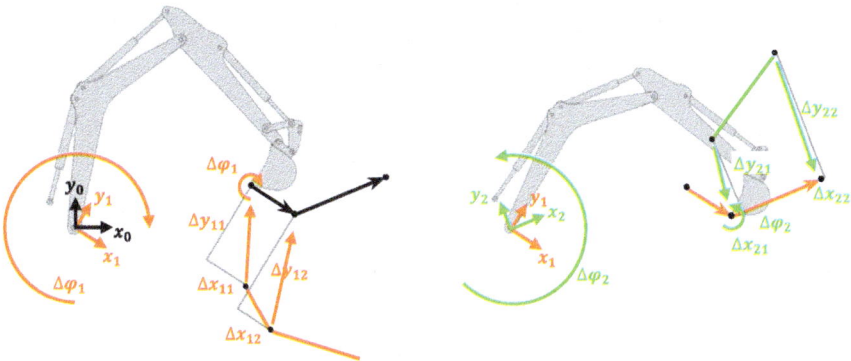

Fig. 6 Waypoint transformation with rotation of the base frame relative to the global frame

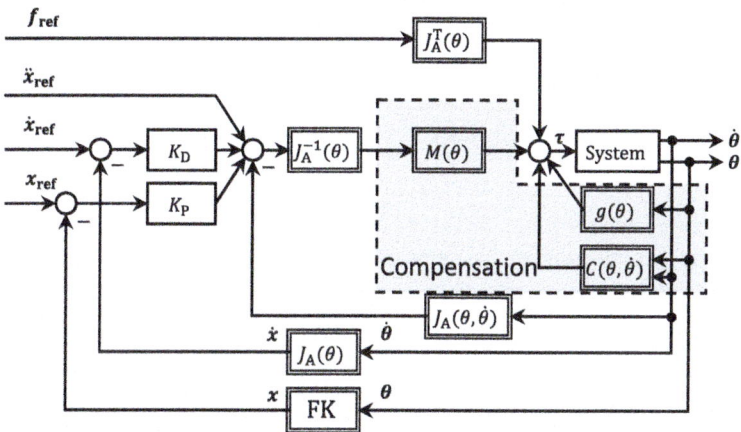

Fig. 7 Structure of the Computed Torque Controller

this is straightforward for a simplified three-linked manipulator and can be obtained by the Lagrange Formalism [2, p. 47 et sqq.]. Unfortunately, the given manipulator has a closed kinematic structure and the inertial properties of the cylinders and bucket linkages are not negligible compared to the main structure. This leads to a situation where even the configuration-independent properties of the compensation matrices are configuration-dependent [2, p. 58 et sqq.]. As an approximate solution we calculated the inertial properties of the manipulator at different configurations using the rigid body model functionalities of the Robotic System Toolbox of MATLAB/SIMULINK [27] (Fig. 7).

The CTC incorporates three proportional and 3 derivative gains which were tuned based on the desired dynamic properties of the controlled system [17, p. 188 et sqq.]. An eigenfrequency $f = 10$ Hz, i.e., $\omega = 20\pi$ rad/s and damping $D = 0.7$ was chosen for all three links (Table 3).

Table 3 Gain parameters based on desired dynamics of the controlled system

$\forall i = 1, \ldots, 3$		
$K_{P,i}$	$= \omega_i^2$	$= 3947 \, (\text{rad/s})^2$
$K_{D,i}$	$= \omega_i \cdot D_i \cdot 2$	$= 88 \, \text{rad/s}$

Independent Metering Valve Controller

Each hydraulic actuator is equipped with three valves as depicted in Fig. 4. The HFPC provides desired cylinder speeds and the CTC supplies the required cylinder forces. The velocities are transformed into flow requirements and the cylinder force is transformed into a load force

$$p_{L,\text{ref}} = F_{\text{cyl,ref}} \cdot A_A. \tag{8}$$

We define the weighted chamber pressure as

$$p_K = \frac{p_A p_B}{p_A + p_B} \tag{9}$$

and use it as a pressure limitation for the cylinder chamber side facing away from the load. In the experiments p_K is set to 5 bar to increase cylinder damping. Along with $p_{L,\text{ref}}$ and the area ratio $\Phi = A_B/A_A$ we can formulate chamber pressure requirements similar to Bachmann et al. [25].

$$p_A^* = \frac{p_{L,\text{ref}} + (1 + \Phi) \, p_K}{2} + \sqrt{\left(\frac{p_{L,\text{ref}} + (1 + \Phi) \, p_K}{2}\right)^2 - p_{L,\text{ref}} \, p_K} \tag{10}$$

$$p_B^* = \frac{-p_{L,\text{ref}} + (1 + \Phi) \, p_K}{2\Phi} + \sqrt{\left(\frac{-p_{L,\text{ref}} + (1 + \Phi) \, p_K}{2\Phi}\right)^2 - \frac{p_{L,\text{ref}} \, p_K}{2\Phi}} \tag{11}$$

Instead of computing the load pressure p_L from the current chamber pressures as a load feedback and defining a reference trajectory for p_K we use the torque commands from the CTC as pressure requirements. Thus, there is no necessity for pressure sensors, because the pressure demands result from the computed dynamics of the system based on a position and force trajectory.

The valve spool position is then calculated based on an inverse flow mapping. For the sake of simplicity and because the valve behavior is not the focus of this paper, we used the flow description Eq. 7 and the corresponding inversion

$$y_A = K_{Qy}^{-1} \left(v \cdot A_A \cdot \sqrt{\rho}\right)/\sqrt{2 \cdot |\Delta p|}, \quad \Delta p = \begin{cases} p_P - p_A^* & v > 0 \\ p_A^* - p_T & v < 0 \end{cases} \tag{12}$$

$$y_B = K_{Qy}^{-1}\left(v \cdot A_B \cdot \sqrt{\rho}\right)/\sqrt{2 \cdot |\Delta p|}, \quad \Delta p = \begin{cases} p_B^* - p_T & v > 0 \\ p_P - p_B^* & v < 0 \end{cases} \quad (13)$$

with the linear, invertible factor K_{Qy}. This can be replaced by more sophisticated flow descriptions if necessary. The switching valve is simply controlled by the sign of the velocity requirement

$$y_R = \text{sign}(v). \quad (14)$$

4 Simulation Study

Implementation and simulation of the excavator model and controller were performed in MATLAB/SIMULINK® R2023b using the Simscape® toolboxes for multi-domain simulations. Contact forces between the excavator bucket and obstacles are computed using *Spatial Contact Force* elements from the Simscape® library which apply smooth spring damper interaction when the convex hulls of the elements collide with each other.

In the simulation study, two use cases of typical work situations of an excavator were analyzed. One task consists of moving the bucket tip along a solid surface with a desired speed and orientation while exerting a defined normal force against the surface. This manoeuvre is required, for example, to sweep up loose bulk material on a hard surface such as asphalt. The force specification ensures that the tip always touches the ground and precise knowledge of the actual ground contour is not required.

The other task represents the collision with an obstacle during the trajectory tracking. The TCP may only exert a specified maximum force and has to stop if the object resists. Applications for this scenario are digging in sensitive or unknown ground conditions like in the vicinity of gas or electricity infrastructure or at the recovery of explosives.

In both cases the unconstrained movement is position controlled and only during interaction with the obstacles, i.e., sliding along the wall and touching the obstacle, one direction of the controller is set to force control. There are three reasons for this decision.

Firstly, force control effectively means acceleration control, and tracing a trajectory with a reference acceleration leads to larger errors in position accuracy compared to a velocity definition. Secondly, in real-world applications, the resistance of the ground during the digging process is not constant and this will lead to fluctuations in speed when there is no force feedback loop. And finally, and most importantly in the case of hydraulic actuation, force control means pressure control. Typically, in hydraulic systems pressure control is very difficult, as small variations in flow can create large fluctuations in pressure. Therefore, the presented IMVC basically incorporates a flow feedforward control with pressure feedback through the CTC. If a trajectory is supposed to be traced tangentially in force control, then technically

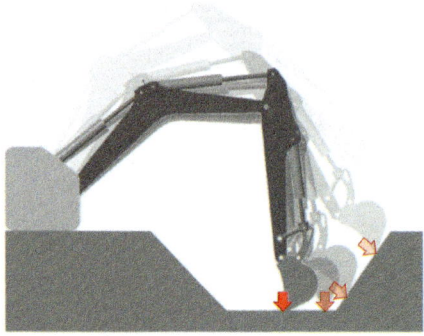
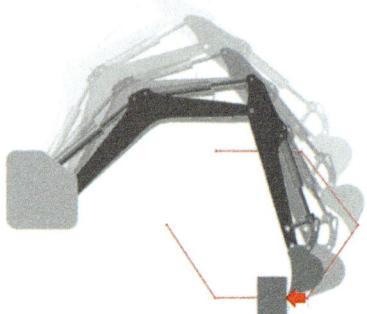

(a) Use Case A: Normal force surface tracing

(b) Use Case B: Obstacle Collision

Fig. 8 Use cases for simulation study

the flow feedforward path is zero and only pressure control is active. This works well for use case A where the movement is constrained in force direction but does not perform satisfactory in unconstrained movement (Fig. 8).

Use Case A The TCP is moved toward the surface with full position control in all three coordinates. Once the TCP is close to the surface, the normal space of the trajectory is set to force control and the TCP accelerates toward the surface and exerts the specified force while the other two coordinates, i.e., tangent trajectory and bucket orientation, are position controlled.

In Fig. 9 the deviation in the normal coordinate y can be observed starting from the contact at $t \approx 6.5$ s. At the same instance, the normal contact force between TCP and wall rises. Due to the fact that the force coordinate has no feedback control, the desired value is not exactly reached. In future research, this will be added by estimating the contact force with the measured chamber pressures. The measured flows and pressures (See Fig. 10) show good tracking during unconstrained movement and deviations in flow once the normal coordinate y is force controlled. The deviations affect all three actuators, because of the transformation between joint coordinates and task space coordinates.

At the onset of the contact, the required pressures are traced with little error, and the desired contact force is approximately reached; as said before, feedback control might enhance this behavior. While sliding along the wall though, the error between desired and actual pressures increases and consequently, the normal force error rises. The explanation for this behavior lies in the controller design, specifically the CTC. The current implementation of the dynamic compensation does not incorporate the internal frictions of the cylinder. Those create a configuration-dependent disturbance force. To overcome this issue, characterization, and mapping of those forces with

Design of a Cartesian Hybrid Force-Position Controller ...

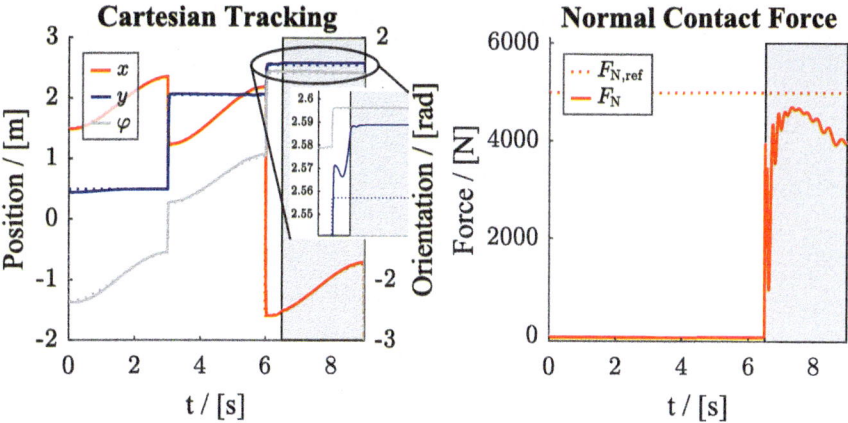

Fig. 9 Position and force trajectory during Use Case A. Reference values in dotted lines

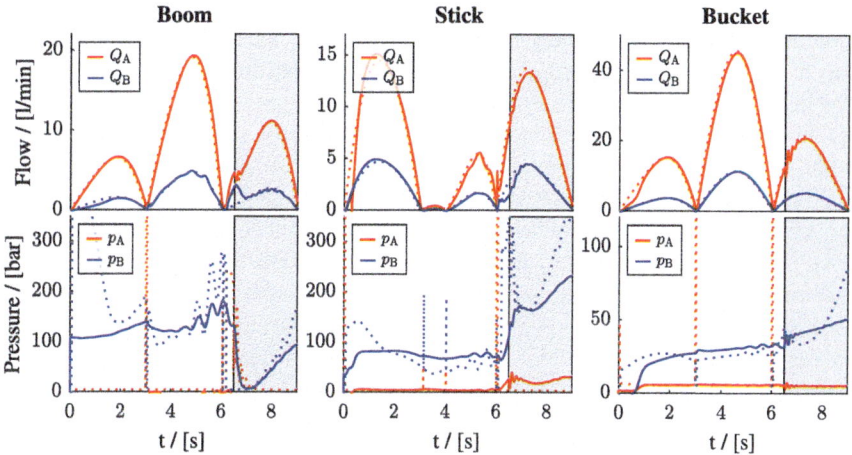

Fig. 10 Flow and pressure measurements during Use Case A. Reference values in dotted lines

respect to the joint configuration is required. Accurate experimental parametrization of such models is complex and was not part of this study.[2,3]

Use Case B The main difference toward use case A is that the force control task is in the tangent space of the trajectory. The specified trajectory is traced using pure position control until a collision is detected. In the current implementation, this collision detection is based on a significant speed deviation combined with a

---

[2] The steps in the coordinates result from the waypoint transformation of the trajectory in the base frame. Due to the counter rotation of θ_1 in the Jacobian matrices of the CTC, both target and measurement values are transformed equally.

[3] Pressure peaks in required pressure signals (dashed) are the result of discontinuities at the instance of the waypoint transformation and are filtered.

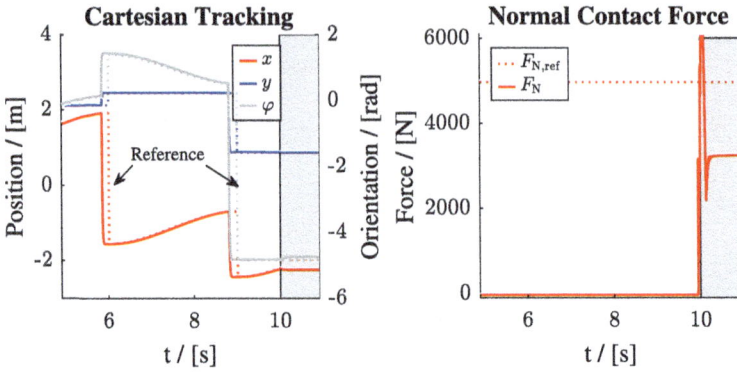

Fig. 11 Position and force trajectory during Use Case B. Reference values in dotted lines

large error between required and actual chamber pressures. If this is detected, the direction of movement, i.e., the tangent of the trajectory, is set to force control and the valves are set to pressure control to limit the interaction force between the obstacle and bucket. If one valve is completely shut during this process, the controller state machines are set to a standstill and the excavator eventually stops completely. At this point, the risk of damage is prevented and evasive actions can be implemented.

Figures 11 and 12 show the system states in use case B. During unconstrained movement, pressure, and flow requirements are met, and consequently the position error in task space is small. At the contact interaction at $t \approx 10$ s the tooltip of the manipulator collides with the object. The object collision is modeled with spatial contact elements in MATLAB Simscape® with a linear increasing spring stiffness over a transition region of 2 cm. At the contact pressure values spike and the controller state machine switches into pressure control of the valves. As the considered system incorporates a constant pressure source the inlet flow control valves are gradually shut while the outlet valves are actuated in order to reduce the pressure on the high-pressure side of the cylinder to limit the interacting force. Once the inlet control valves are shut, the controller state machine is set to standstill. The chamber pressures of boom and stick actuators are traced with little error during the process but overshooting in the stick actuator and deviations in the bucket actuator result in significant error of the desired force. The maximum target force is not exceeded, though. This behavior can be enhanced in future work by incorporating the switching valve in the pressure control task so that not only pressure reduction on the high-pressure side but also pressure incrementation on the low-pressure side to reduce the overall acting cylinder force is possible. Another enhancement would be the application—and inclusion in the control process—of a variable pressure supply in order to accurately control the chamber pressures at the contact with the obstacle.

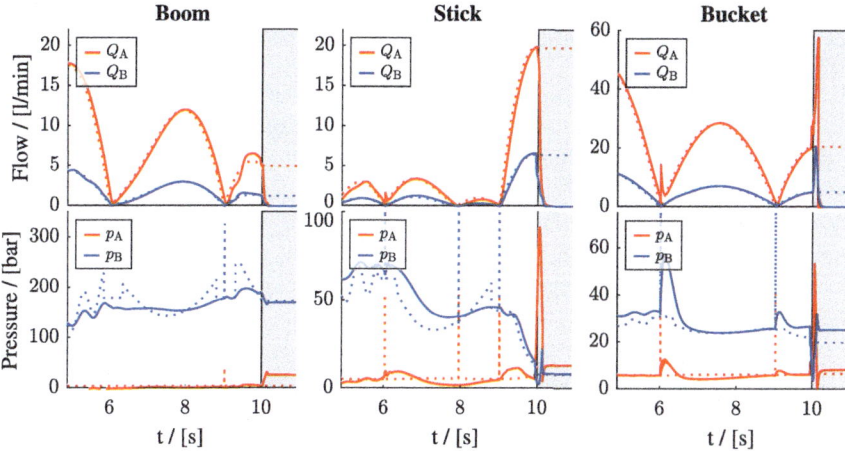

Fig. 12 Flow and pressure measurements during Use Case B. Reference values in dotted lines

Fig. 13 Lab demonstrator on which the simulation was based

5 Conclusion

In this paper, we presented a control strategy for a hydraulic actuated manipulator with hybrid force-position control on arbitrary trajectories. In two use cases, the capabilities of position and force tracing were shown. In general, the position control works satisfactory while force control has to be enhanced using force feedback. For this, a force sensor at the tool tip is not an adequate solution for an earth-moving machinery therefore, pressure measurement in the cylinder chambers can be used to estimate the achieved force. Even though, an accurate characterization of the configuration-dependent friction forces has to be included in the compensation control, and would also be necessary for the force estimation by pressure. This is no straightforward task and would have exceeded the scope of this paper hence, it will be a focus of further research.

This simulation study was performed on the detailed simulation model of a hydraulic excavator which also exists at the institute as a demonstrator in the lab. The demonstrator incorporates many sensors including draw wire sensors to measure cylinder displacement as well as angle sensors and IMUs for orientation measurements. Furthermore, there exists a highly versatile hydraulic system consisting of multiple servo valves with which different independent metering structures can be easily realized without reassembly. Future work will focus on the implementation and testing of the presented approach on this demonstrator.

Acknowledgements The work reported in this paper was performed within the project: "Development and testing of load- and motion-decoupled actuator structures using nonlinear, model-based control algorithms" funded by the DFG (German Research Society GZ: WE 4828/8-1 Transfer Project). The permission for publication is gratefully acknowledged.

References

1. Raibert M, Craig J (1981) Hybrid position/force control of manipulators. J Dyn Syst Meas Control 103(2):126–133. https://doi.org/10.1115/1.3139652
2. Siciliano B, Khatib O (2016) Springer handbook of robotics, 2nd edn. Springer, Berlin, Heidelberg
3. Topcon: New Topcon automatic excavator system featuring fingertip control (2019). https://www.topconpositioning.com/articles/new-topcon-automatic-excavator-system-featuring-fingertip-control
4. Trimble Earthworks Vers 2.0. (2021). https://www.sitech.de/fileadmin//sitech-content/Brochures/Brochures_DE/SITECH_Earthworks_Bagger-Dozer-Grader_2021-DE_web.pdf
5. Leica iCON excavate—The future of excavating (2022). https://leica-geosystems.com/-/media/files/leicageosystems/products/brochures/machine

6. Johns RL, Wermelinger M, Mascaro R, Jud D, Hurkxkens I, Vasey L, Chli M, Garmazio F, Kohler M, Hutter M (2023) A framework for robotic excavation and dry stone construction using on-site materials. Sci Robot (84) (2023). https://doi.org/10.1126/scirobotics.abp9758
7. Luengo O, Singh S, Cannon H (1998) Modeling and identification of soil-tool interaction in automated excavation. In: Proceedings of the 1998 IEEE/RSJ international conference on intelligent robots and systems. Innovations in Theory, Practice and Applications (Cat No 98CH36190). IEEE, Victoria, BC, Canada, pp 1900–1906. https://doi.org/10.1109/IROS.1998.724873
8. Singh S (1995) Learning to predict resistive forces during robotic excavation. In: Proceedings of 1995 IEEE international conference on robotics and automation, vol 2. IEEE, Nagoya, Japan, pp 2102–2107. https://doi.org/10.1109/ROBOT.1995.526025
9. Jud D, Hottiger G, Leemann P, Hutter M (2017) Planning and control for autonomous excavation. IEEE Robot Autom Lett 2(4):2151–2158. https://doi.org/10.1109/LRA.2017.2721551
10. Hogan N (1985) Impedance control: an approach to manipulation: parts I–III. ASME J Dyn Syst Meas Control 107(1):1–24. https://doi.org/10.1115/1.3140702, https://doi.org/10.1115/1.3140713, https://doi.org/10.1115/1.3140701
11. Ha Q, Nguyen Q, Rye D, Durrant-Whyte H (2000) Impedance control of a hydraulically actuated robotic excavator. Autom Constr 9(5–6):421–435. https://doi.org/10.1016/S0926-5805(00)00056-X
12. Heinrichs B, Sepehri N, Thornten-Trump A (1997) Position-based impedance control of an industrial hydraulic manipulator. IEEE Control Syst Mag 17:46–52. https://doi.org/10.1109/37.569715
13. Volpe R, Khosla P (1993) A theoretical and experimental investigation of explicit force control strategies for manipulators. IEEE Trans Autom Control 38(11):1634–1650. https://doi.org/10.1109/9.262033
14. Guangjun L, Goldenberg AA (1991) Robust hybrid impedance control of robot manipulators. In: Proceedings. 1991 IEEE international conference on robotics and automation. IEEE, Sacramento, CA, USA, pp 287–292. https://doi.org/10.1109/ROBOT.1991.131589
15. Guangjun L, Goldenberg AA, Robust hybrid impedance control of robot manipulators via a tracking control method. In: Proceedings of IEEE/RSJ international conference on intelligent robots and systems (IROS'94). IEEE, Munich, Germany, pp 1594–1601 (0012/1994-09-16). https://doi.org/10.1109/IROS.1994.407643
16. Dunnigan M, Lane D, Clegg A, Edwards I (1996) Hybrid position/force control of a hydraulic underwater manipulator. In: IEE Proceedings—control theory and applications 143(2):145–151. https://doi.org/10.1049/ip-cta:19960274
17. Lewis FL, Dawson DM, Abdallah CT (2003) Robot manipulator control—theory and practice, 2 edn, No 15 in Control Engineering. Marcel Dekker Inc, New York, Basel. https://lewisgroup.uta.edu/FL
18. Xia QH, Lim SY, Ang Jr MH, Lim TM (2006) Parallel force and motion control using robust velocity observer. In: IFAC proceedings volumes, 16, vol 39. Elsevier, Heidelberg, Germany, pp 289–294. https://doi.org/10.3182/20060912-3-DE-2911.00052
19. Koivumaki J, Mattila J (2015) Stability-Guaranteed force-sensorless contact force/motion control of heavy-duty hydraulic manipulators. IEEE Trans Robot 31(4):918–935. https://doi.org/10.1109/TRO.2015.2441492
20. Backé W (1974) Systematik der hydraulischen Widerstandsschaltungen in Ventilen und Regelkreisen. Krausskopf-Verlag
21. Jansson A, Palmberg JO (1990) Separate controls of meter-in and meter-out orifices in mobile hyraulic systems. SAE Trans 99:377–383. https://www.jstor.org/stable/44469407
22. Sitte A, Beck B, Weber J (2014) Design of independent metering control systems. In: Proceedings of the 9th international fluid power conference. Aachen
23. Lübbert J, Sitte A, Beck B, Weber J (2016) Load-Force-Adaptive outlet throttling: an easily commissionable independent metering control strategy. In: BATH/ASME 2016 symposium on fluid power and motion control. American Society of Mechanical Engineers. http://proceedings.asmedigitalcollection.asme.org/proceeding.aspx?articleid=2580077

24. Bachmann L, Sitte A, Weber J (2022) Evaluation of nonlinear MIMO controllers for independent metering in mobile hydraulics. In: IEEE global fluid power society PhD symposium. IEEE, Naples, IT (To appear)
25. Bachmann L, Liu Weber J (2023) Investigation of the temperature influence on electrohydraulic valve control and presentation of a novel compensation approach for independent metering valve systems. In: ASME/BATH 2023 symposium on fluid power and motion control. ASME, Sarasota, FL, USA. https://doi.org/10.1115/FPMC2023-111588
26. Denavit J, Hartenberg RS (1955) A kinematic notation for lower-pair mechanisms based on matrices. ASME J Appl Mech 22(2):215–221. https://doi.org/10.1115/1.4011045
27. MathWorks: Rigid Body Tree Robot Model (2024). https://de.mathworks.com/help/robotics/ug/rigid-body-tree-robot-model.html

Open Access This chapter is licensed under the terms of the Creative Commons Attribution 4.0 International License (http://creativecommons.org/licenses/by/4.0/), which permits use, sharing, adaptation, distribution and reproduction in any medium or format, as long as you give appropriate credit to the original author(s) and the source, provide a link to the Creative Commons license and indicate if changes were made.

The images or other third party material in this chapter are included in the chapter's Creative Commons license, unless indicated otherwise in a credit line to the material. If material is not included in the chapter's Creative Commons license and your intended use is not permitted by statutory regulation or exceeds the permitted use, you will need to obtain permission directly from the copyright holder.

Parameter Identification for Optimized Simulation Models in Mobile Hydraulic Applications

Bernhard Sender[ ], Johannes Sprink[ ], and Katharina Schmitz[ ]

1 Introduction

1.1 Digital Twins and Lumped Parameter Simulations

The importance of digital twins for mobile machinery is on the rise, representing a significant shift in how the industry utilizes technology to enhance operations. These virtual replicas offer advantages useful for the machine operators, owners, and manufacturers. Research shows, that operator skills directly correlate with the productivity and efficiency [1–3] of construction machines. Therefore, the use of a digital twin of a construction machine to train operators or to automatically derive operating recommendations would facilitate a more efficient and effective operation on the construction site [1]. Potential other use cases of digital twins of off-highway mobile machinery also include status, performance, and progress reports and analysis or predictive maintenance useful for fleet and site management [4]. Another use case for machine manufacturers is virtual testing and the incorporation of field data from the machines into the machine development processes.

Madni et al. [5] define the concept of digital twin as follows: "A digital twin is a dynamic digital representation of a physical system or device that has bidirectional communication with the physical system and can operate in parallel to the physical system operating in the real world. A digital twin can represent the full physical system, a subsystem, or only specific aspects of the system needed by specific applications. A digital twin is dynamically updated in real time/near real time as the development status of the physical system changes during design or as the state of other physical systems changes during operation; for example, through

---

B. Sender (✉) · J. Sprink · K. Schmitz
RWTH Aachen University, Institute for Fluid Power Drives and Systems (ifas), Aachen, Germany
e-mail: bernhard.sender@ifas.rwth-aachen.de

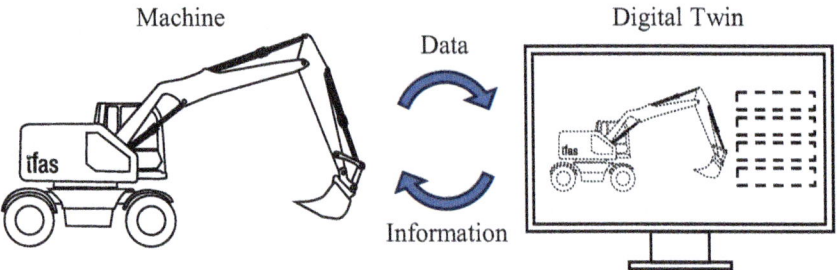

Fig. 1 Digital machine twin

wear or environmental forces." Following this definition, Fig. 1 schematically illustrates the digital twin concept using an excavator as an example. It highlights that the core component of digital twin is the information exchange between the virtual representation and the real system.

Digital twins often include simulation models. Hydraulic systems are an integral part of the drive train of many mobile machines and have a significant contribution to the machines behavior. This means that the simulation of the hydraulic system of the machines should be an integral part of its digital twin. For simulating the dynamics of hydraulic systems, lumped parameter models are often used. To dynamically update the lumped parameter simulation model included in the digital twin, its parameters need to be adjusted to data obtained during the field operation of the machine. This paper focuses on the possible ways of achieving this.

1.2 Model Parametrization as an Optimization Problem

For identifying the parameters of a lumped parameter model as a part of a digital twin of an excavator a method is proposed in the following section. The principle is illustrated in Fig. 2. It consists of the dynamic model, an error metric, and the optimization algorithm. The function of the dynamic model is to predict the behavior of the modeled system based on the given input. For an excavator, this might be to predict the movement of the bucket according to the joystick signals commanded by the operator. The displayed sub-model of the hydraulic system may include differential equations, for example, derived from equations of motion (hydraulic cylinders, valve spools) or pressure build up in hydraulic capacities, accompanied by algebraic equations or empirical data like pump or motor efficiency maps or characteristic curves for valve dynamics [6].

The process of fitting model parameters involves adjusting the values of these parameters to minimize the discrepancy between simulated results and empirical data obtained from the real system under different operating conditions. The goal of parameter fitting is to find the set of parameter values that result in the best match between the simulated behavior of the model and the observed behavior of the real

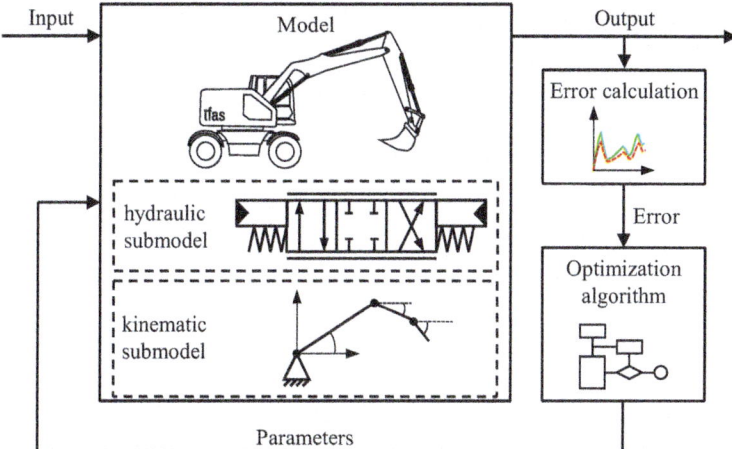

Fig. 2 Schematic representation of the parameter identification process for a lumped parameter simulation model of an excavator

system. This can be achieved by defining a fitness function or error metric that quantifies the difference between simulated and measured data. After this, optimization algorithms can be utilized to perform the iterative search for the set of parameter values that minimize this error metric. This kind of mathematical problem is called inverse problem [7].

Finding the optimization algorithm which performs best for this kind of problem poses a significant challenge. Wolpert et al. [8] developed the *No-free-lunch-theory for optimization,* which argues that averaged over all possible objective functions no optimization algorithm can universally outperform any other [8, 9]. This means that, without additional context or domain knowledge for any application, there is no a priori certainty which algorithm is suited best. However, the theorem acknowledges, that many objective functions imaginable are irrelevant or uninteresting in real-world scenarios [8, 9]. For subsets of problems, which are relevant, it becomes feasible to demonstrate the superiority of one algorithm over another [8, 9]. To make a well-informed a priori choice for an algorithm for the described problem, structured, extensive research is needed. This involves comparing the performance of different optimization algorithms in model parameter fitting applications in the domain of mobile hydraulic machines. The primary goal of the paper is to investigate whether such a tendency can be identified from reviewing preliminary work conducted by other researchers. Therefore, a review of relevant publications regarding this problem within the domain of (mobile) hydraulics is performed in Sect. 3. As an introduction, first an excerpt from the mathematical field optimization is presented in Sect. 2. The proposed strategy to overcome this gap is simply to perform a structured, comparative study of lots of different algorithms on the same excavator simulation model.

2 Optimization Methods

To tune the model parameters to fit the behavior of the model to the real system, a fitness function needs to be maximized. This is equivalent to minimizing an error function which describes the deviation between simulation and measurement in terms of the model parameters [7]. The challenge is to identify the algorithm or group of algorithms from the mathematical field of optimization that are suitable for the specific application. Derivative-based optimization methods like the Gauß-Newton algorithm require the calculation of the gradient or hessian of the fitness/error function in terms of the parameters [10]. To accurately depict reality, lumped parameter models, especially for hydraulic systems, often include non-linearities or discontinuities [6]. This means, explicitly (numerical or analytically) expressing the (multidimensional) error function is a lot of development effort, especially for larger models. Because of this and because derivative-based methods tend to converge prematurely to local minima [10], this paper exclusively focuses on non-derivate methods. An overview about common gradient-free optimization algorithms is shown in Fig. 3.

Since optimization is an essential field of mathematics, there are countless non-derivative optimization methods. Therefore, presenting a larger portion of those would exceed the scope of a paper. For a detailed overview about different optimization algorithms see for example [10]. This paper will thus only present the Particle Swarm Algorithm (PSO), representative for the non-derivative-based algorithms. A related algorithm, also from the field of evolutionary algorithms is the genetic algorithm (GA). It is inspired by the principles of natural selection and genetics and involves evolving a population of candidate solutions through processes such as selection, crossover, and mutation to find optimal solutions to optimization problems [10, 11]. This algorithm is mentioned at this point, but not explained further in this

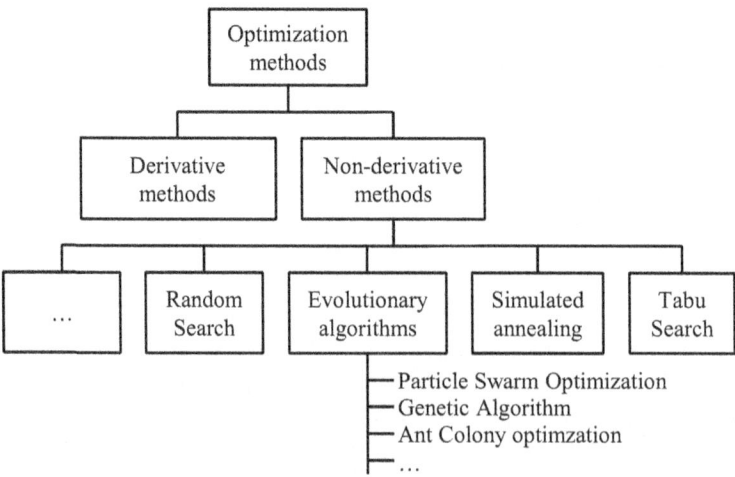

Fig. 3 Optimization methods [10]

paper. Following the argument of the previously introduced No-free-lunch-theorem, the PSO algorithm is not chosen because there exists a prior certainty that it might outperform others but because it is prominent among the field of non-derivative optimization methods and in the relevant research for the intended application.

2.1 Particle Swarm Optimization

The particle swarm optimization algorithm (PSO) was introduced in 1995 by Eberhart and Kennedy [12, 13]. It was inspired by an effect found in nature: the collective (complex) behavior of groups of animals emerging from limited responses from its individuals, such as fish schools or bird flocks [9]. During their research trying to simulate the social interactions leading to group motion, Eberhart and Kennedy discovered that algorithms mimicking cooperative swarm behavior can be applied to solve optimization problems [9]. This led to the development of the particle swarm algorithm [9]. Since then, PSO and its various adaptations have found application across many domains in the field of optimization. Furthermore, numerous related swarm-based algorithms have emerged since the inception of the PSO. They form the category of algorithms called Swarm Intelligence [9, 14].

Working Principle of the Basic PSO Algorithm. PSO algorithms work by searching for an optimum in the search space with a "population" of collision-free search agents, called *particles*, whose positions change with every iteration. Each position represents a potential solution to the function sought to optimize. This means in every iteration the particle positions are the input at which the function is evaluated. The advantage of PSO algorithm is, that the function doesn't need to be explicitly defined. What is needed is a process to assign a value of the degree of optimality to a given input (the positions). In the context of this application, this means to allocate the accuracy of the model to the model parameters used. During the optimization, the best position every particle has visited so far is stored. This value is called *personal best* [9, 12, 13]. The personal best is related to the concept of fitness [11, 13] found in other evolutionary algorithms [9, 12, 13]. Another key component is the information transfer between particles. The best position off all particles within the swarm is called *global best* [9, 12, 13]. This value is also retained and updated during the iteration process. Sharing this information across all particles imitates the social behavior of individuals in swarms. Within one iteration, the new position for every particle is influenced by its personal best and the global best [14, 15].

The following paragraph formalizes and expands on the described particle movement process. In the notation used, $\boldsymbol{p}_i^t$ represents the current position-vector of the particle $i \in \{1, 2, \ldots, N\}$, where N is the total number of particles and t is the iteration count. Subsequently p_{ij}^{t+1} denotes the position component in the dimension $j \in \{1, 2, \ldots, D\}$, with D being the number of dimensions of the search space. Equation (1) describes the particle movement component-wisely.

$$p_{ij}^{t+1} = p_{ij}^t + v_{ij}^{t+1} \qquad (1)$$

In every iteration, the position of every particle is updated with the vector v_{t+1}^i called *velocity*. For the first iteration, every particle is initialized with a random position and velocity. The velocity component v_{ij}^{t+1} of the particle i is calculated according to Eq. (2).

$$v_{ij}^{t+1} = \omega \cdot v_{ij}^t + c_c \cdot r_1 \cdot \left(p_{ij}^{best} - p_{ij}^t \right) + c_s \cdot r_2 \cdot \left(g_j^{best} - p_{ij}^t \right) \qquad (2)$$

The above-described personal best and global best are represented in (2) by $\boldsymbol{p}_{ij}^{best} = \left(p_{i1}^{t,best}, \ldots, p_{iD}^{t,best} \right)$ and $\boldsymbol{g}_j^{best} = \left(g_1^{best}, \ldots, g_D^{best} \right)$, respectively. The two scalars r_1 and r_2 are chosen at random for every component following a continuous uniform distribution bound by 0 and 1. The two scalar parameters c_k and c_s are called *acceleration constants* [14, 16]. The values of these constants determine the particle's magnitude of attraction towards its personal best and the global best. Since the second term in (2) only involves a particle's information of its own trajectory, it is called *cognitive term* [9], whereas the third term is called *social term* [9], because it includes information contributed by other particles. The scalar parameter ω in the first term is called *inertia weight* or *constriction coefficient*. Note that the original algorithm [12, 13] didn't have the inertia term, which was introduced by Shi and Eberhardt [17]. The three parameters ω, c_c, and c_s have great influence on the stability and performance of the algorithm. For example, the influence of these parameters is described in [14, 16, 18]. Note that the complete PSO algorithm requires defining an appropriate termination condition. The specifics regarding this and the initialization of the particles position and velocity are not covered in this paper. For details regarding this, see [12–14, 16].

This relatively simple set of rules, described above, produces the emergent behavior of the particle swarm to converge towards (local or global) optima of the problem fitness function [9]. To evaluate and benchmark algorithms, many hard to minimize test functions have been developed. An early non-convex test function developed by *Rosenbrock* in 1960 [19] is the subsequently called Rosenbrock function, which is often used for benchmarking algorithms. Figure 4 depicts the capability of the PSO algorithm finding the global minimum of the Rosenbrock function. While the upper left graph shows the surface plot of the function, the other three images show the contour plot with the particle positions during iterations 5, 17, and 60. It is visible, that the particles converge to the global minimum at (1, 1). For this application, a PSO variant with 40 particles was used, which needed 60 iterations to find the global minimum.

Variants. The basic algorithm has seen numerous variants since its inception. For example, *Sedighizadeh* and *Masehian* [20] identify 95 different PSO variants. Examples of other publications focusing on describing PSO variants are [21, 22]. To distinguish, the basic, historical version of the PSO described above is referred to as Canonical PSO [16]. There are efforts by *Clerc,* an early PSO researcher [15, 16,

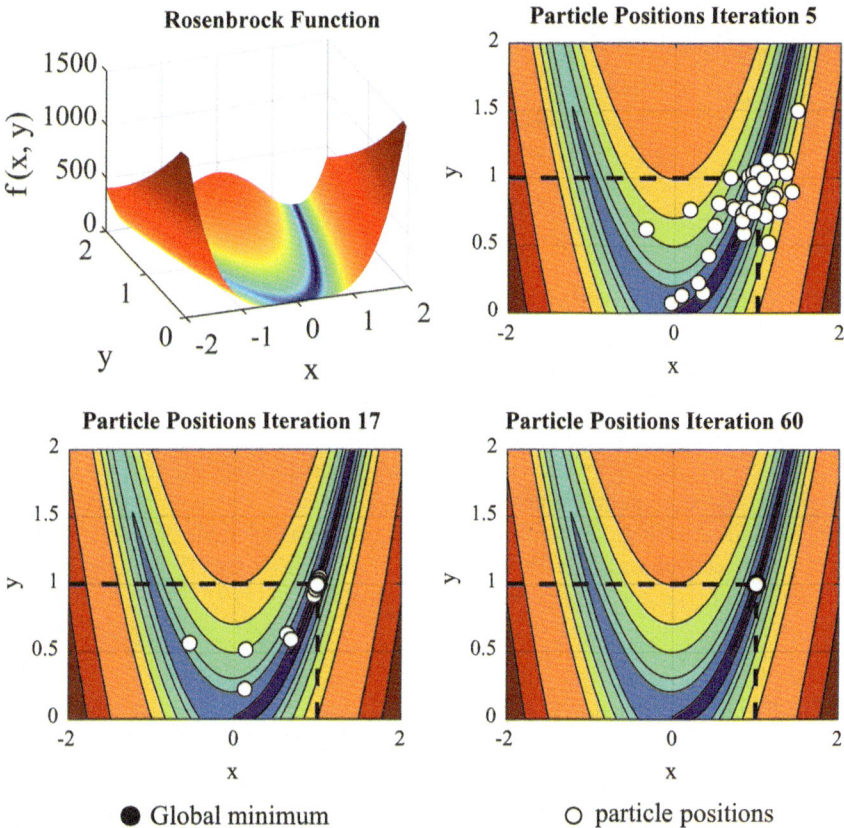

Fig. 4 Surface and contour plot of the Rosenbrock function and particle positions

23], to define a standardized version of the PSO algorithm, called Standard PSO, short SPSO [14, 15].

A common modification is to use the best position of a specific subset of the whole particle swarm as the second attractor for each particle instead of using the global best. This concept is called *neighborhood* [14, 23, 24] and is also part of the Standard PSO. Note that there are several commonly used methods how to define which particles are considered a neighborhood. Another common modification is to decrease ω with increasing iteration count [14, 25] to progressively increase the significance of exploitation over exploration [9]

Applications. There are numerous review papers focusing on listing different applications throughout different scientific or engineering domains, in which PSO has been applied. Examples of these kinds of papers are [20–22, 26]. Details regarding the usage of PSO related to the addressed problem can be found in Sect. 3.

Table 1 Usage of evolutionary algorithms for finding optimal design parameters in (mobile) hydraulic applications

Publication	Application	Number of parameters	Algorithms used
Zhao et al. [29]	Hydraulic steering system (Simulation)	8	PSO
Hui [30]	Hydraulic transmission (Simulation)	6	GA

3 Related and Previous Work

Optimization algorithms have emerged as valuable tools in engineering, offering solutions to complex optimization tasks. These algorithms have found application in various areas. This section, especially focusing on the PSO and GA algorithms, gives an overview of how they can be utilized in the domain of (mobile) hydraulic systems, emphasizing their potential contributions to addressing challenges in system design, control, and modeling. Before the usage in fitting model parameters is described in Sect. 3.3, related applications in design optimization and controller parameter tuning are depicted in Sect. 3.1 and Sect. 3.2, respectively.

3.1 Finding Optimal Design Parameters

The effectiveness of the PSO in engineering design optimization has been showcased by Fourie et al. [27] and Kim et al. [28]. Table 1 lists additional examples for publications utilizing evolutionary algorithms for design optimization in hydraulic applications.

3.2 Controller Parameter Tuning

A related use of non-derivative based optimization algorithms is the tuning of controller parameters. Ou et al. performed tuning of the control parameters of a PID controller with the GA and PSO algorithm. In their tests, two theoretical systems are described by a second-order and third-order transfer functions as the plant [31]. Nagaraj et al. conducted a similar study with two different second-order transfer functions and one fourth-order transfer function, where the PID parameter tuning was performed with a PSO, GA, and two other evolutionary algorithms [32]. Both publications assess the usage of evolutionary algorithms beneficial for control parameter tuning. In both cases the transfer functions are linear. However, hydraulic systems in general are highly non-linear.

Table 2 gives seven examples of publications focusing on the tuning of control parameters with PSO or GA optimization algorithms applied to systems in the field of (mobile) hydraulics.

Table 2 Usage of evolutionary algorithms for tuning controller parameters in (mobile) hydraulic applications

Publication	Application (method of error calculation: simulation or test rig)	Controller (number of control parameters)	Algorithms
Nedic et al. [33]	Position-control of a six-degree-of-freedom parallel robot platform (non-linear model)	Cascade controller (4 parameters)	PSO
Ye et al. [34]	Position-control of a hydraulic cylinder for the bucket actuation of an excavator (non-linear model)	PID controller (3 parameters)	PSO
Karam et al. [35]	Angular position-control of a hydraulic motor (linearized and non-linear model, test rig)	PID controller (3 parameters)	GA
Wonohadidjojo et al. [36]	Position control of an electro-hydraulic actuator system (non-linear model)	Fuzzy logic controller (3 parameters)	PSO
Feng et al. [37]	Trajectory control for a grade-assist-system of a hydraulic excavator (non-linear model, test machine)	PID controller (3 parameters)	PSO
Feng et al. [38]	Trajectory control for a grade-assist-system of a hydraulic excavator with friction compensation (non-linear model, test machine)	PID + feed forward controller (3 + 6 parameters)	PSO
Guo et al. [39]	Position control of a hydraulic actuator (linearized model)	PID controller (3 parameters)	GA

3.3 Parameter Identification

The following section describes the utilization of evolutionary algorithms to find parameters with a physical representation in an application with the intent of improving the accuracy of lumped parameter simulation models. Ye [40] showed with simulations that PSO and GA algorithms are capable of finding known parameters of chaotic systems. In electrical engineering, for fitting the parameters of diode models of photovoltaic cells to measurement data, PSO and other non-derivative-based optimization methods are widespread [41, 42].

Maier et al. use a PSO variant to fit the parameters of a lumped parameter simulation modeling the behavior of hydraulically actuated clutches. The utilized simplified model features 25 parameters to optimize. To implement value ranges, a penalty function is introduced, which adds a penalty value to the error function, when a parameter exceeds its defined value range. A parameter deviation term, which adds another penalty value proportional to the difference between a pre-parameterization (based on an expert educated guess) and the estimates, is also included. The goal with these two additional terms in the cost function is to ensure the physical plausibility of the identified parameters. The drawback of this method is a high dependency on the pre-parameterization [43].

Table 3 provides seven additional examples of publications which use evolutionary algorithms (mainly PSO and GA) to identify model parameters in hydraulic applications.

From the provided overview, two observations can be made. Firstly, while there is plenty of publications focusing on the application of different optimization algorithms in related applications, relevant publications addressing the parameter identification in mobile hydraulic simulation models are sparse. Secondly, the majority of the identified publications focus on one or a limited number of algorithms. Because of this and the fact, that the performance of algorithms highly depends on the specific problem, no compelling argument can be made solely from the literature about the existence of an overperforming algorithm for the intended application. The proposed strategy, depicted in Fig. 5, is to build a lumped parameter model of an excavator and use this as a benchmark to comparatively test multiple optimization algorithms for parameter identification under the same conditions. Also, this can be used to conduct studies about the nature of the underlying error function.

Table 3 Usage of evolutionary algorithms for identifying simulation parameters in hydraulic applications

	Application (model type)	Number of parameters	Algorithms
Alfi et al. [44]	Hydraulic suspension system (estimating known simulation parameters)	3–5	GA, PSO
Chen [45]	Identifying inertia and friction parameters for a six-degree-of-freedom parallel robot platform (fitting a non-linear model to experimental data)	26	PSO
Yu et al. [46]	Hydraulic press (fitting a linear model to experimental data)	6	PSO
Montazeri et al. [47]	Hydraulic robot manipulator (estimating known simulation parameters)	19	GA
Quian et al. [48]	Servo-hydraulic actuator (fitting a linear model to experimental data)	5	GA
Yousefi [49]	Servo-hydraulic actuator (fitting non-linear model to experimental data)	16	Differential evolution
Sarhadi et al. [50]	Servo-hydraulic actuator (fitting a linear model to experimental data)	3	PSO, GA, other evolutionary algorithms

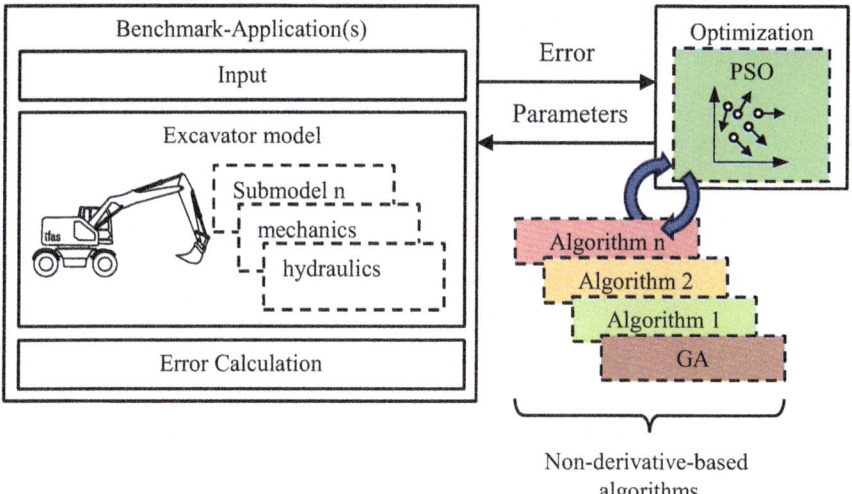

Fig. 5 Proposed algorithm testing strategy

4 Conclusion and Outlook

In conclusion, this paper has provided an introduction to methods for identifying the parameters of simulation models in mobile hydraulic applications, with a focus on non-derivative optimization methods, particularly the particle swarm optimization algorithm. Additionally, a literature review about the relevant publications for the application and related fields has been conducted. From this could be concluded, that for the application of identifying parameters in lumped parameter models for mobile hydraulic machines, without further study no statement which algorithm is suited best, can be made a priori. A strategy to investigate the effectiveness of different methods for the application of a hydraulic excavator has therefore been proposed. Future work is to execute the proposed strategy and analyze the results. An important question is to what extent the method is capable of online parameter identification.

Funded by:

on the basis of a decision
by the German Bundestag

Acknowledgements This research is supported by the Federal Ministry for Digital and Transport of Germany within the research initiative mFUND (project number 19FS2036).

References

1. Lehmann B, Jacobs G, Habermehl C et al (2021) Daten von Land- und Baumaschinen gewinnbringend nutzen. In: Trauth D, Bergs T, Prinz W (eds) MONETARISIERUNG VON TECHNISCHEN DATEN. Springer Berlin Heidelberg, Berlin, Heidelberg, pp. 471–520. https://doi.org/10.1007/978-3-662-62915-4_25
2. Kunze G, Mieth S, Voigt S (2011) Bedienereinfluss auf Leistungszyklen mobiler Arbeitsmaschinen. ATZoffhighway 4(1):70–79. https://doi.org/10.1365/s35746-011-0010-2
3. Frank B, Skogh L, Alaküla M (2012) On wheel loader fuel efficiency difference due to operator behaviour distribution. In: Commercial vehicle technology 2012. Shaker, Aachen
4. Schöberl M, Kalla T, Sauermann T et al (2020) Schöberl Kalla 2020 The Process-oriented Digital Twin. In: Bramberger R, Breitenbach T, Finzel R, Hermann J, Herrmann A, Lüddemann R (eds) 8 Fachtagung Baumaschinentechnik 2020, pp. 203–214
5. Madni AM, Augustine N, Sievers M (2023) Handbook of model-based systems engineering. Springer Nature Switzerland, Cham, Switzerland
6. Hansen AH (2023) Fluid power systems. Springer Nature Switzerland, Cham
7. Aster RC, Borschers B, Cliford HT (2013) Parameter estimation and inverse problems, 2nd edn. Elsevier, Oxford, United Kingdom
8. Wolpert DH, Macready WG (1997) No free lunch theorems for optimization. IEEE Trans Evol Comput 1(1):67–82. https://doi.org/10.1109/4235.585893
9. Kennedy J, Eberhart RC, Shi Y (2001) Swarm intelligence, vol 8. Morgan Kaufmann, San Francisco
10. Persson J, Ölvander J (2020) Engineering optimization
11. Kramer O (2017) Genetic algorithm essentials. Springer International Publishing, Cham, Switzerland
12. Kennedy J, Eberhart R (1995) Particle swarm optimization. In: Proceedings of 1995 IEEE international conference on neural networks. IEEE, Piscataway, NJ, pp. 1942–1948. https://doi.org/10.1109/ICNN.1995.488968
13. Eberhart R, Kennedy J (1995) A new optimizer using particle swarm theory. In: MHS'95. Proceedings of the sixth international symposium on micro machine and human science. IEEE. https://doi.org/10.1109/MHS.1995.494215
14. Parsopoulos KE (2018) Particle swarm methods. In: Martí R, Pardalos PM, Resende MGC (eds) Handbook of Heuristics. Springer International Publishing, Cham, pp. 639–685. https://doi.org/10.1007/978-3-319-07124-4_22
15. Clerc M (2012) Standard particle swarm optimisation
16. Clerc M (2006) Particle swarm optimization, 1st edn. ISTE, London, United Kingdom
17. Shi Y, Eberhart R (1998) A modified particle swarm optimizer. In: 1998 IEEE international conference on evolutionary computation proceedings. IEEE, pp. 69–73. https://doi.org/10.1109/ICEC.1998.699146
18. Bonyadi MR (2019) A theoretical guideline for designing an effective adaptive particle swarm. IEEE Trans Evol Comput 24(1):57–68. https://doi.org/10.1109/TEVC.2019.2906894
19. Rosenbrock HH (1960) An automatic method for finding the greatest or least value of a function. Comput J 3(3):175–184. https://doi.org/10.1093/comjnl/3.3.175
20. Sedighizadeh D, Masehian E (2009) Particle swarm optimization methods, taxonomy and applications. Int J Comput Theory Eng 486–502. https://doi.org/10.7763/IJCTE.2009.V1.80
21. Khan MS, Ali W, Qyyum MA et al (2021) Introduction to particle swarm optimization and its paradigms: a bibliographic survey. In: Malik H, Fatema N, Alzubi JA (eds) AI and machine

learning paradigms for health monitoring system, vol 86. Springer Singapore, Singapore, pp. 105–124. https://doi.org/10.1007/978-981-33-4412-9_6
22. Gad AG (2022) Particle swarm optimization algorithm and its applications: a systematic review. Arch Comput Methods Eng 29(5):2531–2561. https://doi.org/10.1007/s11831-021-09694-4
23. Clerc M, Kennedy J (2002) The particle swarm—explosion, stability, and convergence in a multidimensional complex space. IEEE Trans Evol Comput 6(1):58–73. https://doi.org/10.1109/4235.985692
24. Kennedy J, Mendes R (2002) Population structure and particle swarm performance. In: Proceedings of the 2002 congress on evolutionary computation. CEC'02 (Cat. No.02TH8600). IEEE, pp. 1671–1676. https://doi.org/10.1109/CEC.2002.1004493
25. Arasomwan MA, Adewumi AO (2013) On the performance of linear decreasing inertia weight particle swarm optimization for global optimization. Sci World J 2013:860289. https://doi.org/10.1155/2013/860289
26. Poli R (2008) Analysis of the publications on the applications of particle swarm optimisation. J Artif Evol Appl 2008:1–10. https://doi.org/10.1155/2008/685175
27. Fourie PC, Groenwold AA (2002) The particle swarm optimization algorithm in size and shape optimization. Struct Multidiscip Optim 23(4):259–267. https://doi.org/10.1007/s00158-002-0188-0
28. Kim T-H, Maruta I, Sugie T (2010) A simple and efficient constrained particle swarm optimization and its application to engineering design problems. Proc Inst Mech Eng C J Mech Eng Sci 224(2):389–400. https://doi.org/10.1243/09544062JMES1732
29. Zhao W, Luan Z, Wang C (2018) Parametric optimization of novel electric–hydraulic hybrid steering system based on a shuffled particle swarm optimization algorithm. J Clean Prod 186:865–876. https://doi.org/10.1016/j.jclepro.2018.03.180
30. Hui S (2010) Multi-objective optimization for hydraulic hybrid vehicle based on adaptive simulated annealing genetic algorithm. Eng Appl Artif Intell 23(1):27–33. https://doi.org/10.1016/j.engappai.2009.09.005
31. Ou C, Lin W (2006) Comparison between PSO and GA for parameters optimization of PID controller. In: 2006 International conference on mechatronics and automation. IEEE Operations Center, Piscataway, NJ, pp. 2471–2475. https://doi.org/10.1109/ICMA.2006.257739
32. Nagaraj B, Murugananth N (2010) A comparative study of PID controller tuning using GA, EP, PSO and ACO. In: 2010 International conference on communication control and computing technologies. IEEE, pp. 305–313. https://doi.org/10.1109/ICCCCT.2010.5670571
33. Nedic N, Prsic D, Dubonjic L et al (2014) Optimal cascade hydraulic control for a parallel robot platform by PSO. Int J Adv Manuf Technol 72(5–8):1085–1098. https://doi.org/10.1007/s00170-014-5735-5
34. Ye Y, Yin C-B, Gong Y et al (2017) Position control of nonlinear hydraulic system using an improved PSO based PID controller. Mech Syst Signal Process 83:241–259. https://doi.org/10.1016/j.ymssp.2016.06.010
35. Karam M, Zongxia J, Huaqing Z (2008) PID controller optimization by GA and its performances on the electro-hydraulic servo control system. Chin J Aeronaut 21(4):378–384. https://doi.org/10.1016/S1000-9361(08)60049-7
36. Wonohadidjojo DM, Kothapalli G, Hassan MY (2013) Position control of electro-hydraulic actuator system using fuzzy logic controller optimized by particle swarm optimization. Int J Autom Comput 10(3):181–193. https://doi.org/10.1007/s11633-013-0711-3
37. Feng H, Ma W, Yin C et al (2021) Trajectory control of electro-hydraulic position servo system using improved PSO-PID controller. Autom Constr 127:103722. https://doi.org/10.1016/j.autcon.2021.103722
38. Feng H, Yin C, Cao D (2023) Trajectory tracking of an electro-hydraulic servo system with an new friction model-based compensation. IEEE/ASME Trans Mechatron 28(1):473–482. https://doi.org/10.1109/TMECH.2022.3201283
39. Guo Y-Q, Zha X-M, Shen Y-Y et al (2022) Research on PID position control of a hydraulic servo system based on Kalman genetic optimization. Actuators 11(6):162. https://doi.org/10.3390/act11060162

40. Ye M (2006) Parameter identification of dynamical systems based on improved particle swarm optimization. In: Huang D-S, Li K, Irwin GW (eds) Intelligent control and automation, vol 344. Springer Berlin Heidelberg, pp. 351–360. https://doi.org/10.1007/978-3-540-37256-1_42
41. Khare A, Rangnekar S (2013) A review of particle swarm optimization and its applications in solar photovoltaic system. Appl Soft Comput 13(5):2997–3006. https://doi.org/10.1016/j.asoc.2012.11.033
42. Cuevas E (2020) Recent metaheuristics algorithms for parameter identification. Springer International Publishing AG, Cham
43. Maier CC, Schröders S, Ebner W et al (2019) Modeling and nonlinear parameter identification for hydraulic servo-systems with switching properties. Mechatronics 61:83–95. https://doi.org/10.1016/j.mechatronics.2019.05.005
44. Alfi A, Fateh MM (2011) Identification of nonlinear systems using modified particle swarm optimisation: a hydraulic suspension system. Veh Syst Dyn 49(6):871–887. https://doi.org/10.1080/00423114.2010.497842
45. Chen C-T (2012) Hybrid approach for dynamic model identification of an electro-hydraulic parallel platform. Nonlinear Dyn 67(1):695–711. https://doi.org/10.1007/s11071-011-0020-8
46. Yu Y, Ren X, Du F et al (2012) Application of improved PSO algorithm in hydraulic pressing system identification. J Iron Steel Res Int 19(9):29–35. https://doi.org/10.1016/S1006-706X(13)60005-9
47. Montazeri A, West C, Monk SD et al (2017) Dynamic modelling and parameter estimation of a hydraulic robot manipulator using a multi-objective genetic algorithm. Int J Control 90(4):661–683. https://doi.org/10.1080/00207179.2016.1230231
48. Qian Y, Ou G, Maghareh A et al (2014) Parametric identification of a servo-hydraulic actuator for real-time hybrid simulation. Mech Syst Signal Process 48(1–2):260–273. https://doi.org/10.1016/j.ymssp.2014.03.001
49. Yousefi H, Handroos H, Soleymani A (2008) Application of differential evolution in system identification of a servo-hydraulic system with a flexible load. Mechatronics 18(9):513–528. https://doi.org/10.1016/j.mechatronics.2008.03.005
50. Sarhadi P, Khosravi A, Bijani V (2015) Identification of nonlinear actuators with time delay and rate saturation using meta-heuristic optimization algorithms. Proc Inst Mech Eng Part I J Syst Control Eng 229(9):808–817. https://doi.org/10.1177/0959651815595569

Open Access This chapter is licensed under the terms of the Creative Commons Attribution 4.0 International License (http://creativecommons.org/licenses/by/4.0/), which permits use, sharing, adaptation, distribution and reproduction in any medium or format, as long as you give appropriate credit to the original author(s) and the source, provide a link to the Creative Commons license and indicate if changes were made.

The images or other third party material in this chapter are included in the chapter's Creative Commons license, unless indicated otherwise in a credit line to the material. If material is not included in the chapter's Creative Commons license and your intended use is not permitted by statutory regulation or exceeds the permitted use, you will need to obtain permission directly from the copyright holder.

Intelligent Approach to Enhance Redundancy in Novel Steer-by-Wire for Heavy Earth Moving Machinery

Vinay Partap Singh[ID], Abid Abdul Azeez, and Tatiana Minav

1 Introduction

In recent years, the electrification and automation of Heavy Earth Moving Machinery (HEMM) have emerged as pivotal areas of research, revolutionizing the industry with innovative technologies and intelligent systems. Amid this technological renaissance, Artificial Intelligence (AI) has played a crucial role with different applications in these machines. Apart from autonomous and assisted operations of these machines, integration of Machine Learning (ML) and Deep Learning (DL) methodologies has been used for fault detection in hydrostatic systems, which are integral parts of these machines [1–3].

This paper specifically focuses on one of the most vital and safety-critical aspects of HEMM: the steering operation. Traditionally, HEMM has relied on hydrostatic steering systems primarily because of their reliability and redundancy. The redundancy in hydrostatic steering is usually achieved by the manually operated steering, which entirely depends on the operator for response time and effectiveness. In previous studies, the authors proposed an Electro-Hydrostatic Steering System (EHSS), which operates on the Steer-by-Wire (SbW) principle and has potential to comply with safety standards [4, 5]. In the proposed EHSS, the primary steering operation is by an electric motor-controlled electro-hydrostatic actuator, while redundancy is achieved by a proportional valve system. In scenarios where the steering system encounters a fault, the timely activation of a secondary steering channel is imperative to prevent potentially hazardous outcomes. This research proposes an intelligent approach to swiftly identify faults in the primary steering system and accordingly activate the secondary steering, utilizing steering pressure signals as a key indicator. The overall architecture of the steering system is shown in Fig. 1.

V. P. Singh (✉) · A. A. Azeez · T. Minav
Innovative Hydraulics and Automation, Tampere University, Tampere, Finland
e-mail: vinaypartapsingh@tuni.fi

© The Author(s) 2025
L. Ericson and P. Krus (eds.), *Advancements in Fluid Power Technology: Sustainability, Electrification, and Digitalization*, Lecture Notes in Mechanical Engineering,
https://doi.org/10.1007/978-3-031-84505-5_10

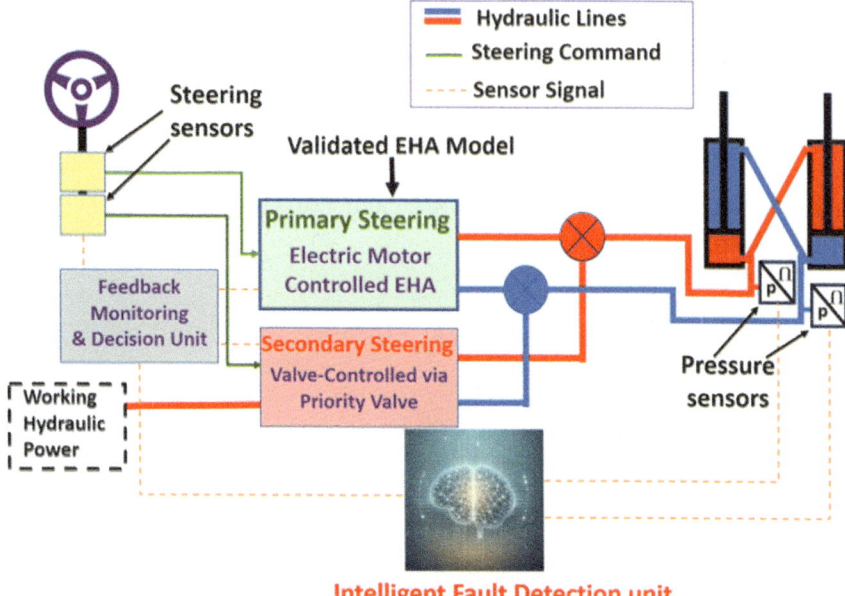

Fig. 1 Overall architecture of the steering system with intelligent fault detection

Addressing the complexity and critical nature of fault detection in steering systems, this paper explores an intelligent approach using machine learning and deep learning algorithms. These classification algorithms are carefully chosen and applied to analyze the intricate patterns of steering pressure signals, enabling the timely detection and response to any anomalies or faults. Algorithms including bagged ensemble of decision trees, neural network-based multi-layer perceptron, and Gaussian kernel-based Naive Bayes classifiers are chosen and tuned among various trained and tested ML and DL classification algorithms to analyze the pressure signal data. This data is generated through a partially validated model of steering, created in a unique real-time interactive co-simulation environment integrating MATLAB/Simulink, Simcenter AMEsim, and Mevea multi-body dynamics, with control using a physical joystick. The study simulates two major fault scenarios identified in previous research [4], ensuring a comprehensive and representative dataset for training the classifiers. Further, an ensemble of these trained classifiers is created and integrated into the co-simulation model for real-time fault detection. This study showcases the potential of artificial intelligence (AI) in enhancing safety through redundancy in safety-critical systems like steering. The implementation of AI in such critical systems underscores the evolving landscape of industrial machinery and the continuous push towards safer, more efficient, and technologically advanced solutions.

The next section in paper describes the real-time interactive co-simulation environment, and steering system operation with experimentally validated EHA model. The data generation, data analysis, and description of different ML and DL models trained and used along with finally selected model hyperparameters and the structure of the final ensemble for fault detection are described in the third section. The fourth section contains results of fault detection in real time using the intelligent fault detection unit, followed by conclusion and future aspects of the study.

2 Real-Time Interactive Co-simulation Environment and Steering Model

2.1 Real-Time Interactive Co-simulation Environment of Wheel Loader

A 16-ton wheel loader's real-time interactive co-simulation model is created using MATLAB/Simulink, Simcenter AMESim, and Mevea [6], each chosen for their unique advantages. Mevea software specializes in digital twin technology and simulation platforms, effectively simulating the mechanics, hydraulics, power transmission, and operating environment of the machine with its physics engine. Siemens' AMESim stands out for its object-oriented programming capabilities, providing extensive libraries for hydraulics, mechanics, and interfaces. Finally, MATLAB/Simulink excels in data management, ML/DL and statistics, controller design, and its ability to interact with multiple software platforms simultaneously.

The model of the wheel loader, incorporating multibody dynamics and interaction with a realistic environment, is developed in Mevea. The design of steering actuators, secondary steering hydraulics, and the connection between primary and secondary steering systems are executed in AMESim. The intelligent fault detection unit using ML and DL algorithms, electric motor-controlled Electro-Hydrostatic Actuation (EHA) model, overall control, steering command, and signal monitoring, is integrated within MATLAB/Simulink.

Control over the wheel loader's operation in co-simulation model is versatile, allowing for a physical joystick, keyboard inputs, or a scripted set of commands. A physical joystick is integrated for this study for more interactive operation. Figure 2 illustrates the comprehensive structure of the environment, with each block representing the components realized in the respective software, and arrows indicating the direction of data flow.

It's important to note that the steering functionality is exclusively powered and managed by MATLAB/Simulink and AMESim. In contrast, all other operations, including driving and handling the working implements of the wheel loader, are controlled using a physical joystick exclusively in Mevea software. Although the steering can also be managed using the joystick, uniformity in command is maintained by inputting steering references in MATLAB/Simulink. The movement of the wheel loader can be visualized in Mevea software along with real-time data monitoring in MATLAB/Simulink.

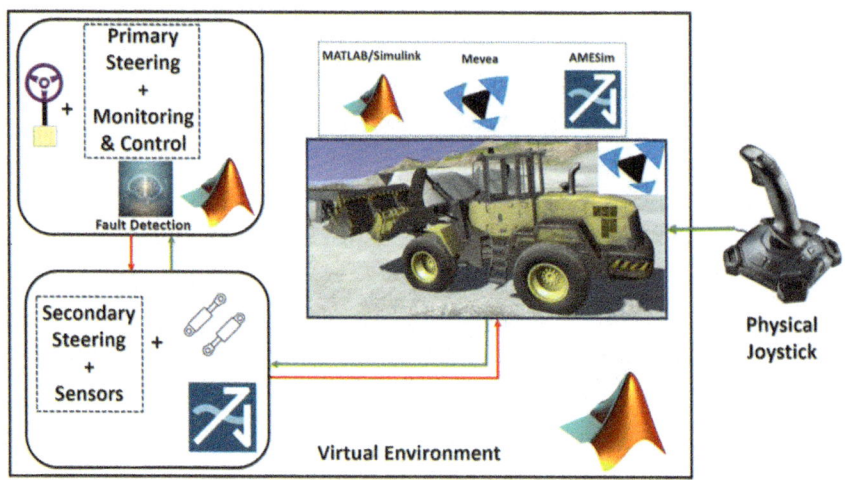

Fig. 2 Real-time interactive co-simulation model with EHSS and intelligent fault detection

2.2 Steering System Model and Operation

The design of the wheel loader's steering mechanism used in the co-simulation model is shown in Fig. 3. There are two independent modes of steering, primary steering function is enabled by an electric motor-controlled electro-hydrostatic actuator (EHA), while the redundancy is achieved using a proportional valve connected through a priority valve. For an in-depth understanding of the steering mechanism, its functionality, and its adherence to specific standards, references [3] and [4] provide comprehensive details. Under standard conditions, the primary steering is managed by the electric motor in the EHA. If a hazardous situation arises, mechanical isolation of the primary steering is achieved via locking valves (4.1, 4.2), allowing the continuation of operation through the redundant path. This system utilizes a priority valve (5) to redirect flow to other hydraulic functions if steering is not required. Upon activation of secondary steering, and when the proportional valve (6) demands flow, it is rerouted to power steering. The EHA model as shown in green dotted boundary of Fig. 3, is validated on an experimental test bench. The key values and parameters for the steering components are listed in Table 1. It shall be noted that more detailed description of all the components with their specification, detailed description of the test bench, model, and validation, which are not considered crucial for this study can be found in [7].

Figure 4 depicts the control block diagram and signal flow for the steering operation under non-hazardous conditions as part of the wheel loader's co-simulation model.

In this study, for consistency, the steering actuator position command (X_{ref}) is inputted in MATLAB/Simulink, while the remaining functions of the wheel loader are

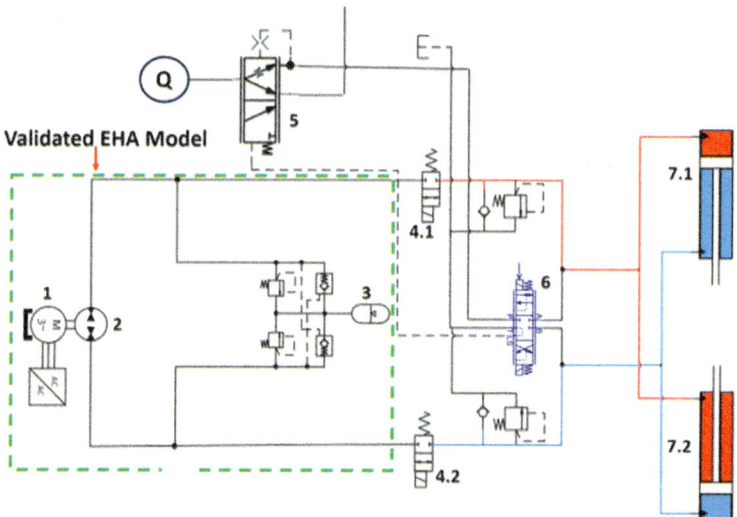

Fig. 3 Schematic of steering system

Table 1 Parameters of steering system components

Component	Main parameters
Electric motor and drive	Rated power 2.54 kW, Rated speed 3000 rpm, Max continuous current 11 A
EHA hydraulic pump	14.53 cc/rev
Steering actuators	80/50, stroke—340 mm
Proportional valve	Nominal flow—50LPM, pressure drop—15 bar, LS port
Constant flow source (Q)	68 LPM
Pressure relief valves	210 bar with anti-cavitation

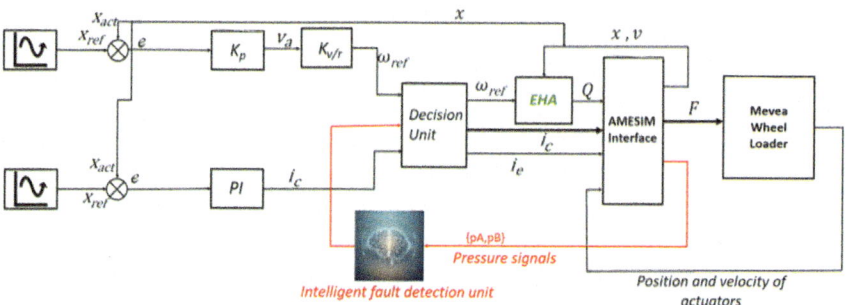

Fig. 4 Control diagram and signal flow in steering

governed in Mevea using joystick. For primary steering, the actual steering actuator position (X_{act}) is derived from AMESim using virtual sensors.

A "P" controller (K_p) generates the rotational speed command (ω_{ref}) for the electric motor. The linear actuator's velocity command (V_a) is translated into the motor's rotational speed command using a coefficient ($K_{v/r}$). The validated EHA model processes the electric motor speed reference, along with the position and velocity of the actuators (x, v) from AMESim, to regulate the hydraulic flow (Q) to the steering actuators in AMESim. Secondary steering involves a PI controller, which adjusts the valve command signal (i_c) to the load-sensing proportional valve in AMESim. An integral element (I) compensates for the non-linearities introduced by the valve's dead zone.

The intelligent fault detection unit receives pressure signals and sends signal in binary form of healthy or faulty to the decision unit. The decision unit receives input from both steering channels and the fault detection unit. In normal functioning, the decision unit transmits the electric motor speed reference to the EHA and adjusts the locking valve signal (i_e) as needed. When it receives the signal for fault, it deactivates the primary steering, shuts down the electric motor, de-energizes the locking valves, and transfers control to the secondary steering system for the rest of the operation.

The force command (F) for steering is sent to the wheel loader in Mevea, which calculates the real-time position and velocity of the actuators. This data is then relayed back to AMESim, where it is converted into the necessary force and pressure requirements. The MATLAB/Simulink environment serves as the central hub, connecting, monitoring, and managing the real-time data exchange between the various software systems.

3 Intelligent Fault Detection Unit

3.1 Data Generation

The data for training and testing of different ML and DL classification algorithms is generated from the previously described interactive co-simulation environment. The wheel loader was operated in digital environment with a time series reference steering command which utilizes steering actuator stroke on both sides. However, the path followed by the wheel loader was kept random for more representative dataset. The movement of the wheel loader, except the steering command, was controlled by the physical joystick. Hence, despite being the same steering command, the steering forces and hence steering pressures are random because of the non-uniform and random terrain the wheel loader is operating, with some level of similarity in the beginning of the operation.

Two different fault scenarios are simulated to generate the data for the training of ML and DL classification models. The simulated fault scenarios are loss of power in primary steering, and unintended/uncommanded steering, they have been chosen as they have been considered the most hazardous scenarios for the present case according to ISO 19014-5 [8]. The faults are introduced by manipulating the speed of electric motor of EHA manually in MATLAB/Simulink. For each fault scenario, eight simulations are performed with failure at different time instances, making it a total of sixteen simulations for training data. Each simulation time ranges from 10 to 25 s depending on the timing of fault, with 1 ms of sample time. The pressure data of both the chamber sides is recorded and labeled as healthy or faulty in binary form. It shall be noted that although both faults are simulated separately, for training of the classifiers they are merged, and data is labeled as healthy or faulty where faulty includes both the fault scenarios.

For testing dataset, the same procedure was followed for two simulations per fault scenario, a total of four simulations with time range varying from 15 to 25 s with 1 ms of sample time. Quantitatively with a number of data samples, it makes it in ratio of 80%–20% for training and testing datasets, respectively.

3.2 Classification Models

As stated earlier, raw pressure signals are used as key features for training and testing of ML and DL classification models, as the final goal of the study is to identify faults using signals from pressure sensors. For this, a wide range of these classification models are trained and tested with a range of respective hyperparameters. The training of the said models and tuning of the hyperparameters is a complicated and computationally demanding process, limiting the ability of further improvement for the current study because of available computational resources. Finally, three models are selected to be used in real-time fault detection based on the accuracy, model size, and compatibility and prediction speed in MATLAB/Simulink as it becomes crucial in real-time simulations.

Bagged Decision Tree Ensemble Classifier

The bagged decision tree ensemble classifier is a machine learning model that uses an ensemble of decision trees to enhance classification accuracy. It employs bootstrap aggregation (bagging) to train each tree on a unique subset of the data, drawn with replacement. This technique helps mitigate overfitting and improves model robustness. The classifier combines predictions from all trees, typically through majority voting, to make final decisions. This method is particularly effective in handling complex datasets with high variance, offering a balanced approach between bias and variance.

The main hyperparameters for this model are the number of learners in the classifier and the maximum number of splits in each learner. The robustness and accuracy can be improved with higher value of both the parameters. But with the increase in the said hyperparameters, the computational power required for training and size of the final ensemble model increases. The size of the ensemble model has a significant

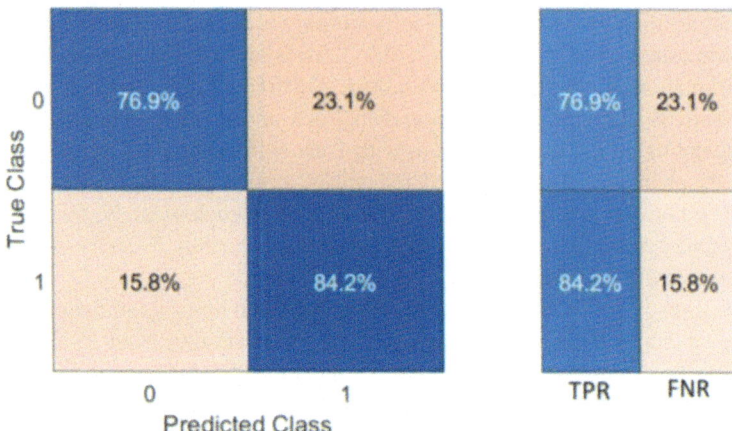

Fig. 5 Confusion matrix of bagged decision tree ensemble classifier

effect in this case as the bigger size model makes the real-time prediction significantly slower. The used hyperparameters are 50 learners, with maximum number of splits allowed as 1000.

Figure 5 shows the confusion matrix of bagged decision tree classifier. Here, "0" represents the healthy case, while "1" corresponds to fault. The True Positive Rate (TPR) is the proportion of actual faulty cases correctly identified as faulty, out of all actual faulty and healthy cases. While, the False Negative Rate (FNR) is the proportion of actual faulty cases incorrectly identified as healthy, out of all actual faulty and healthy cases.

The accuracy of the model can be calculated as

$$Accuracy = \frac{TP + TN}{TP + FN + TN + FP} \quad (1)$$

$$Accuracy = \frac{74.2 + 76.9}{84.2 + 76.9 + 15.8 + 23.1} = 80.55\% \quad (2)$$

In Eq. 1, True Positives (*TP*) is the number of faulty cases correctly identified as faulty, while True Negatives (*TN*) is the number of healthy cases correctly identified as healthy. On the other hand, False Negatives (*FN*) is the number of faulty cases incorrectly identified as healthy while False Positives (*FP*) is the number of healthy cases incorrectly identified as faulty.

Neural Network-Based Multi-layer Perceptron

The neural networks are widely used because of their good accuracy and robustness on complex data in variety of applications. The network is a feedforward, fully connected neural network where each layer has a connection to the previous layer. Each fully connected layer process the input from previous one with a weight matrix and adding a bias vector. There is an activation function following each fully connected layer, and finally output layer produce the predictions of the network.

The important hyperparameters for this type of neural network are number of hidden layers, number of neurons each layer, activation function, and regularization strength. Similar to the previous classification model case, the robustness and accuracy of the neural network can potentially be improved with a greater number of fully connected layers and number of neurons in each layer. Although the size of trained classifier is very compact and prediction speed is fast, the training time for the model increases significantly with increasing the said parameters. Figure 6 shows the confusion chart of the used classification model, the accuracy of model is 81%. The data is standardized for training and a network of three hidden layers with number of neurons as 20 for the first two layers and 15 for the last hidden layer is selected. For the fully connected layers a hyperbolic tangent (tanh) function is used as activation function, it ensures that input to the next layer is bounded between − 1 and 1. Whereas the output is produced by SoftMax activation function, it makes sure that the sum of the probability distributions of all the predicted classes is 1.

The regularization strength in a neural network classifier controls the degree of regularization applied, helping to prevent overfitting by penalizing complex models. The regularization strength applied in this case is selected from a widely applied practice as in the range of inverse of total number of training observations.

Gaussian Kernel-Based Naive Bayes

In the context of Naive Bayes classifiers, it's a simple yet effective approach used for classification tasks, particularly suited for large datasets. This method relies on the assumption of independence between predictor variables. It is highly efficient in training and prediction, capable of handling both continuous and categorical data. The model calculates the probability of different classes based on input features

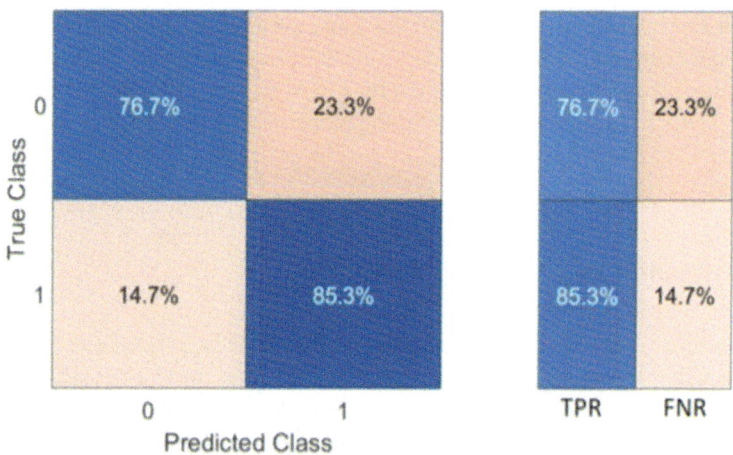

Fig. 6 Confusion matrix of neural network-based multi-layer perceptron

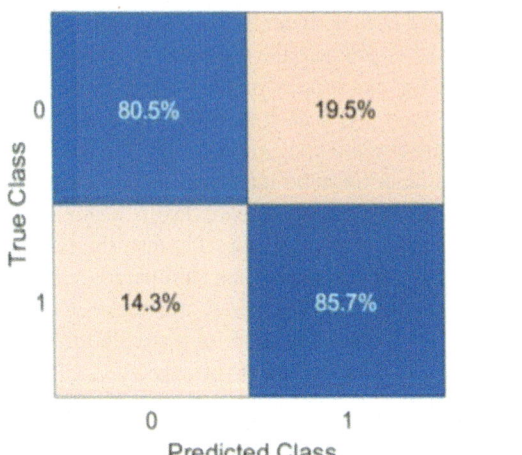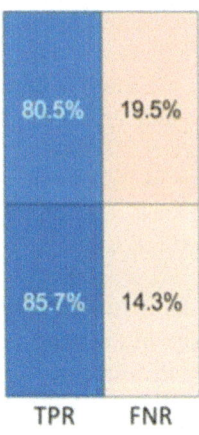

Fig. 7 Confusion matrix of Gaussian kernel-based naive Bayes

and selects the class with the highest probability as the output. This technique is versatile and can be adapted for various types of data distributions. The effectiveness of a Naive Bayes classifier often depends on the degree to which the independence assumption holds true in the given dataset. The main reason for including this in the present study is its good probability score output for classification.

The classifier selected is a Naive Bayes classifier with Gaussian assumptions for the data distribution. The model is configured to handle data that is unbounded and standardizes the data before processing. Figure 7 shows the confusion chart of the used classification model, the accuracy of model is 83.1%.

It is worth noting that prediction speed of this model is significantly slower than the previous two models. Naive Bayes calculates probabilities using all features for each class during prediction, which is time-consuming. Nevertheless, it did not affect significantly in overall fault detection for the present case, as due to limited computational capability the simulation time loop is more than 100 ms in the whole process. Which allows enough time for the model to make predictions and does not slow down the process further.

3.3 Ensemble of Trained Classifiers

The pressure signals in an active and dynamic system like steering can have very random patterns. Any ML or DL models need to be trained on an extensive amount of data to be able to capture the real-life behavior in such cases. Moreover, the models can label the signal wrongly at any given time step in continuous signal despite the majority of signal in that range are identified correctly, resulting in a false alarm. Ensemble of different ML and DL models have been used in different applications to eliminate such possibility and enhance the accuracy [9], and redundancy in real-time monitoring [10].

The three classification models explained beforehand are integrated to form an ensemble for more robust and accurate fault detection. Each model has its unique ability, Naive Bayes for its probabilistic output, neural networks for their ability to capture complex patterns through layers of computation, and bagged trees for their variance reduction and stability. The prediction score of fault from three models has been used instead of the output label of healthy or faulty, and a weighted moving average of scores is taken over a rolling window. This gives the advantage of ignoring any false alarms and wrong prediction in any individual classification model. A threshold value of this mean is defined to classify the signal as healthy or faulty. Figure 8 shows the pictorial representation of the ensemble to make the intelligent fault detection unit. The final output is in binary form, where 0 represents the healthy and 1 represents the fault.

As evident from the accuracy of the trained classification models, the performance of the models is at the similar level in terms of accuracy. Hence, the final weights of the models are kept same for the weighted moving average of score. For the current study, the size of rolling window is 200 points, whereas two threshold values are used for case of active steering command and absence of steering command at 0.8 and 0.5, respectively. The difference in the threshold values represents separate criterion to label the signal faulty, for fault scenarios representing loss of power in primary steering and uncommanded activation of steering. The separate thresholds have been implemented to differentiate whether a sudden change in pressure signals is due to an intentional steering command or a fault. It effectively means that if the rolling

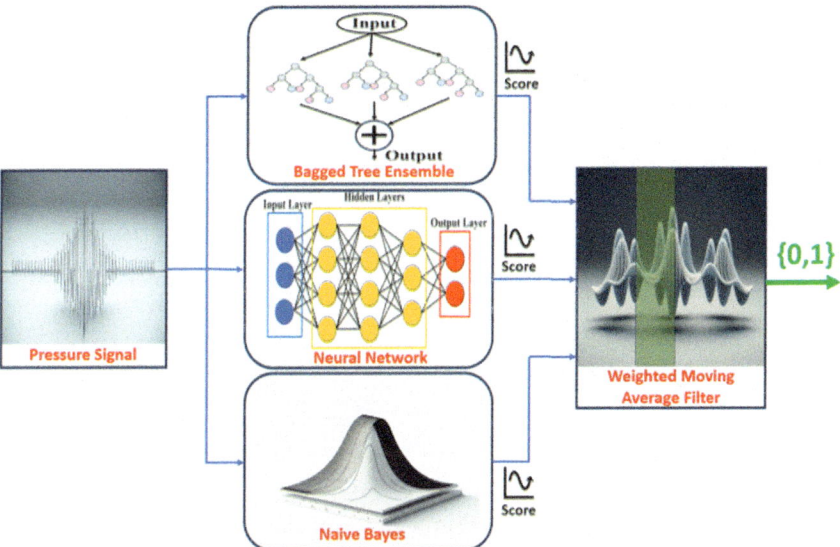

Fig. 8 Intelligent fault classification unit

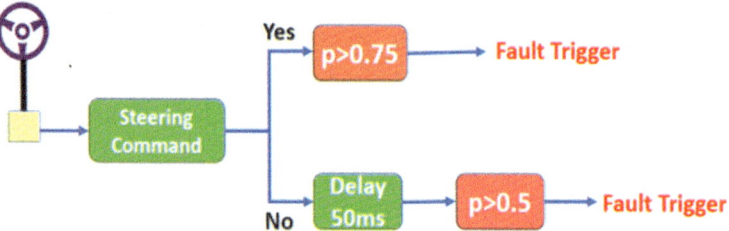

Fig. 9 Flow diagram to label the weighted average signal as faulty

average of weighted average scores for the last 200 ms data have a confidence of more than threshold for signal being faulty, the alarm is raised for a fault in steering system. Figure 9 shows the flow diagram to label the weighted average signal as faulty where "p" is the probability of signal being faulty.

4 Real-Time Fault Detection

The intelligent fault detection unit explained in the previous section is integrated into the real-time interactive co-simulation environment explained in Sect. 2. The two major fault scenarios in steering are injected in simulations in the same way as explained in Sect. 3.1. Figure 10 shows the physical setup for the real-time interactive co-simulation environment.

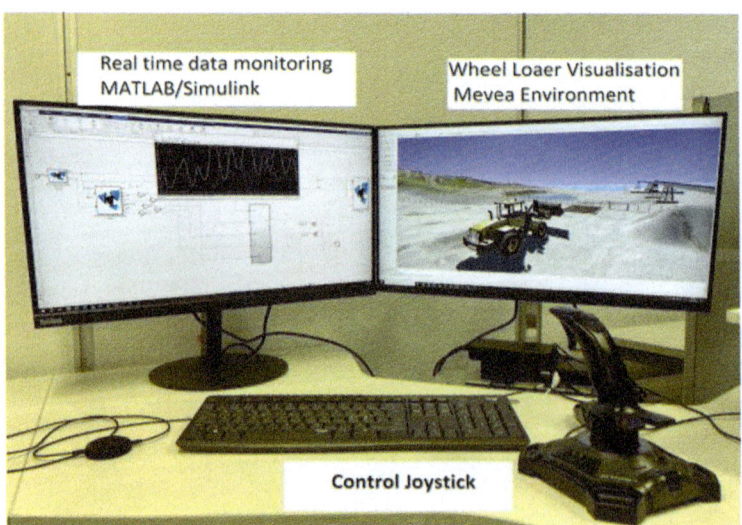

Fig. 10 Setup for real-time interactive co-simulation environment of wheel loader

A total of ten simulations have been performed to analyze the performance of the fault detection unit over different steering commands. Four simulations are performed for each fault scenario, uncommanded steering activation, and loss of power in primary steering, whereas two simulations are performed for no-fault case which include a period for no steering command and active steering, to analyze the performance for false alarms. The motion of the wheel loader is actively controlled using the joystick, which sends analogue signal to move the wheel loader model on a random terrain.

The fault detection unit detected the faults in all the eight fault scenarios; moreover, it did not give any false alarms in two healthy operation cycles. When the fault detection unit detects the fault, it sends a trigger to the decision unit, explained in Sect. 2.2, which then isolates the primary electric motor-controlled EHA steering mechanically/hydraulically and continues the operation with the redundant steering channel.

Figure 11 shows one case of fault for uncommanded activation, where the steering is activated without any steering command present. The fault occurs at t = 2 s when there was no steering command, and the fault is detected in 138 ms. Before t = 2 s, the wheel loader is moving in a straight line with lower pressure values and insignificant pressure peaks. Hence, the fault detection unit is almost certain that it is a healthy signal with near-zero fault probability. As soon as the fault detection unit sends the trigger to decision unit, it isolates the EHA steering and activates the secondary steering. It can be observed from the probability plot that from the time the redundant (secondary) steering takes the control, the probability of signals being faulty goes down consistently. Nevertheless, the probability in this region is fairly below the threshold values which shows the secondary steering is actively bringing the steering to desired value and signifies the redundancy of the fault detection unit. As the steering was activated suddenly resulting in pressure peaks, it takes some time for redundant steering channel to dampen the oscillations which can be improved by better control strategies.

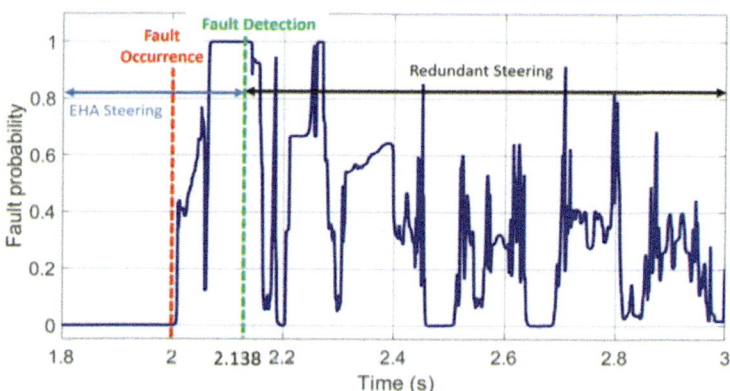

Fig. 11 Fault probability plot for uncommanded steering fault

Figure 12 shows the similar probability plot for the second type of fault, loss of power in primary steering. Before t = 7 s, the steering is healthy with electric motor-controlled EHA, and hence the probability is at significantly less value than threshold. At t = 7 s when there is a steering command present, the electric motor in EHA steering was shut off, leading to no active power source for the steering. This leads to a change in pressure signals, and as evident from the probability curve, the probability of signal being faulty rises and fault is triggered in 337 ms. Once the redundant steering takes over, the probability again falls down after a very short period, indicating the steering being in healthy condition.

Table 2 shows the time taken to detect the fault in each simulation for both the fault cases. The time taken for the detection of fault varies for all the cases, it further depends on the vehicle speed and terrain it is being operated, as the steering pressure is directly affected by these factors. Furthermore, the size of rolling window and threshold values define the sensitivity of the fault detection. As explained beforehand in this section, four simulations are performed for each fault, making it total eight simulations. In the tested cycles explained beforehand in this section, the average time for fault detection is 307 ms for the loss of power in primary steering fault case named Fault B, while 155 ms for uncommanded steering actuation is termed as Fault A.

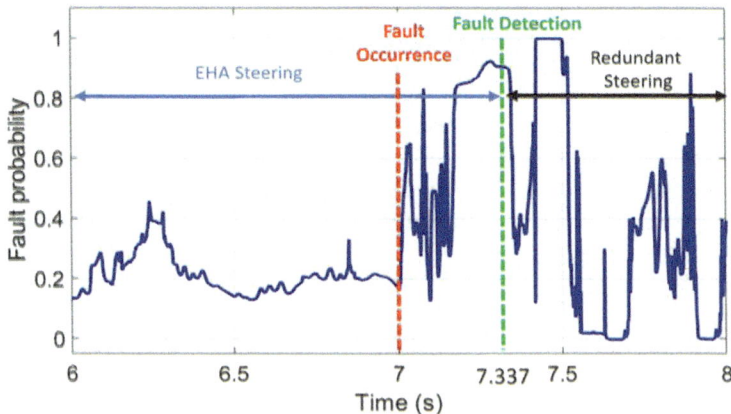

Fig. 12 Fault probability plot for loss of power in steering fault

Table 2 Time of fault detection for each simulation

	1	2	3	4
Fault A (ms)	205	138	138	137
Fault B (ms)	308	242	337	343

5 Conclusion

An intelligent approach has been used to enhance the redundancy in Steer-by-Wire (SbW) for HEMM using ML and DL classification models. A real-time interactive co-simulation environment of a wheel loader integrating the partially validated model of SbW system is used for the purpose. The pressure signal from both sides of the steering actuator is used as main indicator. Three different classification models based on ML and DL are trained on the data for two major steering failure conditions. An ensemble of the trained classification models is created using probabilistic approach to form the intelligent fault classification unit.

The intelligent fault classification unit detected the faults in real-time simulations, leading to timely activation of redundant steering channel. The time of fault detection varies in both the fault scenarios, it took an average of 307 ms to detect loss of power in primary steering, whereas 155 ms for an uncommanded steering. The study demonstrates the potential of using AI to enhance the redundancy of safety-critical systems like steering.

Acknowledgements This research is funded by the Business Finland Project EMMA-2 (2471/31/2021).

References

1. Beck B, Weber J (2017) Enhancing safety of independent metering systems for mobile machines by means of fault detection. In: Proceedings of 15th Scandinavian international conference on fluid power
2. Saeedzadeh A, Habibi S, Alavi M (2021) A model-based FDD approach for an EHA using updated interactive multiple model SVSF. In: Fluid power systems technology, vol 85239. American Society of Mechanical Engineers
3. Abdul Azeez A et al (2021) AI-based condition monitoring of hydraulic valves in zonal hydraulics using simulated electric motor signals. In: Fluid power systems technology, vol 85239. American Society of Mechanical Engineers
4. Singh VP, Huova M, Minav T (2023) Energy efficient steer-by-wire in articulated non-road mobile machines: analysis and proposal. In: Scandinavian international conference on fluid power, SICFP'23
5. Singh VP, Huova M, Minav T (2023) Simulation study of a fail-safe steer-by-wire for heavy earth moving machinery. In: Fluid power systems technology, vol 87431. American Society of Mechanical Engineers
6. Mevea Ltd. (2022) Mevea real-time simulation software | digital twins techology. Mevea. https://mevea.com/solutions/software/. Accessed Aug 2023
7. Singh VP, Raunio V, Niemelä J, Minav T (2024) Hazard-free steer-by-wire in articulated heavy earth moving machinery using co-simulation model. In: Proceedings of the 14th internatinaol fluid power conference. Dresden, Germany

8. SFS-EN ISO 19014-5:2021:en Earth-moving machinery. Functional safety. Part 5: Tables of performance levels (ISO 19014-5:2021)
9. Ghaemi A et al (2022) Accuracy enhance of fault classification and location in a smart distribution network based on stacked ensemble learning. Electr Power Syst Res 205:107766
10. Abdulaal MJ et al (2022) Real-time detection of false readings in smart grid AMI using deep and ensemble learning. IEEE Access 10:47541–47556

Open Access This chapter is licensed under the terms of the Creative Commons Attribution 4.0 International License (http://creativecommons.org/licenses/by/4.0/), which permits use, sharing, adaptation, distribution and reproduction in any medium or format, as long as you give appropriate credit to the original author(s) and the source, provide a link to the Creative Commons license and indicate if changes were made.

The images or other third party material in this chapter are included in the chapter's Creative Commons license, unless indicated otherwise in a credit line to the material. If material is not included in the chapter's Creative Commons license and your intended use is not permitted by statutory regulation or exceeds the permitted use, you will need to obtain permission directly from the copyright holder.

Safety Function-Failure Mode and Effect Analysis a Novel Approach of FMEA for Safety Application in Mobile Working Machinery

Christa Maria Düsing● and Frank Will●

1 Introduction

The Failure Modes and Effects Analysis (FMEA) process, incorporating Safety Functions, is a well-established practice in the automotive industry. Special machines for mobile applications are manufactured in smaller series for both on-road and off-road use. On-road mobile machines adhere to functional safety standards such as ISO 26262 [1] and ISO 13849 [2]. Compliance with local Road Traffic Act regulations may also be necessary for road approval. For mobile off-road machinery, adherence to ISO 13849 [2] and ISO 19014 [3] is essential during operation.

In the development phase, challenges arise in implementing safety architectures (Cat 2, 3 & 4) for on- and off-road machine controls, particularly for (semi-) autonomous machines. Existing standard FMEA structures and software tools often lack the necessary considerations, including process-related data like PFH_D, DC, and PLs. This gap hinders the integration of specifically defined SFs for mobile machinery into the FMEA and impedes the efficiency of FMEA processes.

The current FMEA processes, heavily influenced by the automotive industry, pose a barrier to seamlessly incorporating SFs according to ISO 13849 [2] requirements into professional FMEA SW tools with minimal effort.

The objective of this paper is to provide methods to overcome these challenges by introducing additional safety architectures (CAT 2, 3 & 4), incorporating functions to determine relevant process data and establishing interfaces to integrate all possible

---

C. M. Düsing (✉)
XCMG European Research Center GmbH, Krefeld 47807, Germany
e-mail: duesing@xcmg-erc.com

F. Will
Institute of Mechatronic Engineering, Chair of Construction Machinery, Technische Universität Dresden, Dresden 01069, Germany
e-mail: frank.will@tu-dresden.de

solutions into an FMEA structure and a professional FMEA SW tool. As the approach described focuses on the evaluation of failure modes (ISO 13849), it is also necessary to embed the results of the SF-FMEA in a machine simulation model to approximate real-life conditions for the validation and verification of system behaviour.

Safety standards prescribe various methods to mitigate risks and failures at different stages of the development process for mobile working machinery. The interaction of SFs and safety objectives is outlined through these diverse methods (see Table 1). Designing and developing a product necessitates a multidisciplinary approach integrating individual methods. Each method addresses specific product, system, and component aspects employing various analysis tools with distinct tasks. The challenge lies in combining and integrating these methods effectively. The methods are interconnected to address this, and the results at each level serve as input for the subsequent phase.

The successful execution of FMEA requires the cooperation of an interdisciplinary team of experts who have a deep understanding of the product or system and its safety

Table 1 Interaction of safety goals, hazard and risk assessment, FMEA & FMEDA

Development levels	Risk categories	Specific tools and methods
Product Level 2-way Excavator	Concept optimization risk	Concept optimization, Product-FMEA
	Financial risk	Cash flow forecast
	Strategy risk	SWOT-Analysis
	Time risk	Program evaluation and review technique (Pert)
Vehicle Level 2-way Excavator	Hazard & Risk Assessment	ISO 12100 (Safety of Machinery)
	Safety Specification	Risk graph in accordance with ISO 26262 – ASIL (Automotive Safety Integrity Level)
	Safety Classification	ISO 13849 (Safety of control systems), PL_r (Performance Level required)
System Level Bucket System	Safety-relevant Analysis	Failure modes and effects analysis (System-FMEA, Safety Function-FMEA, MSR-FMEA)
	Systematic Failure Analysis	Fault-based system response analysis (FSR)
Component Level Pressure / Angle Sensor	Mechanical component risk	Failure modes and effects analysis (Design-FMEA, Process-FMEA)
	Analysis tools	FMEDA (failure modes, effects and diagnostic analysis), FTA (fault tree analysis)

aspects. The integration of functional safety considerations and the FMEA process constitutes critical elements in safety management, as requested by ISO 13849 [2] and ISO 19014 [3]. The absence of cohesive processes and functionalities in existing FMEA tools leads to prolonged and failure-prone development of SFs in mobile machinery. Consequently, there is a high need for research to explore more intelligent methodologies for carrying out SFs within FMEA and FMEDA procedures.

2 Conceptual Approach for Integration of SF into FMEA

The specific challenge for functionally safe systems featuring safety-relevant functions for mobile machines, stems from the amalgamation of work sequences and driving functions. This challenge is further complicated by the requirements for swift transport movements on public roads and the flexibility inherent in combinable machine systems. These distinctive characteristics must be considered during the safety-focused development of the system. A tailored development framework based on established guidelines and standards can serve as a foundation to guide this process.

The safety of a machine system is often contingent upon the risk associated with certain system failures or malfunctions. In this context, "risk" is defined as the product of the probability of a specific fault occurring and the resultant level of damage. When developing a machine or machine system, it becomes imperative to ascertain the hazard level of safety–critical functions and identify the necessary measures for reducing partial risks. One quantitative method for risk determination involves risk analysis, hazard and risk assessment (HARA) following ISO 12100 [4]. Another well-established approach for scrutinizing fail-safe hardware (HW) and SW mechanisms, as mandated by functional safety standards, is the combination of FMEA and FMEDA.

Integrating of functional safety and FMEA is an established procedure in the automotive industry. FMEA is prescribed according to ISO 26262 [1], an international standard that deals with the functional safety of electrical and electronic systems and refers to the safety of a system or component in executing a specific function in motor vehicles. This standard specifies requirements for the entire development process to ensure the reliable safety of electronic systems and describes the general reliability and safety of a system or component in vehicles. The FMEA is an integral part of the development process in following ISO 26262 [1]. It is explicitly used in various phases of the safety analysis to identify and evaluate potential faults and their impact on safety. The results of the FMEA are then integrated into other safety-related activities and decisions. Integrating safety functions into simulation tools includes conducting fault injection tests to verify their effectiveness in mitigating identified risks, ensuring functional safety across the development process. The method is discussed in Chap. 2, along with strategies for improving process performance.

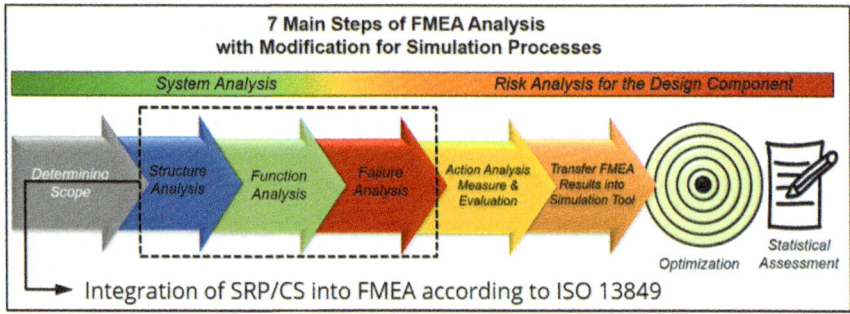

Fig. 1 Integration of SFs into standard FMEA processes with modifications for simulation processes [9, 12, 14, 15]

2.1 Verification of the Requirements

For (semi-)autonomous machines, the proposed SF-FMEA represents a future-orientated approach for the standard FMEA [9, 12] methodology. The SF-FMEA analysis study is based on the conceptual phase processes outlined in FMEA [9, 12, 14, 15] and ISO 13849 [2] and includes methods such as hazard and functional analysis, functional FMEA and system-theoretical process analysis.

Based on the HARA principles in ISO 12100 [4], potential malfunctions are analysed in various scenarios (both safety-relevant and non-safety-relevant) and categorised according to severity, exposure and controllability. Consequently, integrating the standard FMEA risk analysis with functional safety and transferring the FMEA results into a simulation tool for a digital model [16] forms a framework for the development phase of (semi-)autonomous machines (see Fig. 1).

2.2 Analysis of Synergies of FMEA and Safety Function Processes

This study examines the synergies between FMEA processes and functional safety. The main focus is investigating the essential prerequisites for performing FMEA analyses in functional safety methods. The method includes a specific comparison between ISO 26262 [1], which applies to the field of the automotive industry, and ISO 13849 [2], which is tailored to mobile machine applications, particularly in the field of (semi-)autonomous machines. Particular attention is focused on identifying possible overlaps, (see Table 2), in safety–critical processes that may arise when using professional FMEA SW tools per the above-mentioned standards. The primary aim of this targeted investigation is to identify standard interfaces and optimum integration points within the processes of both standards.

Table 2 Main difference between ISO standards for vehicles in the automotive industry and the mobile machinery

Standard Criteria	ISIO 26262	ISO 13849
Implementation of safety functions in the FMEA process	As safety prevention & detection measures which provide intrinsically safe design within the system	As add-on features by integration of new principles
Considered architectures	Functional channel with the structure of input, logic & output	Cat 2, 3, 4 with different channel structure for input, logic & output
Available values for calculation of parameters in the development phase	Parameters e. g.: Single-point fault metric latent-fault metric, failure rate Lambda, FIT, PFH_D, DC, DC_{avg}	Different parameters e. g. failure rate Lambda $MTTF_D$, PFH_D, DC, DC_{avg}
Evaluation of diagnostic coverage	Based on fault behaviour with conventional test procedures and within field tests	Missing evaluation methods for diagnostic coverage
Covered by existing FMEA software tool	Available processes by corresponding FMEA software tool	Missing processes by FMEA software tool

The comparison clarifies that integrating SFs into the FMEA process is provided in ISO 26262 [1] as preventive measures and that the detection measures can be realised via monitoring [5]. In the mobile machinery, the SF is generally regarded as an Add-On. The ISO 13849 [2] uses CAT 1–4 in addition to the required dangerous failure rates. This results in specific HW-Architectures in relation to the sum of the assumed possible failures. The ISO 26262 [1] standard does not provide for specific HW-architectures but aims to regulate them through additional areas of fault metrics. On the one hand, there is the Latent Fault Metric (LFM) and, on the other, the Single Point Fault Metric (SPFM). Table 3 gives an overview of the different calculation methods of both ISO standards.

This difference implies that safety-relevant components enable an intrinsically safe design in the automotive systems. Just the HW-architectures of Cat B and 1 of ISO 13849 [2] are consistent with the structure of ISO 26262 [1], with the input, logic and output.

Both standards, (see Table 3), require a statistical view of a SFs remaining dangerous failure rate and prescribe a specific DC. Statistical information for a Failure Mode Distribution (FMD) comes from the combination of common failure modes listed in ISO 26262 [1]. The DC is determined or calculated using a FMEDA which is already available in FMEA SW tools. Generally, the ISO 13849 [2] specifies that the values for DC_{avg} are estimated or determined by the manufacturer depending on the

Table 3 Main difference between the calculation method of both ISO standards

Parameters	Calculation method ISO 26262	Calculation method ISO 13849
Single-point fault metric	$SPFM = \frac{\sum_{SR,HW}(\lambda_{MPF,DP}+\lambda_S)}{\sum_{SR,HW}\lambda}$	Not available
Latent-fault Metric	$LFM = \frac{\sum_{SR,HW}(\lambda_{MPF}+\lambda_S)}{\sum_{SR,HW}(\lambda-\lambda_{SPF}+\lambda_S)}$	Not available
Failure rate Lambda	$\sum_{SR,HW}\lambda_i[\frac{1}{h}] =$ $\lambda_{SPF} + \lambda_{RF} + \lambda_{MPF} + \lambda_S$	$\lambda_D[\frac{1}{h}] = \frac{0,1 x n_{op}}{B_{10D}*365*24}$
Fit (Failure in Time)	$\sum \lambda_{SPF}[\frac{1}{h}] = 1, 0 FIT$	$FIT = \lambda_D[\frac{1}{h}] * 10^9$
PMHF / PFHD	$PMHF =$ $\lambda_{SPF} + \lambda_{RF} + \lambda_{MPF} + \lambda_{SR} * T_{LT}$	$PFH_D[\frac{1}{h}] = \frac{\lambda_D}{365*24}$
MTTFD	Not available	$MTTF_D[year] = \frac{1}{\lambda_D}$
DC	$DC_{SPF}[\%] = \left(1 - \frac{\lambda_{SPF}}{\lambda_i}\right)*100$	$DC[\%] = \frac{\sum \lambda_{DD}}{\sum \lambda_{DD}+\sum \lambda_{DU}}*100$

values for MTTF$_D$ and measures against common cause failures or from statistical field information.

2.3 Methods to Integrate Missing Safety Function Values into FMEA Process

Statistical methods provide a quantitative measure of the uncertainty and trust in the calculated values of the safety-related functions. This method allows a data-driven approach based on sufficiently available reliable data. To develop a professional FMEA SW tool with functional safety architectures according to ISO 13849 [2], specific calculations and analyses are required to create calculation algorithms that support the FMEA SW tool's functionality and programming. Existing approaches from the automotive industry can be used to determine accurate and reliable methods [14].

Quantitative categorisation is used when data on Failure Rates (λ), Failure Mode Ration (α) (FMR (α)) and Failure Effects are available. This approach is most appropriate when reliable system data can be collected. The following information and criteria are taken from the FMEA results:

- Functional Description
- Failure Modes
- Failure Mechanisms
- Failure Effects
- Severity Classifications / Ranking Numbers

The FMR (α) represents the percentage of parts that fail in a specific way. The majority of the new data collected to support the update of FMD-2013-1 contained known failure quantities, while less than 20% of the data consisted solely of percentage values. Therefore, a data merging algorithm that assigns a weight to each data source based on the total number of reported failures within that source was used. To combine the data in this way, all percentage data had to be converted into quantitative numbers.

The following iterative procedures were used to achieve this goal [7, p. 1–2]:

1. A quantity of "1" was assigned to the lowest percentage failure mode / mechanism (N1)
2. Quantities of all other Failure Modes / mechanisms were calculated:

$$\Sigma_{\text{Failure Modes}} = \left[(\text{Quantity of lowest percentage}) * \left(\frac{N_i}{N_1} \right) \right] \quad (1)$$

where N_i is the percentage associated with the ith Failure Mode / mechanism.

3. The percentages associated with the quantities calculated above were then calculated
4. The difference between the percentages from the original data was compared to the percentage derived from Step 3
5. If the difference of any failure mode/mechanism (between the actual and calculated) was greater than 1.0%, the quantity associated with the lowest percentage was incremented by "1" and Steps 2, 3 & 4 were repeated until all differences were less than "1" [7].

A hypothetical representation of the algorithm and the evaluation of the corresponding failure modes enables the creation of the standardised "FMR", which shows the proportion of the failure rate for each failure mode and the evaluation of the associated events [7, 8, p.6–7].

This method is applied in the Failure Mode, Effects and Criticality Analysis (FMECA) and can be used in addition to the FMEA. The Failure Mode Criticality (FMC) is calculated by multiplying the values "$\beta[1]$", which represent the probability that the failure that has occurred will have an effect. The value Alpha "$\alpha[\%]$" indicates the percentage of the total FMR attributed to each individual failure mode and lambda "$\lambda[\text{failure}/10^6 h]$" indicates the real Failure rate and the time "$t[h]$" gives absolute expected frequency of the specific failure mode per time unit.

The FMC [Failure] is calculated by

$$\text{FMC} = \beta * \alpha * \lambda * t \quad (2)$$

To ensure that all types of faults are recorded and identified, the FMD 2023-1 [7] database follows the fault conditions described in the AIAG & VDA [9] FMEA procedure. This approach ensures comprehensive recording and analysis of possible faults or failures in the analysed systems or processes. By taking the described fault conditions into account, potential problems can be recognised at an early stage of the development and appropriate measures for prevention or detection can be initiated. Typical failure conditions are shown in the following examples:

- Total failure: No operation, Broken, Burnt out, Damaged
- Partial failure: Incorrect pressure, Incorrect capacity, Out of specification, Out of adjustment
- Intermittent failure: Intermittent operation, Sporadic failure, Loose line, Broken spring, Broken seal
- Degrading faults: Deteriorated operation, Wear, Contamination, Corrosion

Additional insights into the fault conditions was provided in detail in the publication [8, pp. 7–9].

By integrating the FMR into the FMEA process, the FMR gains importance as a significant parameter (percentage of time in which or in what way an element fails). The failure mode distribution (FMD-2023–1) catalogue is used here, as a comprehensive model with real data and a structured methodology for determining the DC following ISO 13849 [2]. The use of the FMR and the classification of the listed faults into "Dangerous Detectable" (DD) and "Dangerous Undetectable" (DU) faults enable the determination of the DC and the creation of a forecast of the system behaviour. This approach increases the precision and reliability of the method when evaluating the effectiveness of diagnostic measures. The DC is calculated as the quotient of the Dangerous Failure Rate (λ_D) of the DD failure rates and the sum of all associated failure rates (DD and DU failure rates).

The DC is expressed by [2, 7], and calculated with Eq. (3)

$$DC = \frac{\lambda_{D_{Part}}}{\lambda_{D_{Total}}} = \frac{\lambda_{DD}}{\lambda_{DD} + \lambda_{DU}} = \frac{\sum \lambda_{m_{Detectable}}}{\sum \lambda_{D_{Total}}} \quad (3)$$

In accordance with ISO 13849 [2], the DC is calculated both for each individual component and for the overall function channel. This means that the DC is determined for the safety-related parts of the control system (SRP/CS) for all involved SFs. A calculation example of DC is shown with Table 4.

2.4 Development of a Model to Register Calculated Safety Function Values into FMEA Process

The main process of recording the SF values is then carried out with the assistance of the FMD Catalogue via a plug-in integration in the FMEA SW, (see Fig. 2); this application-specific module of the FMEA SW tool is essential for quantitative inductive safety analysis. The FMD catalogue makes it possible to calculate, validate and analyse the safety of every safety-relevant component within the systems via a pop-up window. The focus here is on inputs, a logic and outputs according to

Table 4 Calculation of diagnostic coverage by failure mode ratio [7, 8]

Failure mode Example Pressure sensor	α Failure Mode Ratio		λ_{D_Part} Calculated Failure rate	λ_m Modal Failure rate	λ_{DD} Failure rate Dangerous Detectable	λ_{UD} Failure rate Dangerous undetectable	DC [%]
Degraded operation	0,56	*	3,6146E-06	2,0242E-06		x	75,0
No Operation	0,19	*	3,6146E-06	6,8677E-07	x		
Functional failure	0,25	*	3,6146E-06	9,0365E-07	x		
Total	**1,00**			**3,6146E-06**			

the HW architectures as per ISO 13849 [2]. For this process, the existing FMD Catalogue following ISO 26262 [1] was adapted to the required performance data of ISO 13849 [2, 14].

Critical parameters such as $MTTF_D$, Dangerous Failure Rate (λ_D), Failure In Time (FIT), PFH_D, Mean time until 10% of the components fail dangerously ($T10_D$), DC, and DC_{avg} are calculated to ensure a thorough assessment of the safety performance of each component in the system of safety-relevant machines. Figure 2 shows the pop-up window for calculating the required safety-related data, which is then displayed

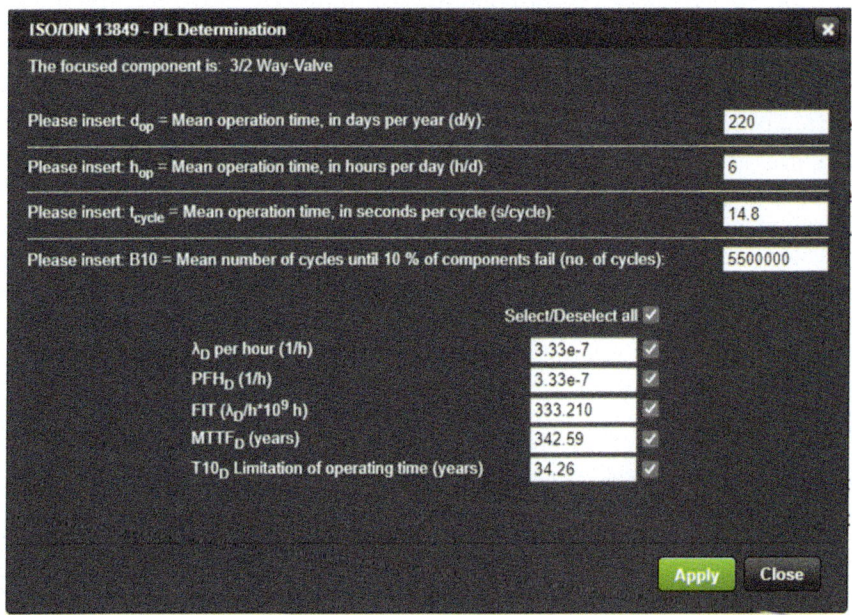

Fig. 2 Pop-up window, calculation of performance parameter of one component [8, 14]

in the FMD catalogue, (see Fig. 3). Several parameters or performance data can be entered or checked in this window to ensure that the safety requirements are fulfilled. The detailed calculation and consideration of these critical safety parameters for each component within the system is essential to ensure the overall safety and reliability of the safety functions. The use of proven and appropriate components, a Failure Mode Catalogue and a special pop-up window, shown with Fig. 2, for calculation underlines the systematic approach to safety assessment [8, 14].

λ_D [1/h] is indicated by [2, p. 82] and calculated for one component by Eq. (4)

$$\lambda_D[1/h] = \frac{0,1 x n_{op}[1/\text{year}]}{B_{10D} * 365[\text{days/year}] * 24[h/\text{day}]} \quad (4)$$

PFH_D[1/h] is a simplified formula given by [2], applies only to one component without considering the diagnostic coverage factor and the fault detection factor. The PFH_D[1/h] is calculated according to Eq. (5)

$$\text{PFH}_D[1/h] = \lambda_D[1/h] = \frac{\lambda_D[1/\text{year}]}{365[\text{days/year}] * 24[h/\text{day}]} \quad (5)$$

An algorithm integrated into SW tool transfers the safety-relevant parameters for the corresponding categories into the FMEDA form sheet to calculate the entire functional channel and check the fulfilment of the specified safety goal.

The FMEDA is a quantitative inductive safety analysis that can be used to compare the safety concept of safety-relevant on-road and off-road vehicle systems. The results of an FMEDA are key indicators of the DC, the proportion of failure rates of DD failures in the sum of all DD failures and DU failures. An FMEDA expands the structural, functional and failure analysis of the FMEA to include information on the reliability of hardware elements (e.g. failure rates for failure modes) and thus makes the effects of random failures quantifiable. The FMEDA can be validated and verified during the development phase of a system's safety–critical components. The essential process of recording the safety function values is then done using the FMEA SW tool with the FMD catalogue, which is significant for performing a quantitative inductive safety analysis. The FMD catalogue is a comprehensive guide to help systematically evaluate the safety performance of all safety-relevant elements. The FMEDA form sheet, (see Fig. 4), then enables the calculation of the entire functional channel and checks the fulfilment of the specified safety target. This allows the safety-relevant parameters for the corresponding categories to be documented and recorded. This approach is an essential module for detecting and recording possible failures and their effects at an early stage of the development phases, the properties are:

- Documentation and evaluation of quantitative parameters
- Calculation of the safety-related parameters with individual procedures and formulas (pop-up window for calculation parameters according to ISO 13849 [2])

Component Type	Category	Combination of Component Type and Operating Condition	Failure Type	Failure Distribution	FIT [$\lambda_D/10^9$]	Recorded FIT [$\lambda_D/10^9$]	T10$_D$ [year]	PFH$_D$ [1/h]	λ_D [1/h]	λ_{DD}	λ_{DU}	MTTF$_D$ [year]	DC	DC$_{avg}$	Description
Temperature Sensor PT100	Cat 1	Temperature Sensor PT100 30°C to +50°C	Signal out of tolerance range (Functional failure)	44,00%	332,80	146,43	34,30	3,33E-07	3,33E-07	x		343,00			
			Degraded Operation (Temperature to high)	40,60%		135,12	34,30	3,33E-07	3,33E-07	x		343,00	91,50%		DC not relevant for CAT B and CAT 1
			No operation	6,30%		20,97	34,30	3,33E-07	3,33E-07	x		343,00			
			Degraded	3,40%		11,32	34,30	3,33E-07	3,33E-07		x	343,00			
			Shorted	0,60%		2,00	34,30	3,33E-07	3,33E-07	x		343,00			
			Unknow	5,10%		16,97	34,30	3,33E-07	3,33E-07		x	343,00			

Fig. 3 Example of failure mode distribution catalogue for one component (FMD)

System element	Component type	Operating condition	Safety function FMEA	Description	Failure Mode FMEA	Failure Mode Component	Failure Mode Distribution	Safety goal	PFH_D [1/h]	λ_D [1/h]	$MTTF_D$ [year]	Recorded FIT [$\lambda_D/10^9$]	$T10_D$ [year]	Prevention Measures	Detection Measures	DC	λ_{DD}	λ_{DU}
SF 13 - Sensor 1	Temperature Sensor PT100	-30°C to +50°C	Generate Sensor Signal for oil temperature	DC not relevant (Distribution not relevant)	Does not use well-tried components and well-tried safety principles	Unknow	5,1%	Ensure hydraulic oil temperature monitoring including tolerances PL = a	3,33E-07	3,33E-07	343,00	16,97	34,30	Specify installation conditions	Operation monitoring Separate evaluation electronics			x
					Does not use well-tried components and well-tried safety principles	Degraded Operation (Temperature to high)	40,6%	Ensure hydraulic oil temperature monitoring including tolerances PL = a	3,33E-07	3,33E-07	343,00	135,12	34,30		Operation monitoring Self-monitoring Watchdog Application		x	
					Does not use well-tried components and well-tried safety principles	Shorted	0,6%	Ensure hydraulic oil temperature monitoring including tolerances PL = a	3,33E-07	3,33E-07	343,00	2	34,30	Regular maintenance	Regular maintenance	90,93%	x	
					Does not use well-tried components and well-tried safety principles	Signal out of tolerance range (Functional failure)	44,0%	Ensure hydraulic oil temperature monitoring including tolerances PL = a	3,33E-07	3,33E-07	343,00	146,43	34,30		Operation monitoring Separate evaluation electronics		x	
					Does not use well-tried components and well-tried safety principles	Degraded	3,4%	Ensure hydraulic oil temperature monitoring including tolerances PL = a	3,33E-07	3,33E-07	343,00	11,32	34,30	Regular maintenance	Operation monitoring Regular maintenance			x

Fig. 4 Detailed section of FMEDA form sheet for one input of a SF [8, 14]

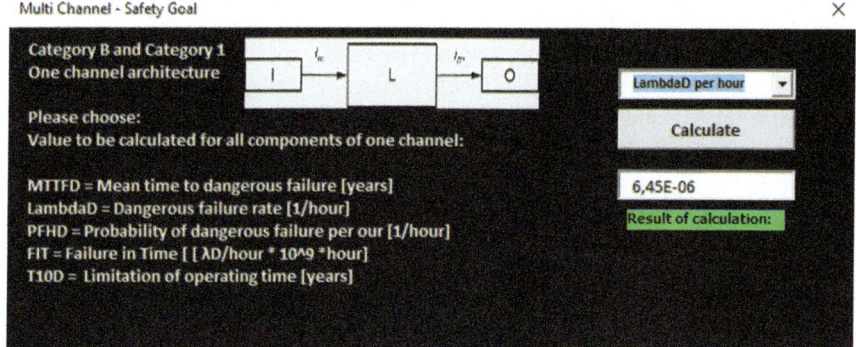

Fig. 5 Pop-up window for calculation of safety parameters of required channel architecture [2, pp. 32–36, 8, 14]

- Value catalogues (FMD Catalogue) for components and information on safety-relevant characteristics from suppliers provide a suitable preparation for practical use
- SFs, with the defined failure modes and the resulting prevention and detection measures from the FMEA are linked here with the diagnostic methods of the FMEDA.

The calculation of functional safety for the function- and the test-channels, (see Fig. 5), in accordance with the HW categories of ISO 13849 [2, pp. 32–36] is summarised by the algorithm in the pop-up window for each SF [8, 14]. A traffic light signal indicates whether the safety objective has been achieved or not. This approach is important for identifying and recording potential failures and their effects early in development.

3 Integration of FMEA Results into Machine Simulation Tool

The simulation tool is used to create a digital model [11, 16] that represents the real mobile working machine, (see Fig. 10) e.g. a (semi-)autonomous mobile working machine equipped with assistance functions and partially autonomous working functions.

A digital model is defined as "an integrated multi-physical, multi-scale, probabilistic simulation of a vehicle or system using the finest available physical models and sensor updates" [11, 16]. A digital model involves manual data transfer using simulation techniques, machine learning and logic operations, so that it can represent a virtual model of a physical object or system over its entire lifetime. It supports engineers in the development phase in mitigating the effects of unpredictable, undesirable events and detecting faults in complex systems [11, 16].

In complement to the conventional development methodology, DIN IEC/ TS 62998–1:2021–10 [10] recommends and requires a specific safety assessment for the performance of simulations. The investigation of complex (multi-)sensor systems places special demands on the performance of a simulation process and requires special sensor simulation methods. The vehicle, the digital model, is equipped with virtual sensors in a three-dimensional environment.

To predict the machine's or system's behaviour in the development phase, the digital model is manipulated with modellable interactions from the identified failure modes of the FMEA. Fault-injection-test are used for this purpose, in which the cause of the fault in the system is brought about in a target manner. In this case, the machine's real control algorithms are retained. The FMEA, in conjunction with the FMEDA, lists the defined failure modes from the analysis and thus provides the input for the manipulator (a SW function block of the simulation tool).

3.1 Results for FMEA and FMEDA

By applying the FMEA in combination with the FMEDA, a comprehensive initial assessment of the potential risks associated with the system or mobile machine under consideration was carried out. The required safety functions for the mobile machine, in particular for (semi-)autonomous machines, were considered in accordance with the RA by ISO 12100 [4] and the Safety Requirement Specification (SRS) [4]. Through integrating additional prevention and detection measures, it is possible to achieve a further reduction in risk. It is important to note that the FMEDA form is an instrument used to evaluate calculated formula and for the performance Level for the safety of electronic systems. The analysed failure modes (see Fig. 6 and Table 5) should be comprehensively documented. The categorization of the listed failures is done by combining FMEA and FMEDA. In the FMEA, (see Fig. 7), potential failures are evaluated based on their impact on safety and reliability. This is usually done using an FMEA form, which typically contains the following columns:

- Failure Mode: description of the potential failure or problem.

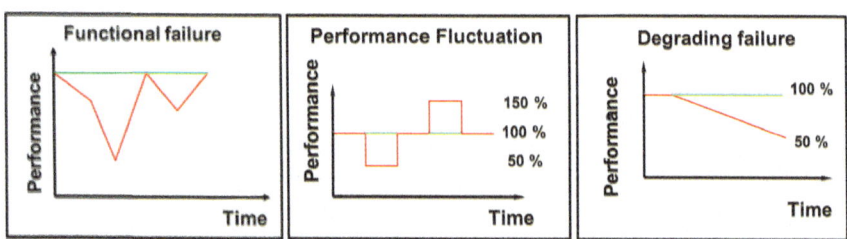

Fig. 6 Failure conditions according to FMEA Procedure from AIAG&VDA [8, 9]

Table 5 Results of failure modes from FMEA process and out of FMD-2023-1 [7, 8]

Functional failure	Performance fluctuation	Degrading failure
Fails to open	Erratic operation	Degraded operation
Does not engage	Spurious operation	Water, Snow, Ice
Out of specification	Sporadically failure	Worn
Out of adjustment	Excessive vibration	Contamination
Drift	Vibration	Corrosion
False operation		Dirt, Dust

- Failure Effect: Evaluation of the potential effect of the failure on the safety or operability of the system, on a scale of 1 to 10, with 10 representing the highest severity.
- Failure Cause: Assessment of the likelihood that the failure will occur, based on various factors such as design, manufacture and operation, also on a scale of 1 to 10, with 10 representing the highest occurrence.
- Failure Detection: Assessment of the occurrence probability of the failure being detected during the development or use of the system, again on a scale of 1 to 10.

In the FMEDA, failures are classified according to their definition into safe failures, DD failures and DU failures. In addition, failures can be further categorised into common cause-failures, systematic failures and random failures.

This optimisation can be visualised using the Failure Criticality Index (FCI), which essentially corresponds to the Risk Priority Number (RPN) [9, pp. 221]. The FCI identifies the top ten failure sources, (see Fig. 8) and additional risk mitigation measures can be implemented correspondingly, particularly in relation to the integration of SFs into the FMEA process [9, 12, 15]. The top ten failures in an FMEA are the most important potential problems or failures identified in a system. These failures are selected based on their assessment of their impact and probability.

The FCI is based on the analysis carried out as part of the FMEA, in which the possible effects of a fault or defect on the overall functionality and safety of the system evaluated. Factors such as the probability of occurrence, the potential damage that the fault could cause and the possibility of recognising and rectifying the fault are considered. The assessment based on the FCI enables effective resource prioritisation and the implementation of targeted measures to prevent, detect, and rectify failures. A higher FCI value indicates a higher failure criticality and usually requires more intensive testing and risk mitigation measures.

In addition, the RA can be continuously monitored using risk matrices with traffic light colours. The example (see Fig. 9) shows a risk matrix with Severity on the X-axis and Occurrence on the Y-axis. This form of holistic analysis can be based either on the solution of the RPN value or on a specific weighting according to the significance of the risks [9, 12, 15].

Potential failure	Potential effect(s) of failure	S	Potential cause(s) of failure	Current preventive action	O	Current detection action	D	RPN	Recommended action for Simulation with Digital Model	R/D	Results: Action taken into Simulation Tool Failure Manipulation	S	O	D	RPN
Function: Move the bucket as commanded															
<Move the bucket as commanded> S (max): 10 Does not move the bucket as commanded	<ensure valve opening / start of movement in desired time> does not ensur valve opening / start of movement in desired time	9	<Move the bucket as commanded> S (max): 10 Clamping valve spool (dirt in the gap, deformed housing of spool part)	System concept chosen according internal standard Protect against internal mechanical impacts Analytical detection methods for robust design Check lubrication concept according internal standard	7	Visual Inspection Monitoring of Sensor Signals (Compare of Target&Actual Value)	3	189	Functional Test Performance Test for the Bucket movement: Test of Bucket start/stop Test of Bucket time sequences, tolerances, ramps	Name: Date:	Hardware: Installation of stroke sensor, increase filter level, inspect production batches, Software-Diagnosis: Virtual flow sensing by comparison set speed (trajectory) vs actual speed value --> no movement after start command, no stopping	9	5	1	45
	<ensure defined accelleration of bucket movement> does not ensure defined accelleration of bucket movement	9	To high leakage in hydraulic actuation line No actuation pressure	Ensure clear interface definition Safety concept reasonable and confirmed System concept chosen according	6	Visual Inspection Monitoring of Sensor Signals (Compare of Target & Actual Value)	3	162	Acceleration Perfomance Test: Test of acceleration cold/hot Test of speed behaviour	Name: Date:	Hardware: Installation of pressure sensors Software-Diagnosis: Virtual flow sensing by comparison set speed (trajectory) vs actual speed value	9	4	1	36
	<ensure defined deccelleration of bucket movement (avoid failure to relief on demand)> does not ensure defined decelleration of bucket movement (avoid failure to relief on demand)	9	Broken Solenoid Hydraulik oil temperature to low Blocked Orifice	Check component local stresses according internal standard Check component performance according internal standard Check housing local stresses according internal standard	5	Visual Inspection Current feedback from solenoid (Monitoring through PLC) Visual Inspection	3	135	Deceleration Test: Test of braking behaviour Test of thermal Stability while Braking Test of speed behaviour while braking	Name: Date:	Software- Diagnosis: Broken Solenoid: Diagnosis of current feedback on PLC Oil Temp: Warm-up function below 40 degree Blocked Orifice - Diagnosis: Actual Reaction time vs. Time for normal system operation (KPI for tracking)	9	3	1	27

Fig. 7 Example section of FMEA form sheet

Safety Function-Failure Mode and Effect Analysis a Novel Approach ... 173

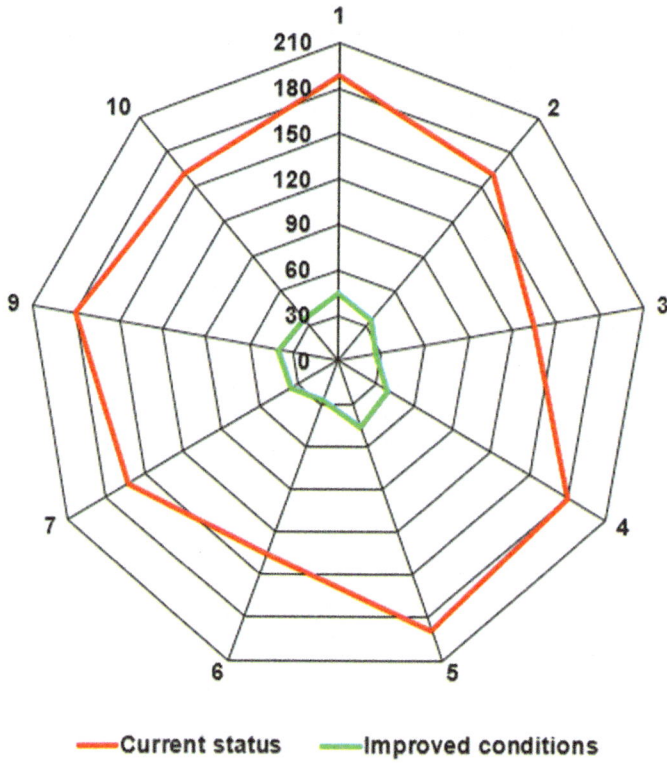

Fig. 8 Visualisation of risk reduction with top ten failures of FCI [9, 12, 15]

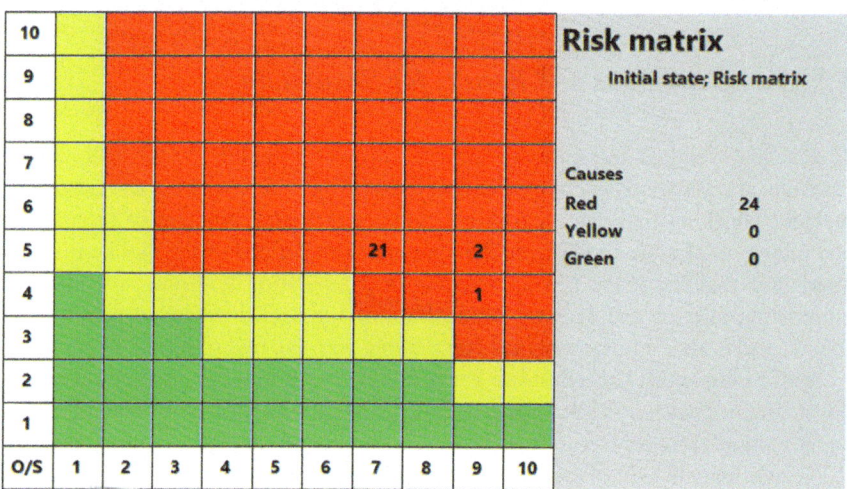

Fig. 9 Detailed decision matrix—example of top ten failures [9, 12, 15]

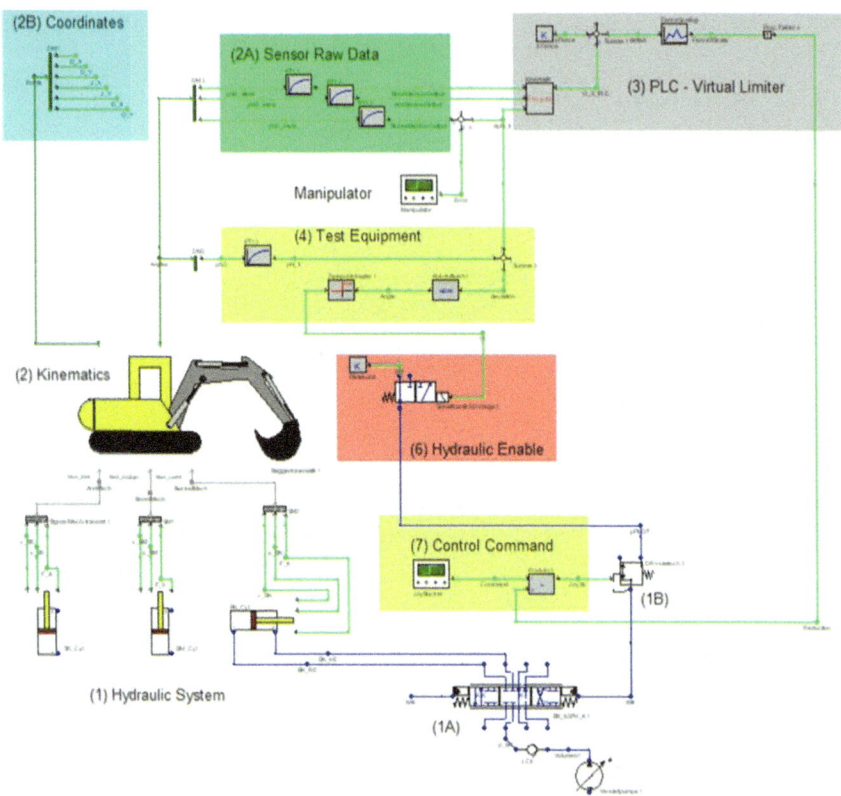

Fig. 10 Simulation example of digital model with manipulator for an excavator [8]

3.2 Test Conditions and Results for Simulation

To test and validate the system's functionality of a mobile (semi-)autonomous machine, specifications and requirements must be met. Verifying the test conditions includes checking the logic, algorithm, functionality and structure of the models, and integrating failure modes into the computer simulations of the mobile machine operations (results of the FMEA analysis in combination with the FMEDA) and eliminating failures and defects. Validation with the digital model ensures that the specifications and requirements are comparable with the real machine.

As the classic test procedures (e.g. endurance test) cannot be verified during the development phase, so-called fault injection tests are conducted in the simulation using a digital model. Here, the failures are specifically induced via a manipulator (SW module) in the simulation tool. The simulation example (see Fig. 10) shows the safety function "Avoid bucket collision with detected obstacle", which is subjected to faults via the manipulator, using the failure modes from the FMEA [8].

Based on the excavator simulation model (see Fig. 10), the hydraulic system consists of main valve spools which are controlled by electro-proportional solenoid valves. The pilot pressure level is activated or deactivated by an electro-hydraulic 3/2-way valve. The movements of the hydraulic cylinders are recorded in a kinematic model of the excavator and used as raw sensor data. These signals are filtered and provided by various sensors for standard Programmable Logic Controller (PLC) and test equipment. One of the sensor signals can be modified by a manipulator implement drifts, sudden signal losses or failure possibilities out of the FMEA tool to simulate faults. The difference between the manipulated and the clean sensor signal is monitored and in case of deviations the hydraulic 3/2-way valve is deactivated to safely stop all functions. This difference also influences the speed reduction to safely stop in front of obstacles. Based on an intelligent excavator simulation model, as outlined in my publication [8, p. 15].

The results of the listed failure modes from the FMEA and FMEDA, in particular, provide information on the test conditions that require a unique modelling approach with the digital model to enable predictions to be made before the first prototype is built. The following list shows examples of results from the FMEA that must be considered in the area of (semi-)autonomous machines for the safety of humans:

- Low and high risk of contamination of sensors for distance measurement, such as camera, lidar and radar
- Installation location / positioning of inclination sensors and gyroscopes (especially the distance from the rotation centre of corresponding parts)
  - Potential position for collisions
  - Signal amplitude varies which distance from the rotation centre
  - Mechanical focuses on cable and connectors includes with distance from the rotation centre
- Consideration of real operating environments, e.g. temperatures, Electromagnetic Combability (EMC), vibrations

In addition to failure manipulation, to the failure injection, supplementary boundary conditions for the simulation tests are essential to ensure that the combination correctly represents the operating conditions and potential failure modes. This ensures that the simulation meets the specification's requirements and objectives and hence, correctly represents the intended functionality. Table 6 illustrates how a degrading fault of a position sensor is generated by the manipulator with different boundary conditions in the digital model.

Figure 11 illustrates the values in Table 6. A higher drift ramp ratio correlates with a faster fault detection and thus, allows a later detection of the drift failure while the attachment is moving towards the object. The right-hand diagram shows, that the bold values, (see Table 6), are exceeding the limit and thus represent a collision caused by the untimely drift failure detection. The results show that the total stop is realized below 7100 mm for any drift ratio and starting point.

When drift failures are caused by common and measurable factors, such as acceleration in another axis or insufficient filters in the software, they are considered

Table 6 Test conditions and results for a sensor drift (Degrading Failure) of the simulation

Parameter	Active	Start time [s]	Ramp [1°/s]	Stop time [s]	Failure [mm]	Stop way [mm]	File
Reference	Yes		No	6,6	3771	*6998*	231,028 Reference V03
Ramp (drift)	No	4	+1	6,6	3228	*6880*	231,028 Ramp 1 4 inactive
	Yes	4	+1	6,2	3228	*6385*	231,028 Ramp 1 4
	Yes	4	−1	6,2	3228	*6509*	231,028 Ramp−1 4
	Yes	4,4	−1	6,6	3615	*6864*	231,028 Ramp−1 44
	Yes	4,5	−1	6,7	3757	*6966*	231,028 Ramp−1 45
	Yes	4,55	−1	6,75	3835	**7018**	231,028 Ramp−1 455
	Yes	4,6	−1	6,8	3915	**7071**	231,028 Ramp−1 46
	Yes	5	−1	7,2	4676	**7088**	231,028 Ramp−1 5
	Yes	4	−0,5	6,6	3228	**7058**	231,028 Ramp−05 4
	Yes	2	−0,5	6,2	3710	*6513*	231,028 Ramp−05 2
	Yes	2,2	0,5	6,7	3598	*6978*	231,028 Ramp−05 25
	Yes	2,6	−0,5	6,8	3570	**7085**	231,028 Ramp−05 26
	Yes	5	+2	6,2	4676	*6500*	231,028 Ramp−2 5
	Yes	5,5	+2	6,7	5755	*6941*	231,028 Ramp−2 55

(continued)

Table 6 (continued)

Parameter	Active	Start time [s]	Ramp [1°/s]	Stop time [s]	Failure [mm]	Stop way [mm]	File
	Yes	5,55	+2	6,75	5847	6992	231,028 Ramp–2 555
	Yes	5,6	+2	6,8	5935	**7044**	231,028 Ramp–2 56

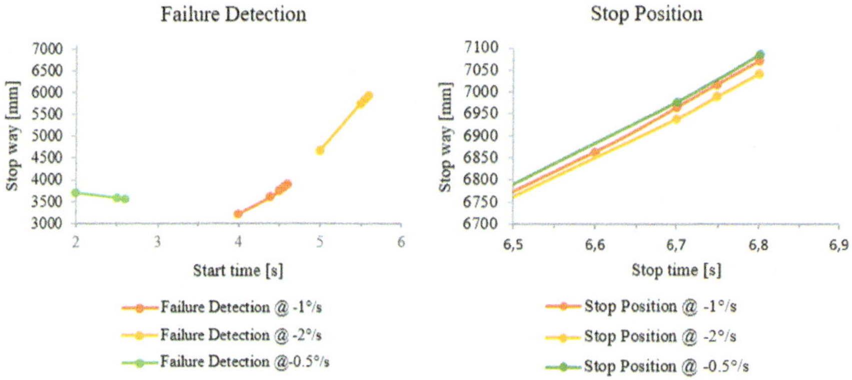

Fig. 11 Graphical representation of test condition values for drift (Degrading Failure) pointed out from Table 6

systematic failures. However, if the drift is unforeseeable and cannot be explained by known factors, it is a random failure. Consequently, a drift failure represents a high risk for the (semi-)autonomous machine and determining the correction factor is essential, which can only be done by simulation before the prototype failed in test operation.

To illustrate the results, Fig. 12 from my previous publication [8, p. 17] shows the behaviour of the bucket in the simulation tool and compares this with the behaviour in regular operation. The characteristic curves in the diagrams for the unexpected but also unrecognised failure show that the system behaves similarly to normal operation. The failure is not recognised in the simulation, so that the system is brought to a standstill by the test conditions set. However, as the signal feeds the algorithms with incorrect values (fault injection test), the expected fault occurs later compared to the drift fault scenario [8].

The analysis of the measurement data, (see Table 6), indicates that a safety distance of 10 cm from the obstacle is suitable and enables the calculation of the Diagnostic Coverage. Equation 3 can be used to calculate the DC. Table 7 shows sample data for three measurement results taken from Table 6 and is used to illustrate the calculation of the DC using the measurement data from Eq. 4:

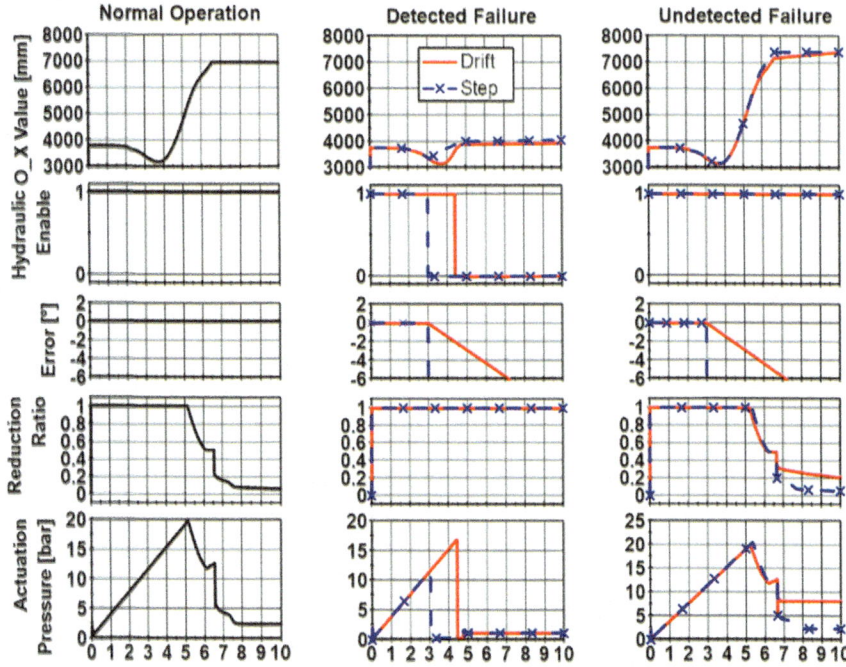

Fig. 12 Simulation results for drift and step response [8, p. 17]

Table 7 Calculation of diagnostic coverage out of measurement data (see Table 6)

Failure mode example sensor	Ramp [1 /s]	Safe failure detection time [s]	Reference cycle time [s]	DC [%]
Drift	0,5	2,6	6,6	**39,4**
Drift	1	4,55	6,6	**68,9**
Drift	2	5,6	6,6	**84,8**

$$DC = \frac{DD}{DD + DU} * 100 = \frac{\text{safe failure detection time}}{\text{reference cycle time}} * 100 \qquad (6)$$

The diagnostic coverage for the drift (degrading fault) with a ramp of 2°/s could be calculated with a value of 84%. The DC of 84% implies that the system is 84% capable of correctly recognising and diagnosing this fault. This partial result focuses exclusively on the diagnostic capability of the system in relation to a specific fault with specific parameters considering the possibility of random failures. It clearly shows how effectively the system could recognise and diagnose this fault with the corresponding setting in the software on the real machine.

4 Discussion and Conclusions

As the complexity of functions in automated mobile machinery increases, especially across the five distinct levels of automation (manual, assisted, partially automated, conditionally automated, and highly automated), and considering various operational contexts (public spaces, restricted-access closed areas) or additional guidance modes (lane-bound, lane-guided, direction-bound, direction-independent, and free navigation within a driving field), appropriate measures need to be chosen for environmental sensing [8]. In the event of a malfunction, the mobile machine must be transitioned to a safe state at any given time. This safe state should align with established standards for functional safety, such as ISO 26262 [1], ISO 13849 [2], ISO 19014 [3], DIN EN 62998 [10], IEC 60812 [12] and DIN EN 61508 [13]. Autonomous systems must be capable of achieving a $PL_r = d$ according to ISO 13849 [2] or, at the very least, an equivalent level of safety [8].

The linking of both methods, namely the FMEA and FMEDA, such as the DC, provides an effective means for risk mitigation in the development and operation of (semi-)autonomous mobile machines [8]. The evaluation in Chap. 4 shows that the information from the FMD-2023-1 [7] database on the FMR can be used in connection with FMEA during the development phase when carrying out reliability analyses and assessments, thus enabling a valuable and cost-saving assessment of risk reduction. The results of the FMEA and the simulation tool show an attractive but effective alternative to the conventional development method. The parallelism of design, calculation, risk reduction via the FMEA, simulation and the linking of the FMEDA with the calculation and the simulation reduces the effort in the development phase, also reduces the probability of failure occurrence and, at the same time, increases the probability of detection to a considerable extent [8].

Integrating the safety functions according to ISO 13849 [2] into the FMEA combined with the FMEDA and the resulting knowledge offers several advantages for implementation in the simulation tool with a defined fault-injection test. The iterations of test bench and vehicle tests can be effectively reduced, which leads to an acceleration of development and research phase. At the same time, the change of components and system elements is minimised, which increases the stability and reliability of the development process. The integration of these methods also helps to reduce the repetition of functional tests, which not only increases efficiency but also helps to save resources. In addition, efficient safety procedures enable effective RA and reduction while simultaneously optimising the process steps by simplifying them. Overall, the integration of FMEA, FMEDA, and fault injection tests leads to a holistic improvement of the development process, contributing to reduced project and development costs [8, p. 18].

Nomenclatures

CAT	Categories
DD	Dangerous Detectable
DC	Diagnostic Coverage
DC_{avg}	Average of Diagnostic Coverage
DC_{SPF}	Diagnostic Coverage for Single Point Fault
DU	Dangerous Undetectable
EMC	Electromagnetic Compatibility
FCI	Failure Criticality Index
FIT	Failure in Time
FMD	Failure Mode Distribution
FMEA	Failure Mode and Effect Analysis
FMECA	Failure Mode, Effects and Criticality Analysis
FMEDA	Failure Mode Effects & Diagnostic Analysis
HARA	Hazard and Risk Assessment
λ	Failure Rate Lambda
λ_D	Dangerous Failure Rate Lambda
λ_{DD}	Failure Rate Lambda Dangerous Detectable
λ_{DU}	Failure Rate Lambda Dangerous Undetectable
λ_{SPF}	Failure Rate Lambda for Single Point Fault
λ_{MPF}	Failure Rate Lambda for Multiple Point Fault
$\lambda_{MPF,DP}$	Failure Rate Lambda for Multiple Point Faults, Design Phase
λ_{RF}	Failure Rate Lambda for Random Failure Rate
λ_{SR}	Failure Rate Lambda for Systematic Reasoning
λ_S	Failure Rate Lambda for Safe Fault
λ_{SPF}	Failure Rate Lambda for Single Point Fault
LFM	Latent Fault Metric
$MTTF_D$	Mean Time to dangerous Failure
MPF, DP	Multiple Point Fault,
PLC	Programmable Logic Controller
PL_r	Required Performance Level
PLs	Performance Levels
PFH_D	Probability dangerous Failure per Hour
PMHF	Predicted Mean Hazard Frequency
RA	Risk Assessment
RPN	Risk Priority Number
SRP/CS	Safety-Related Parts of the Control System
SFs	Safety Functions
SF-FMEA	Safety Function-FMEA
SPFM	Sigle Point Fault Metric
SPF	Single Point Fault
SR, HW	Safety-related Hardware

SW   Software Tools
T_{LT} Lifetime of the System

References

1. International Standard ISO 26262-5:2018-12 (2018) Road vehicles—Functional Safety—Part 5: Product development at the hardware level. Beuth, Berlin
2. International Standard ISO 13849-1:2023-04 (2023) Safety of machinery—Safety related parts of control systems— Part 1: General principles of design. Beuth, Berlin
3. International Standard ISO 19014-05:2021-12 (2021) Earth moving machinery—Functional safety—Part 5: Tables of performance Levels. Beuth, Berlin
4. International Standard ISO 12100:2010-11 (2011) Safety of machinery—General principles for design—Risk assessment and risk reduction. Beuth, Berlin
5. Düsing Ch, Prust D (2020) Supplementary failure mode and effect analysis (FMEA) for safety application standards DIN EN ISO 13849 Safety Function-FMEA. In: 12th International fluid power conference, pp 349–358. Technische Universität Dresden, Dresden
6. Geimer M (2020) Mobile working machines. SAE Intenational. 1st edn. SAE International, Warrendale, USA
7. Prediction, Analysis, reliable system (2023) failure mode distribution—FMD-2023–1. Quanterion Solutions Incorporated, New York, USA
8. Düsing Ch, Inderelst M (2022) Evaluation of the diagnostic coverage for safety-relevant components in automated drive systems for mobile construction machinery. In: 9th Symposium construction machinery, pp 199–217. Technische Universität Dresden, Dresden
9. AIAG & VDA (2019) FMEA Handbook, failure mode and effect analysis, design FMEA, process FMEA, supplemental FMEA for monitoring & system response. 1st edn. VDA QMC, Berlin
10. International Standard DIN IEC/TS 62998-1:2021-10 (2021) Safety of machinery—Safety related sensors used for the protection of persons. Beuth, Berlin
11. Xue YY (2022) A digital twin model-based system engineering approach to failure analysis for an engine system. In: Thesis, Naval Postgraduate School Monterey, California, USA
12. International Standard IEC 60812:2018-08 (2018) Failure modes and effects analysis (FMEA and FMECA). VDE, Berlin
13. International Standard DIN EN ISO 61508-5:2011-02 (2011) Functional safety of electrical/electronic/programmable electronic safety-related systems—Part 5: Examples of methods for the determination of safety integrity levels. Beuth, Berlin
14. Düsing Ch (2023) Integration of a safety function-FMEA into a professional FMEA software tool to meet the new draft of ISO 13849. In: 13th International fluid power conference F – Systems I (2022) RWTH-Aachen, Aachen pp 297–308
15. Carlson CS (2012) Understanding the fundamental definitions and Concepts of FMEAs. John Wiley and Sons, Hoboken, New Jersey, USA
16. Kritzinger W et al (2018) Digital twin in manufacturing. a categorical literature review and classification IFAC Papers online, pp 51–11: 1016–1022

Open Access This chapter is licensed under the terms of the Creative Commons Attribution 4.0 International License (http://creativecommons.org/licenses/by/4.0/), which permits use, sharing, adaptation, distribution and reproduction in any medium or format, as long as you give appropriate credit to the original author(s) and the source, provide a link to the Creative Commons license and indicate if changes were made.

The images or other third party material in this chapter are included in the chapter's Creative Commons license, unless indicated otherwise in a credit line to the material. If material is not included in the chapter's Creative Commons license and your intended use is not permitted by statutory regulation or exceeds the permitted use, you will need to obtain permission directly from the copyright holder.

Improvement of Mobile Crusher Energy Efficiency Through Hybridization and Electrification

Jesse Backman and Tatiana Minav

Abstract The energy efficiency of non-road mobile machines (NRMM) requires significant enhancement to mitigate harmful emissions. Mobile crushers, akin to numerous other NRMMs, encounter a challenge due to the absence of standardized load cycles, impeding efficiency assessments and system-level developments. This paper addresses the issue by utilizing measured load cycles, establishing a foundation for comprehending mobile crusher power requirements. This understanding forms the basis for augmenting efficiency and performance of the selected machine. To achieve these improvements, this paper proposes and studies the applicability of a hybrid powertrain to the selected machine. In this work, a simulation model of a series-hybrid powertrain for a mobile crushing machine is developed, and its efficiency is compared to that of a conventional diesel-electric powertrain using the measured load cycles. The simulation results demonstrate that the hybrid powertrain enhances the mobile crusher's efficiency, concurrently improving crushing performance by increasing available peak power.

Keywords Non-road mobile machines · Mobile crusher · Hybridization · Electrification · Efficiency

1 Introduction

With the increasing electrification of road vehicles, non-road mobile machines (NRMM) have emerged as significant contributors to mobile pollution [6]. Consequently, there has been an increase in the enforcement of stricter emission regulations on internal combustion engines (ICEs) powering NRMMs [2, 4], necessitat-

---

J. Backman (✉)
Metso, Tampere, Finland
e-mail: jesse.backman@metso.com

T. Minav
Tampere University, Tampere, Finland

ing a reduction in their environmental impact and harmful emissions. Despite the diverse range of tasks performed by different NRMMs, ranging from agricultural to construction activities, traditional reliance on ICEs to power, e.g. hydraulics or direct-driven work functions has been ubiquitous across machine types. In an effort to limit NRMM environmental impact, recent developments in both literature and practice have been trending toward hybrid powertrains and electrification of different work functions. These developments have lead to an increased variety in powertrain configurations due to the inherently variable powertrain requirements of NRMMs. Noteworthy examples include a parallel hybrid electric tractor [12], battery-electric excavator [1], hybrid hydraulic excavator [3], series-hybrid wheel loader [15], underground mining loader [9], and others [8]. Nevertheless, identifying an appropriate powertrain configuration necessitates a thorough understanding of NRMM operating conditions and load cycles, underscoring the need for specific studies of the machine under investigation. Although at least one commercially available mobile crusher exists [13], the scholarly exploration of mobile crusher energy efficiency is lacking.

With current mobile crusher powertrain configurations, the diesel engine is either directly mechanically connected to the load in diesel-hydraulic arrangement, or limited to a very narrow range of allowed rotational speeds in diesel-electric configuration, as stable frequency for the generated alternating current is required. Diesel engines are dimensioned to the peak loads, resulting in poor operating areas in average load points. Despite diesel-electric setups utilizing external power sources to circumvent onboard genset losses, mobile crushers often lack access to such sources. Consequently, they remain reliant on diesel fuel. Furthermore, the diesel-electric machines still deploy many hydraulic functions via pumps driven by electric motors; thus high systemic losses of hydraulic systems [5] are still present in these machines. However, this paper concentrates on enhancing the efficiency of the powertrain, deferring the examination of work system enhancements to future studies. Unlike standardized load cycles of the automotive industry, the area of mobile crushers lacks such benchmarks, posing a significant challenge for efficiency assessments and system developments. To address this, this study is grounded in measured load cycles from the mobile crushing machine performing typical crushing, ensuring a robust foundation for analysis. Comprehending the load cycles serves as a foundation for proposing a hybrid powertrain for the machine, with a twofold goal: to elevate both the performance and efficiency of the mobile crusher.

In summary, the motivation behind this study is fuelled by the absence of standardized load cycles, the pursuit of increased machine efficiency and performance, and a dedicated effort to mitigate the environmental impact of non-road mobile machines. Through these objectives, a more sustainable and technologically advanced future in mobile crushing is pursued. The structure of this paper is organized as follows: the existing powertrain configuration and measured load cycles are presented in Sect. 2. Section 3 introduces the hybrid crusher structure, simulation model architecture and control principles, and simulations are performed in Sect. 4. Finally, conclusions and discussions are given in Sect. 5.

2 Mobile Crusher and Load Cycles

This section discusses the specifics of the load cycles of this study, measured from a track-mounted mobile crushing machine. The machine is powered by a diesel-electric powertrain, and the crusher is driven with an electric motor. Powertrain arrangement and data are presented in Fig. 1 and Table 1, respectively.

Load cycles were obtained from the machine to gather and comprehend the performance requirements for the powertrain, to establish a reference point for performance and fuel consumption, and to create relevant load cycles for the simulation study. Crusher motor load was of specific interest, as the crusher is the primary work function component and main energy consumer of the machine. In addition to crusher motor data, engine speed, load and torque, machine parameters and command messages, and many other data were recorded. This comprehensive dataset is the foundation of this study, and provides insights into the machine's behavior under different operating conditions. In total the machine's performance was recorded in three different crushing processes, in one of which an external power source was

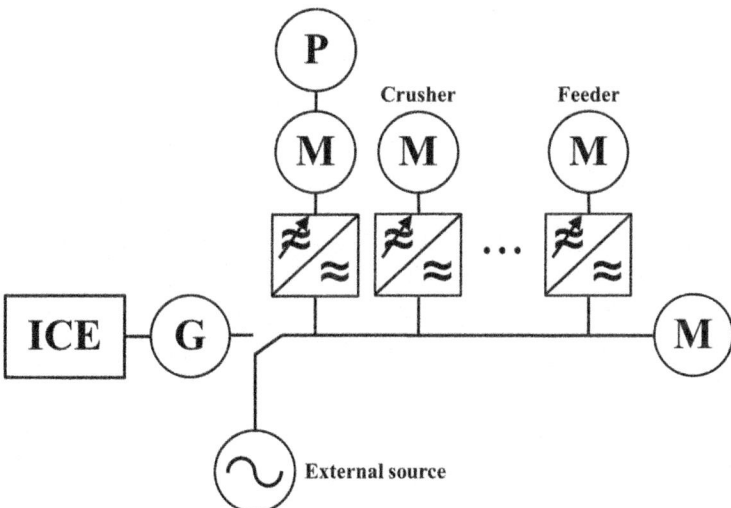

Fig. 1 Diesel-electric powertrain configuration with a genset providing stable frequency alternating current for work functions. Most motors in the configuration are operated with a converter for speed control, but a direct connection is also utilized for a constant-speed motor

Table 1 Diesel-electric mobile crusher data

Crusher motor type and nominal power	Electric 250 kW
Engine power	310 kW
Generator power rating	420 kVA
Total weight	55 tons

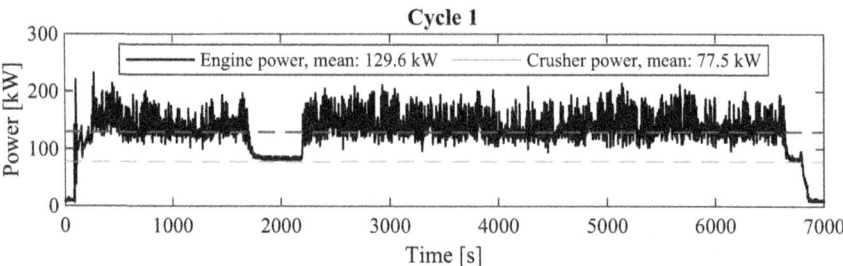

Fig. 2 Engine and crusher power during load cycle 1

introduced to evaluate the crusher load and machine performance without the constraints of available on-board power. A fourth load cycle was synthesized from the externally powered load cycle to create an extremely demanding crushing profile, augmenting the dataset.

In subsequent sections each of these load cycles will be dissected, providing an examination of the machine's performance during different operational conditions. This analysis is performed through the lens of hybrid powertrain suitability, and aims to contribute insights to the specific domain of mobile crushing machinery.

2.1 First Load Cycle

The first load cycle was derived from a waste crushing process of a demolished building. The primary constituent of the feed material comprised concrete blocks, including a significant concentration of fine-grained particles along with a minor quantity of more resilient natural rocks. Notably, concrete exhibits relatively undemanding characteristics for crushers, as evidenced by the modest average power requirements of 130 kW for a 310 kW engine, depicted in Fig. 2.

The average crusher motor power recorded was 78 kW, with intermittent peaks reaching 180 kW, resulting in an average-to-peak ratio of 2.3. Concurrently, the engine exhibited an average power of 130 kW, with sporadic peaks reaching 210 kW, yielding an average-to-peak ratio of 1.6. It is noteworthy that the engine operated within an inefficient zone in this particular application, attributed to the low load conditions and the consistent 1500 rpm operation of the genset for generating 50 Hz electricity. Additionally, the data reveals a considerable idle period lasting 15 min amidst the approximately 110 min of the recorded crushing process.

2.2 Second and Third Load Cycles

The second load cycle investigated in this study was conducted using an external 700 kVA generator to power the machine. This approach aimed to evaluate the machine's performance in a scenario where the crusher motor's power is not restricted by

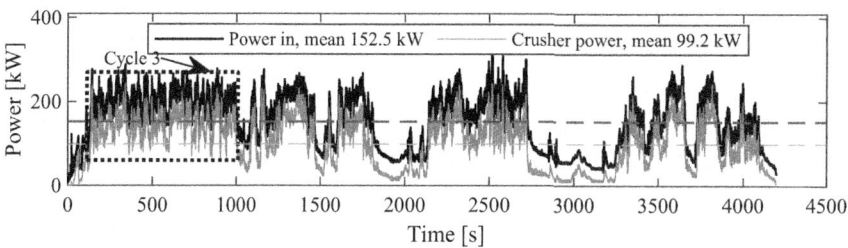

Fig. 3 Crusher and engine power during load cycle 2. A third load cycle was synthesized from the highlighted crushing segment

available input power. The feed material for this cycle consisted of recycled asphalt, presenting a distinct challenge to the crusher. Recycled asphalt was fed to the machine in large pieces, and the breakdown of asphalt introduces tarmac and other adhesive materials, which in combination required a considerate amount of power from the crusher.

As illustrated in Fig. 3, the input power surpassed 300 kW during peak load times. Despite intermittent idle periods in the dataset, the average input power remained above 150 kW. The observed average-to-peak power ratio for the input power reached 1.5, and an average-to-peak ratio of 2.3 for the crusher motor peaking at over 230 kW and averaging 100 kW. This second load cycle required, on average, more power than the first. It is noteworthy that the machine's peak power demand is nearly double the average power demand. Notably, this cycle also exhibited significant idle time between crushing periods.

A third, exceptionally demanding load cycle was synthesized from the available data by replicating the isolated high-power crushing process emphasized in Fig. 3 and eliminating longer idle periods from the load cycle. This synthesized load cycle represents an optimal crushing scenario characterized by a consistent feed rate, allowing the crusher to operate continuously.

2.3 Fourth Load Cycle

The fourth and final load cycle investigated in this study involved concrete crushing, although the composition of the feed material differed slightly from the first cycle, with fewer non-concrete materials present. A visual examination of the data presented in Fig. 4 reveals less variable engine and crusher operations compared to the preceding load cycles.

Despite the observed operational stability, the average-to-peak ratios for this cycle were the highest among all measured cycles. The engine power exhibited a ratio of 1.7, with peaks reaching 230 kW and an average of 136 kW. Similarly, the crusher motor power displayed a ratio of 2.3, peaking at 230 kW with an average of 100 kW.

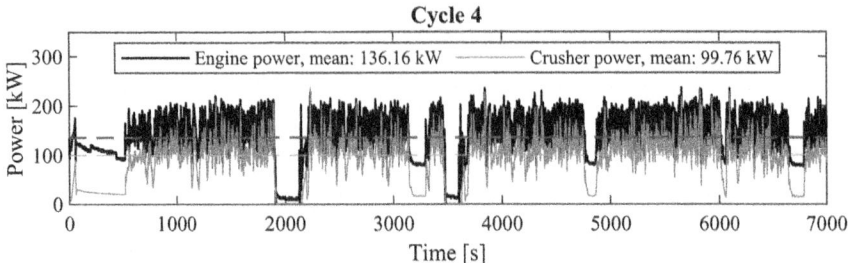

Fig. 4 Crusher and engine power during load cycle 4

Once again, the data indicates periods of idleness between crushing activities, and emphasizes the influence of the crusher motor's power demand on the overall load dynamics.

2.4 General Observations

Upon scrutinizing the analyzed load cycles and associated crushing processes, the following observations are made:

1. The crusher frequently enters an idle state during mobile crushing processes.
2. The engine average power is below 50% of its rated power, causing it to operate in inefficient operating zones.
3. The average-to-peak ratio for the engine falls within the range of 1.5–1.7, while for the crusher motor, it is 2.3.
4. The crusher motor accounts for approximately 60% of the machine's total power consumption, and engine load dynamics are predominantly driven by the crusher's power demand.

Crusher idling occurrences can stem from diverse factors, including blockages within the machine that necessitate manual clearance, pauses in the feed due to alterations in the feed pile, or breaks taken by operators. Throughout these idle periods, the engine remains operational, and the machine is generally maintained in its operating state. Such idle intervals present an opportunity for battery recharge within a hybrid powertrain. Furthermore, an intelligent control system could detect these idle periods and transition the machine into a standby state by either reducing the speed or completely halting the actuators, conserving energy while preserving the machine's readiness to resume operations promptly.

The average-to-peak ratio for the engine emerges as a pivotal metric, advocating the potential viability of a hybrid powertrain for the machine. In the existing diesel-electric powertrain configuration, the engine must be sized to meet peak power demand, resulting in inefficient operation at a suboptimal efficiency zone, given the substantial disparity between average and peak loads. In contrast, a hybrid powertrain allows for engine sizing based on average load requirements, provided peak loads can be accommodated in combination with the energy storage system. The analyzed load cycles suggest a noteworthy opportunity for a considerable downsizing of the diesel engine.

Recognizing that load dynamics are predominantly influenced by the crusher facilitates formulating the powertrain's requisites as a sum of base load and crusher load. The base load comprises the static operational load of the machine, encompassing actuators other than the crusher—such as loads of feeder, conveyors, and hydraulic system. These components exhibit static load profiles, or their contribution to load dynamics is insignificant compared to that of the crusher.

3 Hybrid Mobile Crusher Structure

The powertrain requirements were based on the previously analyzed load cycles. The fundamental technical requirement was that a machine equipped with a hybrid powertrain should be capable of performing the same cycles. Additionally, the overarching objectives aim to achieve fuel consumption reduction and an increase in available power. These objectives were set to pursue an improved technical solution, concurrently enhancing both performance and efficiency of the machine.

The prevalent hybrid powertrain configurations suitable for non-road mobile machines generally fall into series, parallel, and series-parallel categories. While these hybrid configurations comprise similar components, each configuration has benefits and drawbacks. For the mobile crushing machine under consideration, a series-hybrid configuration was chosen due to its advantages in the light of the technical requirements. Series-hybrid configuration features mechanical disconnection of the diesel engine from the load, is adaptable to diverse load profiles, and system control of a series-hybrid is straightforward [7, 10]. Figure 5 illustrates the series-hybrid topology and the principal components. In this configuration, the diesel engine is mechanically connected to a generator, producing alternating current that is subsequently converted to direct current by an AC/DC converter. The core of the electric system within the powertrain is the DC circuit. The genset and energy storage contribute energy to the circuit, which is then consumed by various functions of the machine. Individual work functions are propelled by their dedicated inverters and electric machines, delivering mechanical power to the actuators.

3.1 Simulation Model

The powertrain simulation model was developed using Simcenter Amesim software, while the controller model was created in MATLAB Simulink. A model capable of representing the dynamics of the powertrain and efficiencies of its primary components was required. However, dynamic models for the machine's actuators (such as

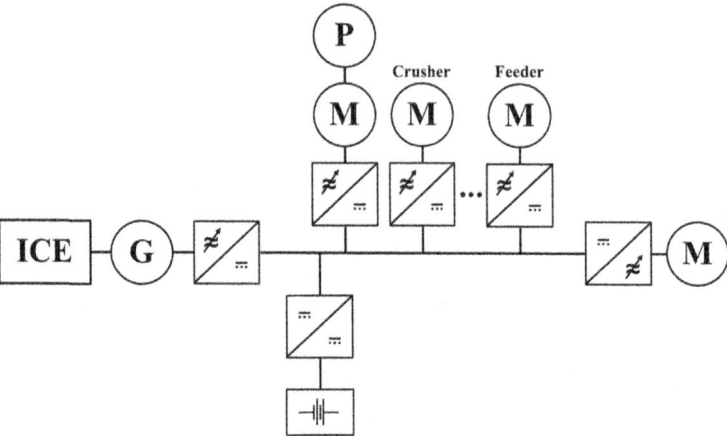

Fig. 5 Series-hybrid powertrain configuration. Alternating current generated by the genset is converted to direct current at a set voltage point. Work function motors and battery are connected to the direct current circuit via converters

feeder, crusher, and conveyors) were deemed unnecessary, as the primary focus was on their power demand for the system. Recorded load cycles and established values for selected actuators were employed to represent their power requirements for the system. The powertrain simulation model was created with configurable model blocks, configured and enriched with data from real components, and was governed by the control model. Given that the load cycles extend for a minimum duration of 100 min, the system model's fidelity was constrained to a level that ensures a reasonable computational efficiency while maintaining accuracy. The model architecture follows that of series-hybrid powertrain configuration displayed in Fig. 5, with an external controller. The controller and its interface with the powertrain model are discussed in Sect. 3.3.

3.2 Battery Modeling

The battery pack is modeled through an equivalent circuit model (ECM) featuring two RC-pairs, as illustrated in Fig. 6. Parameters open-circuit voltage OCV, ohmic resistance R_{ohm}, diffusion resistance R_{diff}, and diffusion capacitance C_{diff} have been configured as functions of battery state of charge SOC, temperature T, and current I. This dynamic approach ensures a comprehensive and dynamic portrayal of the battery's characteristics across diverse operational points. Overall battery pack technical details are listed in Table 2. The battery pack was dimensioned to supply the power difference between average and peak loads, i.e. peak shaving for the ICE.

The model facilitates the computation of energy losses, subsequently manifesting as heat dissipation across multiple material layers encompassing the battery cell,

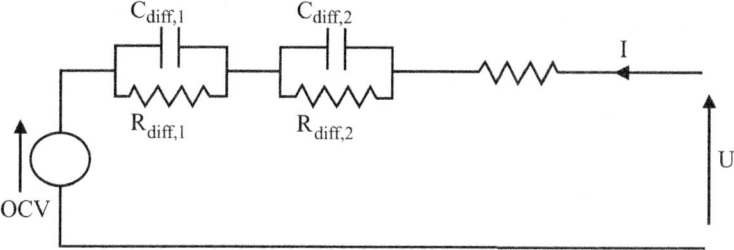

Fig. 6 Equivalent circuit model (ECM) of two-RC-pair battery module model

Table 2 Theoretical battery pack technical details

Modules	12
Voltage	552 V
Max. discharge power (pulse)	440 kW
Max. charge power	240 kW
Energy	11 kWh

module case, and cooling plate. The battery model extends to simulate heat transfer interactions between the cooling plate and a liquid cooling system. Rather than employing a singular block to model the entire battery pack, the model is constructed by connecting twelve configurable battery blocks in series, representing battery modules. Each block is configured with module-level parameters sourced from a battery manufacturer, which have been validated through laboratory testing. This model construction aims to facilitate an in-depth exploration of the battery's thermal characteristics and temperature dynamics during load cycles. The following equations describe the key calculations involved with the battery pack model [14].

State of charge is determined by

$$\frac{dSOC}{dt} = 100 \cdot \frac{I}{Q'}, \tag{1}$$

where I is the input current and Q' is the battery capacity. Total voltage drop of the battery pack is calculated with the equation:

$$\Delta U_{total} = \Delta U_{ohm} + \Delta U_{diff}^{total}, \tag{2}$$

where U_{ohm} is the ohmic resistance voltage drop, and U_{diff}^{total} is the total diffusion voltage drop. These are calculated as

$$U_{ohm} = I \cdot R_{ohm} \tag{3}$$

$$\Delta U_{diff}^{total} = \sum_{1}^{N_{RC}} \Delta U_{diff,i} \quad (4)$$

Diffusion voltage drop for each RC pair is obtained by

$$\frac{\mathrm{d}\Delta U_{diff,i}}{\mathrm{d}t} = \frac{I + \frac{\Delta U_{diff,i}}{R_{diff,i}}}{C_{diff,i}} \quad (5)$$

Total heat flow rate of the battery is a sum of ohmic resistance loss P_{ohm} and diffusion loss P_{diff}:

$$P_{total} = P_{ohm} + P_{diff}, \quad (6)$$

where ohmic resistance and diffusion losses are formulated as

$$P_{ohm} = I \cdot \Delta U_{ohm} \quad (7)$$

$$P_{diff} = I \cdot \Delta U_{diff}^{total} \quad (8)$$

3.3 Control Strategy

The series-hybrid powertrain system model was controlled by a separate controller model, which employed a specific control strategy to optimize the powertrain's fuel consumption and performance, while ensuring the system was operating within the dynamic constraints of its components, such as current limits for power electronics or maximum torque for diesel engine. An overview of the controller interface and strategy is illustrated in Fig. 7. Here, the main principles of the control strategy are examined.

A fundamental component of the control strategy is the selection of the diesel engine's operating point based on required power. Although the battery can supply power for a finite duration, the diesel engine will ultimately provide all power for the machine over time. Consequently, efficient control of the diesel engine is crucial for enhancing overall efficiency. The battery serves as a buffer or a low-pass filter for the engine's power demand, enabling the diesel engine to be controlled with a smooth control curve rather than a highly transient load profile. Furthermore, the diesel engine's operating speed and torque can be freely selected to generate the required power, allowing it to operate in the most efficient areas. It is essential to consider not only the diesel engine's efficiency map but also the generator's efficiency, as it operates at the same points as the diesel engine in this direct connection configuration. With the efficiency maps of both the diesel engine and the generator known, the most efficient operating areas for all available power levels can be determined. These operating areas are determined as a table of rotational speed-torque value pairs for

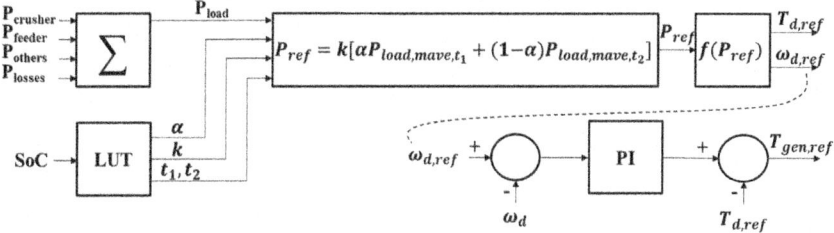

Fig. 7 Control principle of series-hybrid powertrain power generation. Inputs to the controller are defined on the left, and outputs on the right. The controller optimizes ICE-generator combination operating point based on load. A low-pass filter is applied to P_{ref} to ensure smooth ICE operation. Adapted from [7]

a set of power values, and linear interpolation is performed between the explicitly determined values. This control method reduces the transients of the diesel engine load profile, thus reducing its fuel consumption [11].

The reference power for power generation P_{ref} is determined as

$$P_{ref} = k[\alpha P_{load,mave,t_1} + (1-\alpha) P_{load,mave,t_2}], \tag{9}$$

where k is a gain variable, α is the weight coefficient ($\alpha \epsilon [0, 1]$), $P_{load,mave,t_1}$ is moving average of instantaneous power over t_1, and $P_{load,mave,t_2}$ is moving average of instantaneous power over t_2, where t_1 and t_2 are set to a short (e.g. 5 s) and long (e.g. 100 s) values, respectively. Instantaneous power is determined as a sum of measured power from VFDs and estimated or known losses. Coefficient α controls how rapidly the engine reference power responds to a change in overall power demand, and it is dynamically adjusted based on battery SOC, and gain variable k controls whether reference power should be increased, maintained, or lowered based on battery SOC.

The control system defines three distinct states for the battery, which define the values for coefficients α and k. The states are relative to a set target range for the battery state of charge. The first state is entered when battery is below lower bound of target range, and reference power will be closely governed by short average power, and a gain is applied to increase the battery SOC by setting $\alpha > 0.5$ and $k > 1$. In the second state the battery is within its target SOC bounds, aiming to maintain the battery SOC by setting $\alpha < 0.5$ and $k = 1$. In the third state the battery SOC is above its upper bound, and the battery SOC is reduced by setting $\alpha < 0.2$ and $k < 1$.

4 Simulation Analysis

First, the diesel engine model's consumption calculation was validated by comparing its simulated consumption over the load cycles to measured values. As fuel consumption is one of the key evaluation metrics for the system, this task was crucial in order to establish the trustworthiness of the simulation results. As displayed in Fig. 8, the simulated and measured consumptions are very close over the cycle. Similar accuracy was reached over all four load cycles.

The hybrid simulation model was configured to run with parameters displayed in Table 3, and the key simulation results are presented in Table 4. In order to be able to compare the conventional and hybrid powertrains' performance, specific fuel consumption (g/kWh) was calculated instead of volumetric fuel consumption over each cycle. This was done in order to negate the effect of variance in battery SOC over the start and end of the cycle, which impacts the overall work performed by the ICE of hybrid powertrain. From the results it is obvious that the hybrid powertrain can significantly reduce the fuel consumption of the machine. Over all load cycles the fuel consumption was reduced. It should also be noted that load cycles 2 and 3 could not be performed with a conventional powertrain, unless the peak power levels were limited to a level where the diesel engine would not stall. Therefore it is fair to say that the hybrid powertrain also outperformed the conventional configuration in these load cycles.

An example of the hybrid powertrain's performance under load during load cycle 2 is presented in Fig. 9. From this snapshot the control logic and relative smoothness of the ICE's operating curve compared to machine power demand can be observed. At the start of the cycle, the machine is exiting an idle period and begins crushing, which can be seen as a decrease in battery charge and an increase in diesel engine power. As the battery charge reaches its lower limit, the diesel power curve begins to follow instantaneous power more closely, which introduces some transients. As the battery charge is quickly recovered to target levels, the diesel engine's power curve is smoothened again. The machine load is suddenly decreased by approximately 75 kW at $t = 2300s$, but due to the relatively high setting of $t_2 = 100s$, the reference power is not reduced as quickly. Once the battery charge reaches its upper bound, the reference power is also adjusted faster. Once again, as the machine enters a high power sequence, the battery charge is lowered and the engine power is adjusted smoothly with a slight delay.

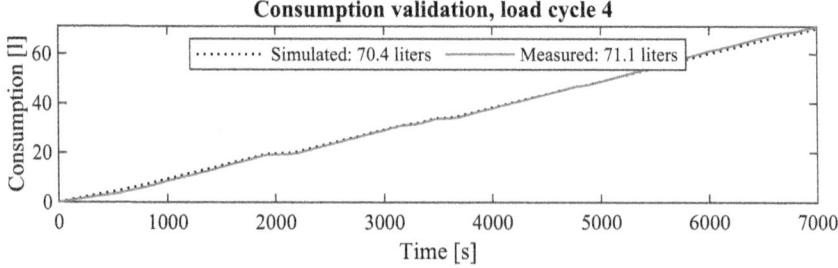

Fig. 8 Engine model fuel consumption calculation validation with load cycle 4 against measured data

Table 3 Simulation parameters

Ambient temperature	25 °C
Cooling liquid temperature	20 °C
Battery initial temperature	20 °C
Battery initial SOC	45%
Battery target SOC range	40–50%
Control parameter t_1	5 s
Control parameter t_2	100 s

Table 4 Simulated powertrain performance over four load cycles for conventional and hybrid powertrains. *The peak loads of cycles 2 and 3 for conventional powertrain were limited to prevent engine model from stalling

	Cycle 1		Cycle 2		Cycle 3		Cycle 4	
	Conv	Hybrid	Conv	Hybrid	Conv	Hybrid	Conv	Hybrid
Fuel consumption (g/kWh)	240	204	238*	204	224*	204	237	204
Consumption reduction (%)	15,2		14,1		8,9		14,2	
Battery max discharge power (kW)	169		215		180		169	
Battery max charge power (kW)	184,3		151,7		135		184	
Battery avg charge power (kW)	27,3		30,4		33,9		25,3	
Battery avg discharge power (kW)	25,3		31,2		32,6		27,3	
Battery avg temperature (°C)	26,6		27,9		28		26,6	
Battery avg SOC (%)	44		44,2		43,4		44	

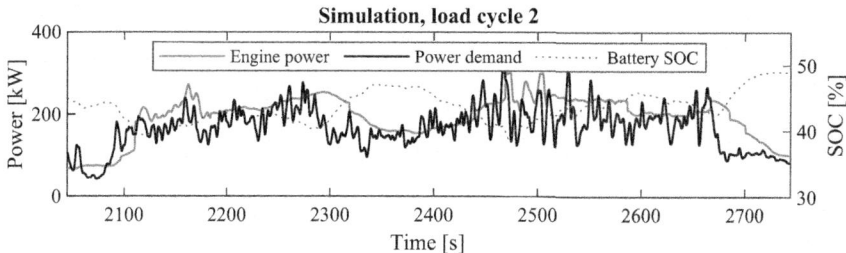

Fig. 9 Total power measurement, engine power, and battery state of charge during load cycle 2

During crushing the diesel engine operation is relatively smooth and satisfactory, but as the machine enters and exits idle periods between crushing work, the battery SOC reaches its bounds, which in turn affects the smoothness of the diesel engine's power curve as reference power is adjusted to bring battery charge to wanted levels. This can introduce large discharge and charge currents to the battery, which can deteriorate its health and performance over time. While the currents observed in the selected load cycles were in an acceptable range for the candidate battery, the controller performance could be improved by, e.g. identifying large, sudden changes in the instantaneous power and adjusting the parameters dynamically to react to large changes faster. Furthermore, a larger target SOC range for the battery would allow for a longer reaction time for the diesel engine. However, the results indicate that overall performance and efficiency of the machine could be improved by introducing a hybrid powertrain.

5 Conclusions

Efficiency of non-road mobile machines must be improved to reduce their harmful emissions. Fully electric and hybrid powertrains are researched and implemented in many NRMMs, but studies in mobile crushers are under-represented in the literature. Unlike some industries with standardized load cycles, the domain of mobile crushers lacks such benchmarks. In this study, this challenge was overcome by utilizing measured load cycles. Additionally, this paper introduced the conventional configuration of a diesel-electric mobile crusher and its load cycles in different crushing applications, providing basis for efficiency analysis. A series-hybrid powertrain simulation model for a mobile crusher was developed, aimed toward improving the energy efficiency and performance of the machine. Performance of the hybrid powertrain in the load cycles was assessed and compared to the conventional solution, revealing an improvement in both efficiency and performance. The key findings and conclusions of this study are as follows:

1. **Load Cycle Analysis**: The analysis of load cycles from different crushing processes highlighted the high power peaks of the crushing processes compared to the average power. Over the load cycles, the average-to-peak ratio for the engine was within the range of 1.5–1.7, while for the crusher, it was 2.3. Notably, the average engine power was in the range of 42–49% of its rated power over the measured cycles, indicating inefficient operating zones. The crusher accounts for approximately 60% of the machine's total power consumption, and engine load dynamics are predominantly driven by the crusher's power demand. Furthermore, the analysis revealed intermittent idle periods in all measured load cycles.
2. **Efficiency Improvement**: The series-hybrid powertrain demonstrated a significant improvement in efficiency compared to the conventional diesel-electric powertrain. The fuel consumption was reduced, on average, by 15% over the simulated load cycles, indicating the potential for substantial environmental and economic benefits.
3. **Hybrid Performance**: The hybrid powertrain effectively utilized the battery as a buffer for the diesel engine's power demand. The battery played a crucial role in smoothing out the engine's operating curve, reducing transients, and enhancing overall system efficiency. The control system effectively adjusted the power distribution between the diesel engine and the battery based on load demands.

Hybrid powertrain was able to perform an exceptionally demanding, synthesized load cycle which was not achievable with the conventional powertrain.

4. **Future Directions**: Subsequent research endeavors may center on refining the hybrid powertrain control strategy, taking into account variables such as battery durability and health. Furthermore, experimental validation of the simulated outcomes on an operational mobile crusher would yield more precise observations regarding the practical efficacy of the hybrid powertrain. Moreover, optimizing the sizing of components, particularly downsizing the diesel engine, could yield both economic and technological advantages. Additionally, exploring load cycles from diverse applications and crusher variations could offer deeper insights. Extensive investigation into the efficiency of work systems within mobile crushers could unveil additional avenues for further improvement.

In conclusion, the presented load cycles reveal the nature of mobile crushing machine work cycles, providing insights into their operational characteristics and requirements they pose for the powertrain. Additionally, the simulation study contributes insights into the potential benefits of hybridizing mobile crushers. The series-hybrid powertrain demonstrated notable improvements in efficiency and performance, paving the way for more sustainable and environmentally friendly operation of mobile crushing machines.

Acknowledgements This work has been conducted under the Digiecoquarry project, funded by the European Union's Horizon 2020 research and innovation program under grant agreement no. 101003750.

References

1. Caterpillar Electric Excavator. https://electrekco/2019/01/29/caterpillar-electric-excavator-giant-battery-pack/. Accessed 21 Jan 2024
2. Council of European Union (2016) Council regulation (EU) no 2016/1628. https://eur-lex.europa.eu/eli/reg/2016/1628/oj
3. Do TC et al (2023) Innovative powertrain and advanced energy man-agement strategy for hybrid hydraulic excavators. Energy 282. issn: 0360-5442
4. European Stage V Non-Road Emission Standards Policy Update. Interna-tional Council on Clean Transportation (ICCT), Nov 2016
5. Fassbender D (2023) Towards energy-efficient electrified mobile hydraulics: considering varying application conditions. Tampere University Disser-tations 821. PhD thesis. Tampere University, Tampere, 2023
6. Hagan R et al (2022) Non-road mobile machinery emissions and regula-tions: a review. In: Air 1.1, Nov 2022, pp 14–36. issn: 2813-4168. https://doi.org/10.3390/air1010002
7. Immonen P (2013) Energy efficiency of a diesel-electric mobile working machine. Acta Universitatis Lappeenrantaensis 518. PhD thesis. Lappeenranta University of Technology, Lappeenranta, 2013
8. Kalociński T (2022) Modern trends in development of alternative pow-ertrain systems for non-road machinery. In: Combustion Engines, Feb 2022, vol 188.1, pp 42–54. issn: 2658-1442. https://doi.org/10.19206/ce-141358

9. Lajunen A (2014) Improving the energy efficiency and operating perfor-mance of heavy vehicles by powertrain electrification. Aalto University Publication Series Doctoral Dissertations, 125/2014. PhD thesis. Aalto University, Helsinki, 2014
10. Lajunen A et al (2016) Electric and hybrid electric non-road mobile machin-ery–present situation and future trends. World Electr Veh J 8.1:172–183
11. Lindgren M, Hansson P-A (2004) Effects of transient conditions on ex-haust emissions from two non-road diesel engines. Biosyst Eng 87.1:57–66
12. Mocera F, Som A (2020) Analysis of a parallel hybrid electric tractor for agricultural applications. Energies 13.12:3055
13. Rockster presents first parallel hybrid crusher. https://www.at-inerals.com/en/artikel/atRocksterpresentsfirstparallelhybridcrusher-1793643.html. Accessed 15 Jan 2024
14. Siemens Simcenter Amesim (2024) Help documentation, IFP Drive Library, Bat-tery modeling. Accessed 05 Feb 2024
15. Wang F et al (2024) A free piston engine generator powered hybrid wheel loader with independent electric drive. Energy (Oxf) 286.129473

Open Access This chapter is licensed under the terms of the Creative Commons Attribution 4.0 International License (http://creativecommons.org/licenses/by/4.0/), which permits use, sharing, adaptation, distribution and reproduction in any medium or format, as long as you give appropriate credit to the original author(s) and the source, provide a link to the Creative Commons license and indicate if changes were made.

The images or other third party material in this chapter are included in the chapter's Creative Commons license, unless indicated otherwise in a credit line to the material. If material is not included in the chapter's Creative Commons license and your intended use is not permitted by statutory regulation or exceeds the permitted use, you will need to obtain permission directly from the copyright holder.

Effect of Electrification on the Energy Efficiency of Boom Trajectories of Semi-Autonomous Mobile Cranes

Timofei Komarov, Victor Zhidchenko, and Heikki Handroos

1 Introduction

The development of mobile machines, widely represented in construction, mining, logistics, and forestry, is carried out in several directions. First of all, electrification increased an overall machine efficiency [1]. Various systems of the machines are being electrified: internal combustion engines (ICE) are being replaced by electric motors [2]; valve-controlled hydraulic actuation systems are substituted with electro-hydraulic (EHA) [3] and electro-mechanical (EMA) actuators [4].

The second development direction is automation. With 5G technologies, fully autonomous operation can be achieved in a controlled and limited environment, such as construction site [5], mine [6], and timber terminal [7]. However, harsh environments and remoteness pose additional difficulties for forestry automation. Therefore, different solutions for semi-automation enhance operator's training and operation productivity: the coordinate boom control [8]; the cut-to-length method [9]; and remote operation in forestry [7], construction [10] and mining [11].

Motion control is another necessary part of automation. It can be considered from two perspectives: work function and traction drive control. This paper studies the control of machine work functions, using a mobile crane as an example. Application of automated movement requires solution of three main issues: observation of the system state, path planning, and position control. Methods of position control are studied in-depth in the field of robotics. Several main approaches are considered

---

T. Komarov (✉) · V. Zhidchenko · H. Handroos
LUT University, Lappeenranta, Finland
e-mail: timofei.komarov@lut.fi

V. Zhidchenko
e-mail: victor.zhidchenko@lut.fi

H. Handroos
e-mail: heikki.handroos@lut.fi

© The Author(s) 2025
L. Ericson and P. Krus (eds.), *Advancements in Fluid Power Technology: Sustainability, Electrification, and Digitalization*, Lecture Notes in Mechanical Engineering,
https://doi.org/10.1007/978-3-031-84505-5_13

for implementation in mobile machines: closed-loop position PI control [12]; feed-forward with a closed-loop position proportional control [13]; and reinforcement learning control [14].

Lastly, path planning is another area that was well-developed in various cases in robotics and logistics. Regarding the cranes of forestry machines, the emphasis was placed on oscillation damping [15, 16]; and on intellectual positioning, such as learning-based grasping positioning [17] and reinforcement learning motion control [18].

Recently, the research team of the Laboratory of Intelligent Machines at LUT University developed a method for building trajectories of mobile hydraulic crane booms from the perspective of energy efficiency [19, 20]. The approach was implemented on conventional hydraulic actuation systems controlled using pressure-compensated direction control valves (DCV). In this article, the effect of using pump-controlled EHAs for mobile crane actuation on the energy efficiency of boom trajectories is investigated.

2 Description of the System Under Study

The mobile machine under consideration is a hydraulic log crane PATU 655. Its mechanical structure is shown in Fig. 1 and is widely presented in the field of small forwarders. The crane can be considered as a four-link manipulator with rotational and linear joints. It includes a pillar, a lift boom, a jib (tilt) boom, and an extension boom [19].

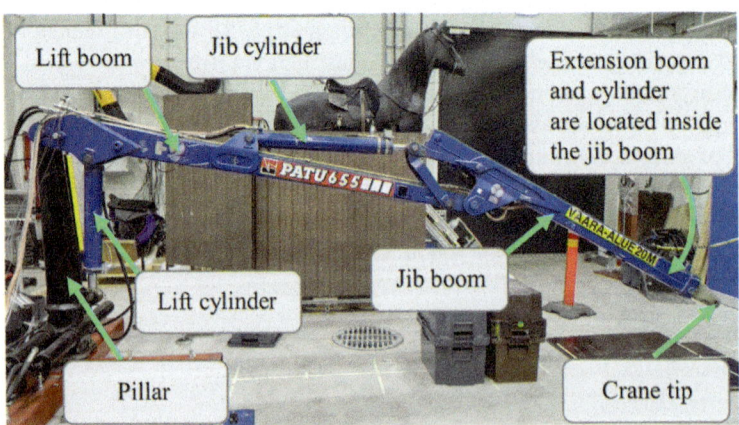

Fig. 1 Mobile hydraulic log crane PATU 655, Laboratory of Intelligent Machines at LUT University, 2023 [19]

2.1 Electro-Hydraulic Actuation System

The considered actuation system of the crane consists of independent closed-loop EHA systems. Three identical configurations were considered to actuate lift, jib, and extension cylinders. The principle diagram of the studied EHA system is depicted in Fig. 2. A study on implementation in mobile machines proved the circuit to be efficient and productive [21].

The electro-hydraulic converter 1 comprises an axial piston pump Parker F11-019 and an electric motor GVM210-050 [22], and directly drives the cylinder 2. An accumulator 3 and pilot-operated check-valves 4 are necessary for the volume flow difference compensation due to the asymmetry of the cylinder. Pressure relief valves 5 protect the system from overload, and externally controlled on/off valves 6 perform load-holding when the machine is inactive or in an emergency. System parameters are described in previous studies [23].

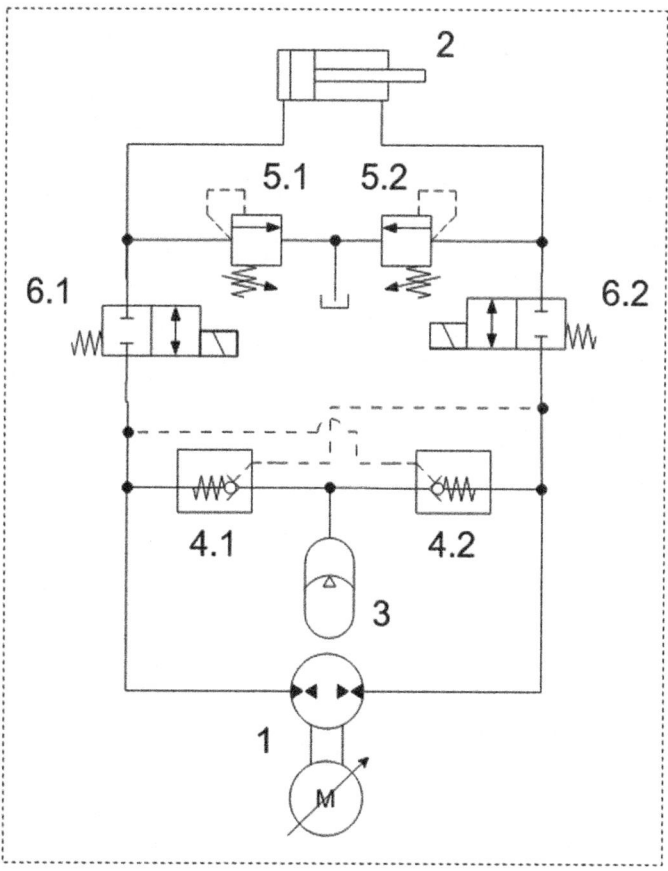

Fig. 2 Diagram of the EHA system

3 Motion Control System

The implementation of the path planning algorithm implies automatic crane motion along the built trajectories. Various approaches to position control of mobile hydraulic cranes have been studied [12–14]. In the presented study, a straightforward closed-loop control system for the position of the cylinders was formulated (Fig. 3) and implemented.

The control algorithm consists of three main parts: the feed-forward proportional control provides the necessary system dynamics; the closed-loop proportional control compensates for position error; and acceleration-based feedback damping attenuates oscillations caused by the high system inertia and flexibility of the hydraulic systems. It can also be expressed in a mathematical form:

$$\omega = \dot{x}_{ref} \cdot \frac{A}{V_{pump}} + K_{FB} \cdot (x_{ref} - x_{meas}) + K_{AD} \cdot \ddot{x}_{meas} \qquad (1)$$

where ω is reference rotation velocity of the motor, rad/s; x_{ref} is the reference cylinder position, m; x_{meas} is the measured cylinder position, m; A is a piston area, m^2; V_{pump} is the pump volume, m^3/rad; and K_{FB} and K_{AD} are proportional coefficients for feedback control and acceleration-based damping, respectively, which were found through manual tuning.

Due to the 4 quadrant operation of the tilt cylinder, the piston area A is chosen based on the pressure in the cylinder's chambers: piston-side area A_{tilt} is chosen if $P_A > P_B$, and rod-side area a_{tilt} if $P_A \leq P_B$. It is so because of the connection of externally controlled on/off valves 4, which are opened if the opposite-line pressure

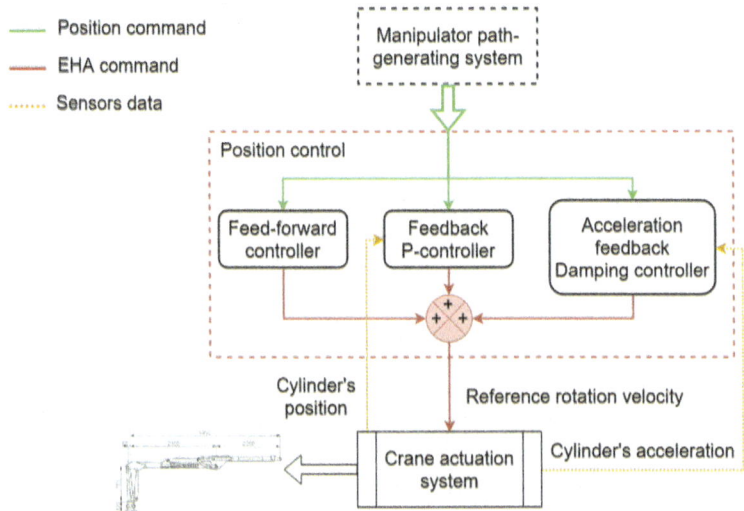

Fig. 3 Diagram of the position control system

is higher. In the case of the lift cylinder, the accumulator is constantly connected to the rod-side line due to the exceeding pressure in the piston-side chamber.

4 Path Planning Algorithm with an Emphasis on Energy Efficiency

The present study considers the path planning algorithm for building trajectories of the crane tip. Yet, during real operation, the crane performance is focused on moving logs. Therefore, the principle point to be traced should be placed within the grapple, so it would represent the movement of the payload. However, due to the lack of the grapple in the considered crane, the study neglects the load swing and considers the tip for following the trajectory. In other words, the current solution assumes a constant 0.5 m offset of the grappling point below the tip.

A successfully built trajectory should allow the crane to move the payload between the end-points, avoiding obstacles. The working condition implies the end-points to be where logs are grappled and released during the work cycle, whereas the main obstacles are the log bunks of a carrier. Other obstacles might occur, but detection of them is not a part of the current study.

A modified version of the A* search algorithm [24] is the base of the studied trajectory-building method. This heuristic algorithm is direct and clear, yet its graph-based nature implies discretization of the crane workspace. As problem space expands, both computational and memory demands are increasing exponentially, posing challenges for the algorithm's implementation on large-scale solutions. However, the considered log crane has a limited workspace, constrained by its mechanical configuration. In this study, the area within the crane reach was divided into 0.25×0.25 m square cells, which is twice smaller than in the previous research.

The A* algorithm uses a starting point and a goal as input. Then, at each iteration, it checks eight adjacent cells corresponding to the main possible movement directions. The considered version of the A* algorithm uses the shortest distance as a heuristic and the normalized consumed or recuperated motor energy as the cost of the path.

The motor energy consumed or recuperated is calculated as follows:

$$E_{motor} = \int_{T_1}^{T_2} \frac{T(t)\omega(t)}{\eta[T(t), \omega(t)]} dt, \qquad (2)$$

where $T(t)$ is the motor torque calculated from the pump supply pressure at time t, N; $\omega(t)$ is the motor rotation velocity at time t, rad/s; $\eta[(T(t), \omega(t)]$ is the motor efficiency at the given work conditions $T(t)$ and $\omega(t)$ according to the data-sheet [22]; T_1 and T_2 are the moments in time when the movement starts and finishes, s.

Hence, a weighted graph for the A* algorithm is created based on MATLAB Simulink simulation experiments by going through all $i, k = 1, \ldots, 20$, $j, r = 1, \ldots, 30$ cells (excluding the cells outside the crane range) and all directions. The first stage for each cell is finding the equilibrium system state at the given tip position. It is done by running a MATLAB/Simulink simulation in the given crane position for 5 s. During the simulation, pressure oscillations are mitigated through active damping using EHA in closed-loop position and damping control. If pressure deviates for more than 2 bar during the last second, the simulation is repeated with duration increased by 5 sec.

The second stage is a simulation of the tip movement between the adjacent cells in MATLAB/Simulink. Therefore, a matrix element $E_{K,R}[i, j]$ is introduced, which contains the sum of the values of motors' energy when the crane tip moved from a position corresponding to cell $C_{i,j}$ to an adjacent cell $C_{i+K,j+R}$, $K, R \in \{-1, 0, 1\}$.

As a result, eight matrices $E_{K,R} \in M_{20 \times 30}$, $K, R \in \{-1, 0, 1\}$ are obtained. In terms of the A* algorithm, $E_{K,R}$ sets the cost of movement between the graph's nodes. The values are normalized to be comparable to the range of the heuristic function, which is distance, as in the previous research [19]:

$$E_{K,R}[i, j] = \frac{E_{K,R}[i, j] + |\min_{\forall K,R}(E_{K,R})|}{|\min_{\forall K,R}(E_{K,R})| + \max_{\forall K,R}(E_{K,R})} \quad (3)$$

The described phase has to be executed once for adaptation of the method for the particular crane accounting for its geometry, kinematics, and actuation system. Further, the A* algorithm performs path planning considering energy efficiency, and the next chapter describes simulation experiments conducted based on the obtained data.

5 Simulation Experiments

The crane multibody model was developed in previous studies [25] using Simscape Multibody software and was verified using the crane installed at LUT University (Fig. 1). The logs were not included in the model because the reference crane from the laboratory doesn't have a grapple, so the swing motion of the load could not be studied. Therefore, the load was represented as a point mass attached to the crane tip. The load mass was 178.52 kg, which corresponds to an approximate weight of a single log. The extension boom cylinder was fixed to a fully retracted position during movements because of the redundant crane kinematics.

The electro-hydraulic actuation system was modeled according to the parameters of the crane cylinders and electro-hydraulic converters. Pressure relief valves and on/off valves were not considered because the present study does not consider crane performance during overload or load-holding. The dynamic model of the electric motor is simplified to a first-order transfer function with a 50 ms rise time. The leakages in the pump are neglected. The models of pilot-operated check valves were developed during previous research [23]. The Simulink crane model was extended by the previously described control system, and the trajectory-building algorithm was implemented by a MATLAB script.

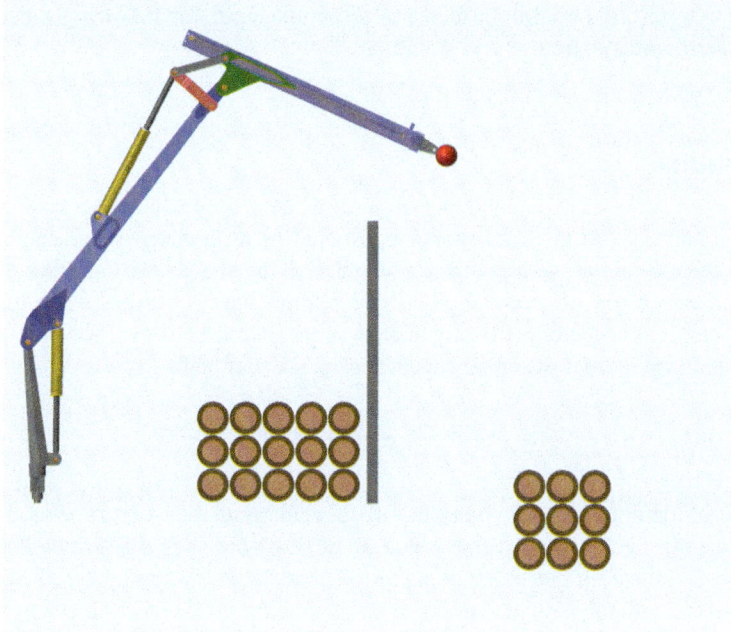

Fig. 4 Simulink crane model. The gray vertical line depicts the log trailer bunks. The brawn circles represent initial and final positions of the logs

The simulation experiments imitated two types of work cycles: firstly, loading the logs from a pile into the trailer, and secondly, unloading the logs to the ground. The used set of points represented a typical layout of a work environment, such as in [26]. The model included a single vertical obstacle representing the log trailer bunks. Its model expanded by 0.5 m in both directions to represent the safety boundaries for the tip movement. Figure 4 illustrates the crane model, the obstacle, and the location of the logs. The latter is done in a simplified way: the circles designate the positions of the centers of mass of the logs. It is justified by the fact that the grapple is supposed to grab a log close to its center of mass to maintain balance, even though the logs are commonly positioned orthogonally to the road.

In total, 9 positions of the logs in a pile and 15 positions in the trailer were considered resulting in 135 combinations of start and final points. For each combination, two path planning options were studied: the shortest path and an energy-efficient one found by the algorithm.

The generated paths were formulated as arrays of reference tip coordinates X_{ref}^L and Z_{ref}^L, where L is the number of steps in a path. The vectors X_{ref}^L and Z_{ref}^L then transformed into arrays $S_{ref_N}^L$ setting the positions of each cylinder N at any step. The mean value of movement duration was 15.96 s, which corresponded to a typical operator performance [26]. Even though the actuation system can provide higher operation speed, the control system has to be enhanced for following trajectories with maximum velocities.

6 Results

Let E_{tot}^s and E_{tot}^e be the total energy consumed or recuperated by all the electric motors during the whole crane movement following the shortest and the energy-efficient trajectories, respectively. Let η be relative energy savings when using an energy-efficient trajectory. It is calculated as a relation between the spent energy during both cycles and the energy spent for the shortest path:

$$\eta = \frac{E_{tot}^s - E_{tot}^e}{|E_{tot}^s|} \cdot 100\% \qquad (4)$$

Figure 5 demonstrates the box plot of η for all simulation experiments. For the logs' loading cycles, the median value is 1.16%, with the 25th percentile being −0.34% and the 75th percentile being 5.01%. For the unloading cycles, the median value is 0%, with the 25th percentile being 0% and the 75th percentile being 3.37%.

Figure 6 demonstrates the paths obtained using the described A* method in one of the simulation experiments. The green and pink dots depict the start and final points correspondingly. Figure 7 shows actual crane tip movement when following the built trajectories. The example demonstrates distinctive abrupt alternate movement during load lowering. Following such an energy-efficient trajectory allows the system to recuperate up to 2.45% more energy in comparison to the shortest path.

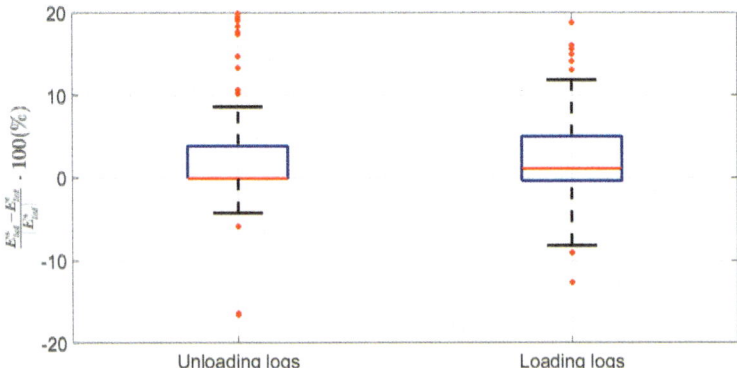

Fig. 5 The ratio of energy savings when following energy-efficient paths $E_{tot}^s - E_{tot}^e$ to the total energy E_{tot}^s consumed or recuperated when moving along the shortest paths

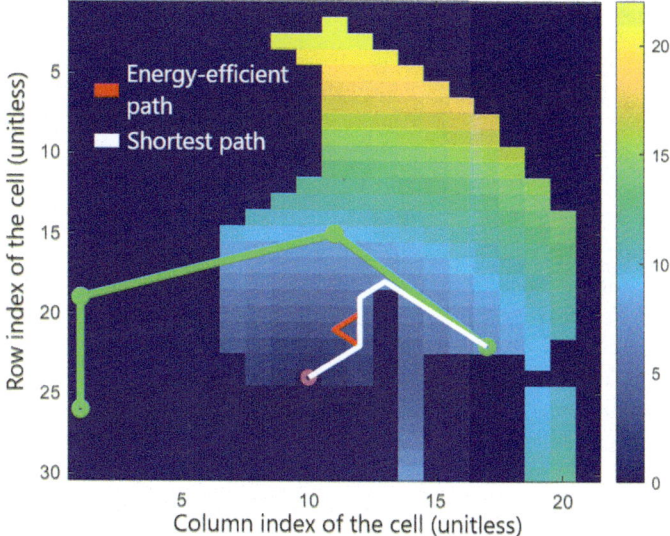

Fig. 6 An example of the paths found using A* method during one of the simulation experiments

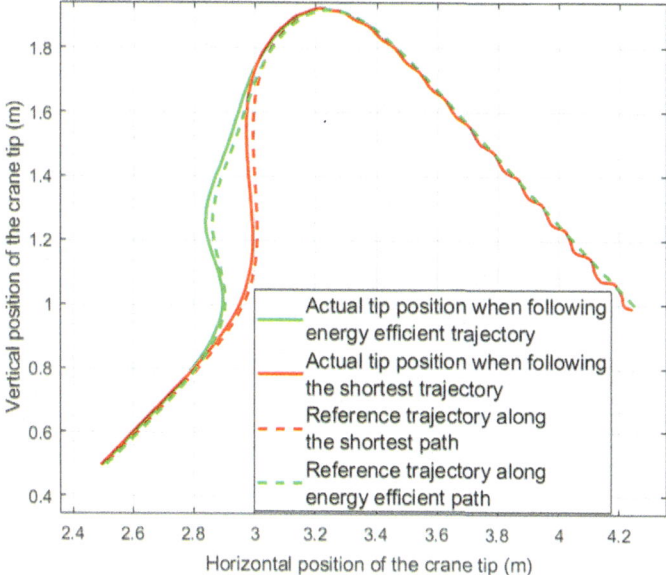

Fig. 7 Smoothed trajectory based on the paths generated by the modified A* method, and actual crane tip movements when following them

7 Conclusion

The method of path planning for building energy-efficient trajectories of the crane tip was implemented in the crane model with EHA-based actuation systems and recuperation capabilities. Simulation experiments demonstrate energy savings with a median value of 1.16% compared to using the shortest path when loading logs into a trailer. For unloading the logs, a median value of 0% resulted from the fact that most of the energy-efficient trajectories coincided the shortest paths. Since the proposed path planning algorithm is implemented in real time, it can be used as a basic method to build trajectories for automatic boom movement. While many of the built trajectories will represent the shortest paths, some of them will allow additional energy saving taking into account specifics of the mechanical structure and hydraulic circuit of the machine.

Electrification of the mobile machines and the implementation of EHA open the door for finding trajectories for optimal energy recuperation. The proposed algorithm provides up to a 2.45% increase in energy recuperation during load lowering. Since the EHA efficiency depends on the load conditions, their speed should be optimized during crane operation to ensure the highest level of efficiency.

The presented study was conducted with a simplified approach to EHA recuperation without considering the machine's electric system. However, further investigation into the detailed capabilities of machine recuperation is of interest for future studies.

The implemented position control system proved to accurately follow the reference trajectories. Nevertheless, the development of robust position control algorithms for mobile hydraulic cranes remains a substantial research task.

In this study, the extension boom was not considered due to the redundancy of the crane kinematics, which poses challenges on finding unique solutions with inverse kinematics. At the same, the opportunity of energy saving by combining joints' movements is a topic of future research.

Acknowledgements Research reported in this publication was financially supported by Business Finland through the Project "EMMA2".

References

1. Lajunen A, Sainio P, Laurila L, Pippuri-Mäkeläinen J, Tammi K (2018) Overview of powertrain electrification and future scenarios for non-road mobile machinery. Energies 11(5)
2. Tianliang L, Qingfeng W, Hu B, Wen G (2010) Development of hybrid powered hydraulic construction machinery. Autom Const 19(1):11–19
3. Minav TA, Bonato C, Sainio P, Pietola M (2014) Efficiency of direct driven hydraulic drive for non-road mobile working machines. In: 2014 international conference on electrical machines (ICEM), pp 2431–2435

4. Shufei Q, Yunxiao H, Long Q, Lei G, Lianpeng X (2022) A novel electro-hydraulic compound driving system with potential energy regeneration capability for lifting device. IEEE Access 10:18248–18256
5. Waurich V, Beck B, Weber J, Will F (2024) Levels of digitalization for construction machinery on the connected and automated construction site. In: Fottner J, Nübel K, Matt D (eds), Construction logistics, equipment, and robotics, Cham. Springer Nature Switzerland, pp 157–165
6. Sandvik (2023) Automine core for loading and hauling
7. Mahmood A, Abedin SF, O'Nils M, Bergman M, Gidlund M (2023) Remote-timber: An outlook for teleoperated forestry with first 5g measurements. IEEE Ind Electron Mag 17(3):42–53
8. Jussi M, Olle G, Anders M, Martin E (2017) Forwarder crane's boom tip control system and beginner-level operators. Silva Fennica 51:01
9. The perfect cut. ctl-the cut-to-length method. Ponsse Plc, Ponssentie 22, 74200 Vieremä (2023)
10. Hayashi K, Tamura T (2009) Teleoperation performance using excavator with tactile feedback. In: 2009 international conference on mechatronics and automation, pp 2759–2764
11. Sandvik (2023) Automine tele-remote for loading and hauling
12. Fodor S, Vázquez C, Freidovich L (2015) Automation of slewing motions for forestry cranes. In: 2015 15th international conference on control, automation and systems (ICCAS), pp 796–801 (2015)
13. Ruderman M (2018) Minimal-model for robust control design of large-scale hydraulic machines. In: 2018 IEEE 15th international workshop on advanced motion control (AMC), pp 397–401
14. Andersson J, Bodin K, Lindmark D, Servin M, Wallin E (2021) Reinforcement learning control of a forestry crane manipulator. In: 2021 IEEE/RSJ international conference on intelligent robots and systems (IROS), pp 2121–2126
15. Jebellat I, Sharf I (2023) Trajectory generation with dynamic programming for end-effector sway damping of forestry machine. In: 2023 IEEE international conference on robotics and automation (ICRA), pp 8134–8140
16. Fodor S, Freidovich L, Vazquez C (2016) Practical trajectory designs for semi-automation of forestry cranes. In: Proceedings of ISR 2016: 47st international symposium on robotics, pp 1–8
17. Gietler H, Böhm C, Ainetter S, Schöffmann C, Fraundorfer F, Weiss S, Zangl H (2022) Forestry crane automation using learning-based visual grasping point prediction. In: 2022 IEEE sensors applications symposium (SAS), pp 1–6
18. Andersson J, Bodin K, Lindmark D, Servin M, Wallin E (2021) Reinforcement learning control of a forestry crane manipulator. In: 2021 IEEE/RSJ international conference on intelligent robots and systems (IROS), pp 2121–2126
19. Zhidchenko V, Komarov T, Williot A, Bauer N, Handroos H (2023) A method for planning the trajectory of mobile hydraulic crane booms with a focus on energy efficiency. In: 2023 IEEE 21st international conference on industrial informatics (INDIN), pp 1–7
20. Zhidchenko V, Komarov T, Handroos H (2023) Towards energy-efficient semi-autonomous operation of hydraulic mobile cranes. Volume ASME/BATH 2023 symposium on fluid power and motion control of Fluid Power Systems Technology, Oct 2023, pp V001T01A068
21. Jensen KJ, Ebbesen MK, Hansen MR (2021) Novel concept for electro-hydrostatic actuators for motion control of hydraulic manipulators. Energies 14(20)
22. Parker Hannifin Corporation (2022) Configured ePumps electro-hydraulic pumps (EHPs) for construction. Truck, Mining, Material Handling, Agriculture and Forestry Applications
23. Shevchuk D (2020) Electric-hydraulic converter simulation and testing. Master's thesis, Lappeenranta-Lahti University of Technology, LUT School of Technology, LUT Mechanical Engineering
24. Hart Peter E, Nilsson Nils J, Raphael Bertram (1968) A formal basis for the heuristic determination of minimum cost paths. IEEE Trans Syst Sci Cybern 4(2):100–107

25. Malysheva I, Handroos H, Zhidchenko V, Kovartsev A (2018) Faster than real-time simulation of a hydraulically actuated log crane. In: 2018 global fluid power society PhD symposium (GFPS), pp 1–6
26. Hera PL, Morales DO (2019) What do we observe when we equip a forestry crane with motion sensors? Croat J For Eng 40(2):259–280

Open Access This chapter is licensed under the terms of the Creative Commons Attribution 4.0 International License (http://creativecommons.org/licenses/by/4.0/), which permits use, sharing, adaptation, distribution and reproduction in any medium or format, as long as you give appropriate credit to the original author(s) and the source, provide a link to the Creative Commons license and indicate if changes were made.

The images or other third party material in this chapter are included in the chapter's Creative Commons license, unless indicated otherwise in a credit line to the material. If material is not included in the chapter's Creative Commons license and your intended use is not permitted by statutory regulation or exceeds the permitted use, you will need to obtain permission directly from the copyright holder.

Study of Cavitation Conditions Inside a Proportional Spool Valve by Means of Modal Analysis on Sound Pressure Level

Luca Romagnuolo, Emma Frosina, Carmela Galdi, Maurizio De Bisceglie, and Adolfo Senatore

1 Introduction

Proportional directional control valves are extensively used in many hydraulic applications, since they can perform fluid control and regulation in a wide range of flow rate with relatively low complexity of the system [1]. Flow metering is realised by changing the position of a moving element, usually a spool, with an external controller (electrically, hydraulically, or manualally activated). The spool position with respect to valve body throttles the flow rate by realising different openings, which correspond to different cross-sectional areas. The geometry of the valve spool is properly realised in order to have precise control of the flow rate with the valve opening, especially at small spool strokes. It is therefore a common use to realise notches on the valve spool in the spool lands, which allow a precise and repeatable control of the cross-sectional area at small valve openings [2].

However, when the spool valve is subjected to working conditions that involve elaborating a high flow with small valve openings, cavitation may occur, specifically near and/or inside the spool notches. As well known, cavitation is an unwanted phenomenon that arises in hydraulic components, which brings several problems, among them performance reduction, vibration, material erosion, and noise. Therefore, this phenomenon is the subject of many studies and research activities that aim for its reduction. Several researches performed experimental activities and numerical simulations in order to reduce cavitation occurrence by acting on the geometry of the valve itself [3–8]. Other studies focused the attention on cavitation detection techniques. Martin et al. [9] showed that the signal of high-frequency response transducers can be used to detect cavitation inception and intensity inside spool valves. He et al. [10]

L. Romagnuolo (✉) · E. Frosina · C. Galdi · M. De Bisceglie
Department of Engineering, University of Sannio, Benevento, Italy
e-mail: lromagnuolo@unisannio.it

A. Senatore
Department of Industrial Engineering, University of Naples "Federico II", Napoli, Italy

© The Author(s) 2025
L. Ericson and P. Krus (eds.), *Advancements in Fluid Power Technology: Sustainability, Electrification, and Digitalization*, Lecture Notes in Mechanical Engineering,
https://doi.org/10.1007/978-3-031-84505-5_14

used a high-speed camera on a plexiglass-body spool valve to detect the formation of cavitation bubbles. Fu et al. [11, 12] studied the noise generated by cavitation in spool valves with different notch shapes, by analysing the noise at specific frequency bands. Osterman et al. [13] used hydrophone signals to detect the incipient cavitation inside a spool valve made of plexiglass, in order to compare results with visualisation of cavitation bubble formation.

Statistical and spectral techniques are widely used to analyse signals from different acquisition methods [14, 15]. However, very few studies adopted acoustic signals to detect cavitation inception and intensity, because of the complexity of the noise signal derived from cavitation itself [16]. Moreover, this analysis often requires to know in advance specific frequency bands to detect the intensity of the phenomenon [17], which is often spread along a wide range of frequencies [18]. Nevertheless, cavitation detection through acoustic measurement can be very powerful since it does not require the system to be heavily modified, for instance, by adding pressure transducers and hydrophones, nor even to have direct access to the component itself, as required by accelerometer installation.

In this paper, cavitation has been studied by performing a modal analysis on the acoustic signal recorded during valve operation. Non-cavitation and cavitation conditions have been imposed on a 2-way 2-position proportional spool valve, and sound pressure level signals have been acquired. Then, a Proper Orthogonal Decomposition has been applied on the frequency spectrum of the acquired acoustic signal, in order to reduce the dimensionality of the problem. The target of this study is to find a unique parameter, which can be derived from acoustic analysis, that allows to detect cavitation inception inside the spool valve. Due to the mentioned effects of cavitation phenomenon on noise, performance, and erosion, cavitation detection is extremely important, since it can be used as a signal to a feedback control mechanism which can change the operating conditions of the component itself to avoid the critical conditions.

2 Materials and Methods

A 2-way, 2-position proportional spool valve has been the subject of the presented analysis. As shown in Fig. 1, the body of the valve has been realised in plexiglass, which allows for direct visualisation of the cavitation phenomenon. Geometrical dimensions of the spool and the cavities of the valve are described in Fig. 2, where it is also possible to notice the presence of a double U-notch machined on the valve spool. Adding a notch on the valve spool edge between the two chambers of the valve is a common use for this type of proportional valves, since it allows a more precise control of the flow rate with the valve stroke [19]. The valve position is manually controlled by the use of a micrometre (Borletti VMW 0–25 mm, precision of 0.01 mm), which pushes the valve spool against a spring placed on the other side of it, that keeps the spool itself in a fixed position during each test. The valve opening X is controlled and measured, for each condition, by comparing the reading of the

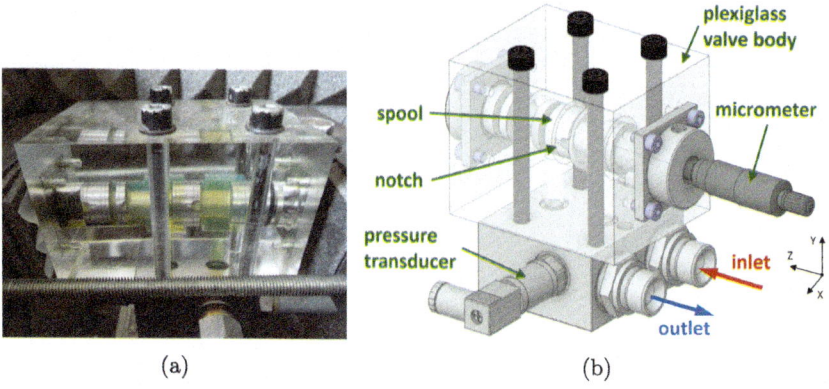

Fig. 1 Plexiglass spool valve (**a**) and its CAD equivalent (**b**)

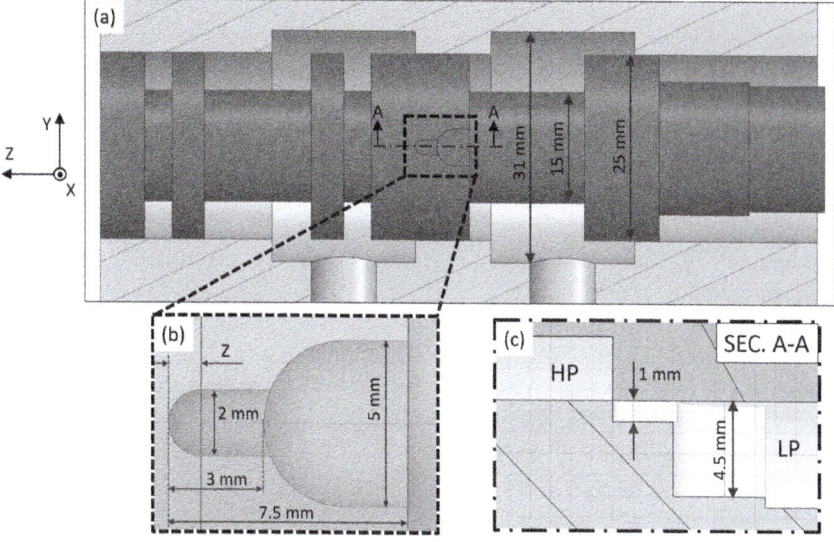

Fig. 2 Detail of the valve spool geometry (**a**), with U-notches dimensions (**b**) and (**c**)

micrometre to a reference position of the spool, that is the fully closed condition, in which the spool is tangent to the lateral wall of the high-pressure chamber (HP) of the valve. If X is less than 3 mm, the low-pressure chamber (LP) is connected to the previous one via a small, controlled section, while if X is between 3 mm and 7.5 mm, the larger notch connects the two chambers, allowing for a higher flow rate and a reduced pressure drop.

Experimental tests presented in this work have been performed in the Fluid Power Laboratories of University of Naples Federico II [20]. The plexiglass spool valve has been mounted on a test rig specifically designed to acquire acoustic signals

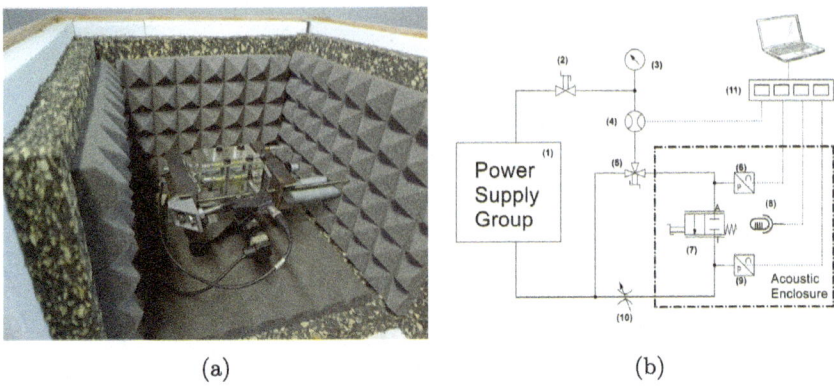

Fig. 3 Plexiglass spool valve (**a**) and its CAD equivalent (**b**)

derived from valve vibration (Fig. 3). A schematic representation of the test rig is given in Fig. 3b. Hydraulic oil (type ISO VG46) is delivered from a power supply group (1), in which an external gear pump model Hesper HP2 (displacement of 11.3 cm^3/rev), connected to an electrical asynchronous motor (model Bronzoni), delivers a maximum flow rate of 17 L/min. A pressure relief valve (Duplomatic PRM5) limits the maximum pressure delivered to the circuit at 20 bar, due to safety concerns about the maximum pressure that the valve can withstand, because of its plexiglass body. Then, the delivered flow rate is accurately controlled by a pressure compensated flow control valve (Duplomatic RPC2). A manual two-way ball valve (2) isolates the power supply group from the rest of the system, for safety. Flow rate is measured by a Coriolis flowmeter (model Endress Proline 83F15 (6), measuring range: 0 kg/h to 6500 kg/h), which also delivers s information of oil density and temperature. The plexiglass spool valve (7) is placed inside an acoustic enclosure, made of different layers of sound-adsorbing material (Fig. 3a), to dampen noise from the motor-pump group and the rest of the laboratory environment. The valve is connected to a manifold that includes slots for two pressure sensors to measure upstream and downstream pressure, respectively, model STS 8370, 0 bar to 60 bar, accuracy 0.2% of full scale (6), and model Gesensing PMP1400, 0 bar to 4 bar, accuracy 0.15% of full scale (9). The manifold is mounted on dampers, to reduce the effects of external vibrations. The valve circuit can be completely bypassed by using a manual three-way valve (5). A microphone preamplifier (8), model Brüel & Kjær DeltaTron type 2671, accuracy of ±0.3 dB, is placed inside the acoustic enclosure, in horizontal position at 2.5 cm from the vertical side of the valve body, in correspondence with the spool notch, where cavitation occurs. Downstream pressure can be regulated by acting on a manual chocking valve (10), however, since the target of this work is to generate cavitation, the downstream pressure has been kept as low as possible by completely opening this valve (the variations recorded from pressure sensor (9) from the atmospheric pressure are only due to the pressure losses inside the delivery branch of the circuit, which can be considered negligible). A data acquisition

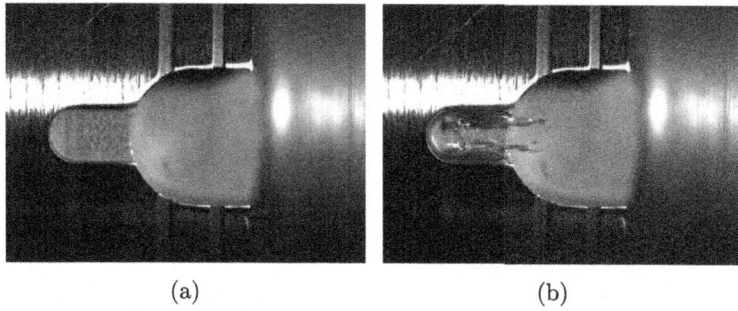

Fig. 4 Cavitation bubbles formation inside the valve spool at opening 1 mm: **a** upstream pressure 4 bar, **b** upstream pressure 5 bar

system, model Siemens SCADAS Mobile, with a high sampling frequency (up to 204.8 kHz) has been used to acquire signals of upstream and downstream pressure from (6) and (9), flow rate from (4), and sound pressure level (SPL) from (8). The acquisition system is managed via Siemens Simcenter Testlab Software.

Tests have been executed in stationary conditions of pressure and flow rate. The valve spool is placed in a position to realise the desired valve opening X by using the micrometre, as previously described, then the spool valve is put in communication with the rest of the system by opening the bypass valve, and the flow is regulated from the power supply group in order to reach the desired value of pressure upstream the valve (pressure difference is generated by the pressure loss realised from the valve itself). Tests have been carried out for 8 different openings of the spool valve (from 0.5 mm to 4.0 mm, with a step of 0.5 mm) and for 23 different values of upstream pressure (p_U (bar): 1, 2, 3, 3.5, 4, 4.25, 4.5, 4.75, 5, 5.25, 5.5, 5.75, 6, 6.5, 7, 8, 9, 10, 12, 14, 16, 18, 20).[1] Each test lasts 20 s, during which values of SPL have been acquired with a frequency of 102.4 kHz. During the test campaign, oil temperature (recorded from the Coriolis flowmeter) has been kept between 30 °C to 40 °C, in order not to have significant variations in terms of viscosity. Cavitation occurrence has been checked by directly visualise cavitation bubbles inside the spool notch, as seen in Fig. 4 and in previous works of the authors [18, 20–22]. The measured $Q - \Delta p$ curves for each condition of valve opening are presented in Fig. 5.

3 Analysis and Results

Cavitation number has been used to consider the pressure levels and pressure drops recorded for each case analysed; it is calculated according to Eq. 1 [23]:

---

[1] For openings at 3.5 mm and 4.0 mm, not all the pressures have been tested, since the valve opening is too big to oppose enough resistance to the flux to generate pressures higher than a certain value (for opening 3.5 mm the highest pressure reached was 19 bar, while for opening 4.0 mm it was 10 bar, as shown in Fig. 5).

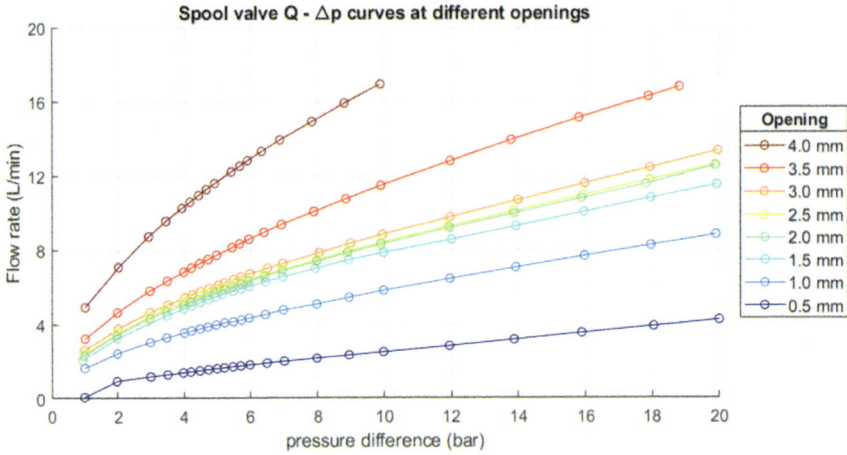

Fig. 5 Flow rate versus pressure loss curves for each tested opening of the spool valve

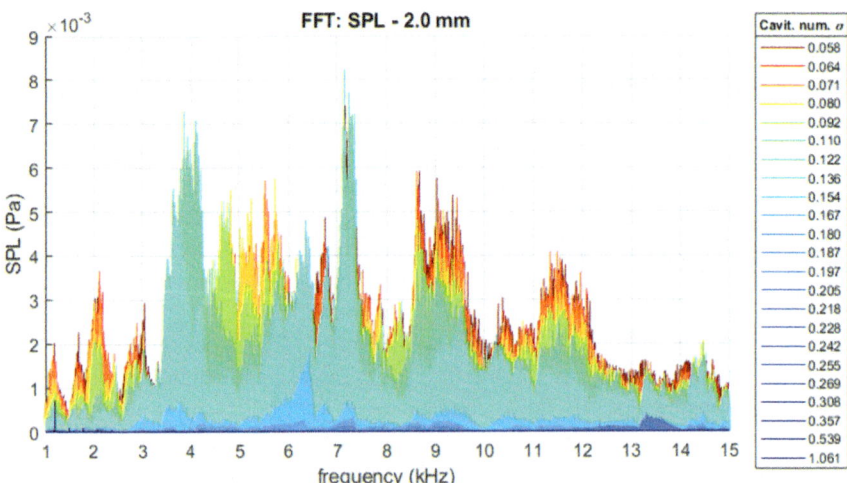

Fig. 6 FFT on SPL signals, for each cavitation number, at valve opening $X = 2$ mm

$$\sigma = \frac{p_D - p_V}{p_U - p_D} \qquad (1)$$

where p_U and p_D are, respectively, the upstream and the downstream pressures recorded in the manifold attached to the valve, whereas p_V is the vapour pressure, which can be considered equal to zero, for hydraulic oils.

Previous analyses [18] showed that SPL recorded during valve operation can be indicative of the presence of cavitation. Indeed, as shown in Fig. 6 (FFT on SPL signals derived at opening 2 mm), the amplitude of the signal generally rises as the

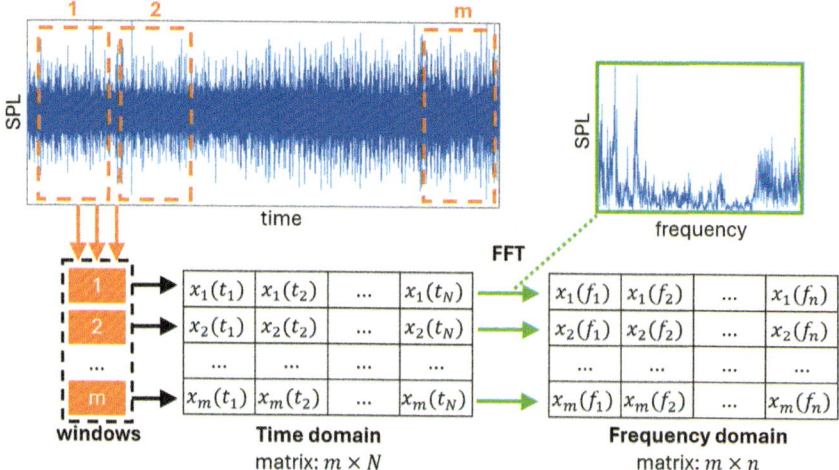

Fig. 7 Matrix M composition: process description

cavitation number decreases (i.e. the pressure difference increases and the cavitation occurrence is more likely), for all the frequencies analysed. However, as can be seen from the same figure, the determination of a specific relevant frequency that can be used to distinguish the state of cavitation of the system is not straightforward, since different opening conditions usually correspond to different frequencies that display amplitude peaks.

To overcome this problem, in this work a Proper Orthogonal Decomposition (POD) [24] has been applied to the frequency spectrum of the signal. Signals from all the openings and upstream pressure conditions have been elaborated in MATLAB® environment (v. R2023b). The total signal is first decomposed in m windows of fixed length N,[2] with an overlap of 50%. For each window, a normalised, single-sided Fast Fourier Transform (FFT) has been computed, which is composed of $n = N/2 + 1$ elements. Then, a $m \times n$ matrix M has been composed, which has on each row the FFT values computed for each window. This process is clarified in Fig. 7. A principal component analysis (PCA) has been applied to the matrix M, considering the n frequency bins, each one composed of m elements, one from each window, as the dimensions of the problem to be reduced. The PCA has been executed via Singular Value Decomposition (SVD), by decomposing the matrix M as follows:

$$M = U \Sigma V' \qquad (2)$$

where U is an $m \times m$ orthogonal matrix, V is a $n \times n$ matrix, that contains the principal components of the matrix M, while Σ is a $m \times n$ matrix with all elements equal to 0 but the ones on the main diagonal, which are the singular values of M

---

[2] By changing the window length, no significant differences have been found in the trend of the results.

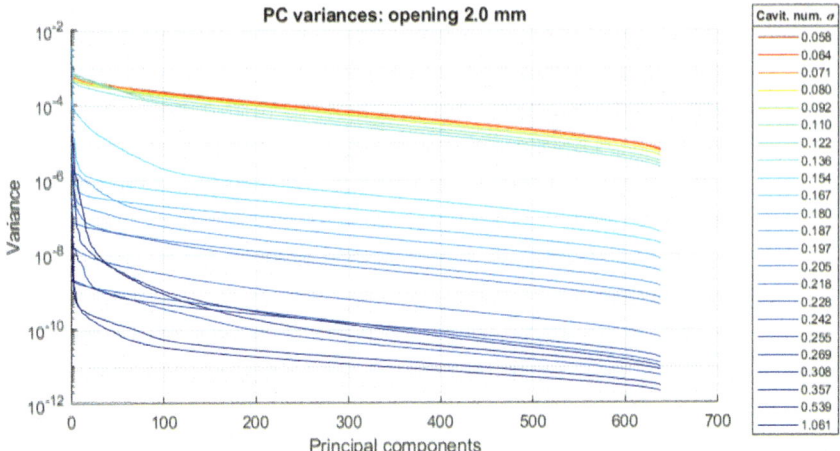

Fig. 8 Principal component variance against number of principal components, for each cavitation number, at valve opening $X = 2$ mm

in decreasing order. The j-th singular value ρ_j represents the variance of the signal described by the j-th principal component.

In Fig. 8, these values have been plotted against the number of principal components (which, as said, are arranged in order of higher variance), for each value of cavitation number acquired at the valve opening $X = 2$ mm. It is possible to notice that, after a rapid decreasing trend for the first principal components, the variance reaches a plateau, which happens to be at increasing levels for decreasing cavitation numbers (i.e. cavitation occurrence increasing).

Therefore, for increasing pressure difference realised by the spool valve, the variance of the principal components that describe the phenomenon increases on average. In Fig. 9, an average of the variances of the principal components has been reported for each case. Plots have been reported against cavitation number and upstream pressure, in order to be comparable with results shown in previous works [18, 20]. In particular, from Fig. 9b it can be seen that the average variance generally increases in correspondence of $p_U = 4$ bar to 6 bar, which is the threshold value for cavitation occurrence for this valve, as shown in [22]. Another behaviour can be noticed in Fig. 9b, that is the trend inversion of the considered value after the aforementioned pressure interval. This is more evident for smaller openings (i.e. 0.5 and 1.0 mm), but it is present also in the other cases. This trend is a manifestation of a reduction in the variability of the signal, after the formation of incipient cavitation. This behaviour indicates the presence of a more stable condition of the cavitation phenomenon at higher pressure differences than the incipient cavitation. Then, after a certain pressure difference, instability rises again in the case of severe cavitation. These observations confirm what was found in a previous work [21], in which the increasing-decreasing trend was found by means of image analysis.

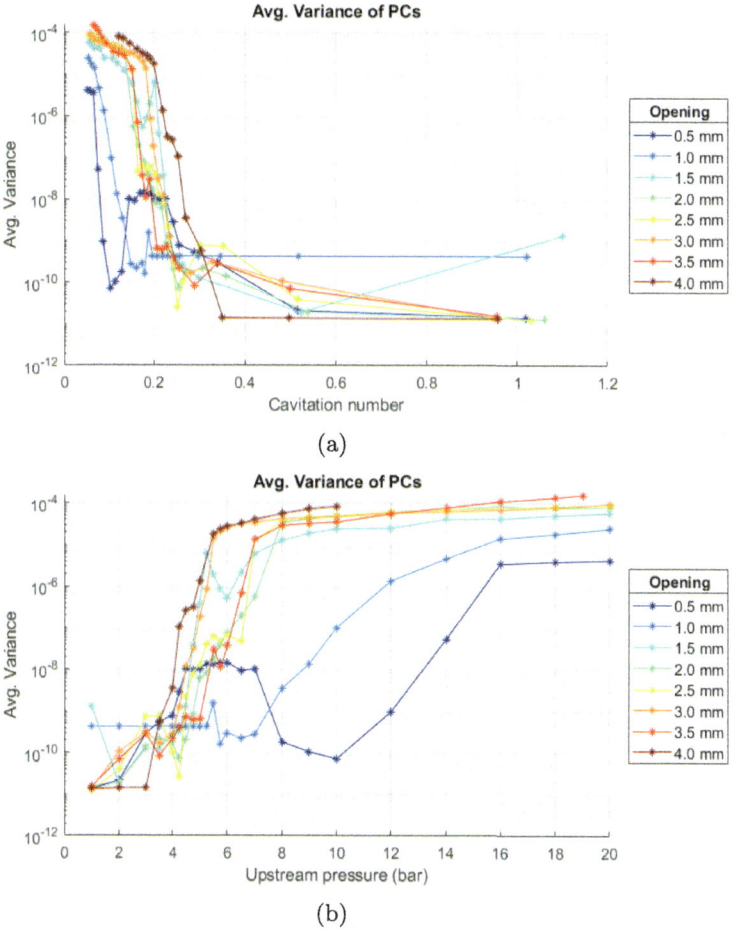

Fig. 9 Average value of principal components variance, for each cavitation number (upstream pressure) and each opening studied

Furthermore, it is also noticed that for high cavitation number the variance decreases more rapidly with the number of principal components. This behaviour can be interpreted as the symptom of the presence of high dispersion of the signal on the frequency domain, i.e. as cavitation number decreases, in order to properly describe the phenomenon, more principal components are needed. This means that the dimensionality of the signal increases, as expected, with cavitation phenomenon. Being R the number of the principal components, it is possible to derive a relative variance $\bar{\rho}_j$ for the j-th principal component. Therefore, a set of J principal components that describe a predefined percentage of the variance of the phenomenon (γ) can be defined as follows:

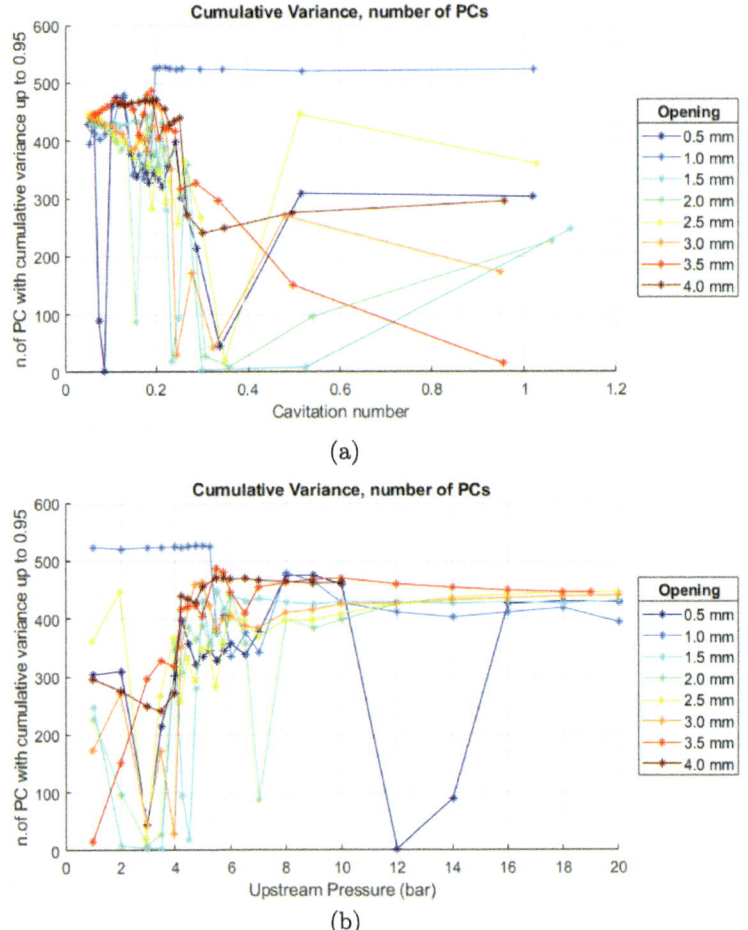

Fig. 10 Number of principal components needed to have a cumulative relative variance of the phenomenon of 95%, for each cavitation number (upstream pressure) and each opening studied

$$\frac{\rho_1 + \rho_2 + \cdots + \rho_J}{\rho_1 + \rho_2 + \cdots + \rho_{N-1}} \geq \gamma \qquad (3)$$

Figure 10 reports the number of principal components that are needed to describe the 95% of the total variance of the phenomenon ($\gamma = 0.95$)[3]. Here it is also possible to see a similar trend to the one shown before, though it is less evident. In particular, the curve that corresponds to the opening of 0.5 mm and 1.0 mm presents a different trend with respect to the other cases. However, these results do not seem to reflect a behaviour of the physical system but are mainly due to the way in which the

---

[3] A similar trend has been found by changing the value of γ to lower values.

significant principal components have been selected. Therefore, a comparison of the cumulative relative variance with a fixed value may not be the best solution to find the correct dimensionality of the problem and other techniques may be required.

4 Conclusions

Cavitation inside a 2-way, 2-position proportional spool valve has been studied by analysing the acoustic signal derived from the phenomenon. Tests have been performed in the Fluid Power laboratories of the University of Naples Federico II. The spool valve has been tested under 179 different conditions of opening and pressure difference. SPL signals have been recorded under stationary conditions by means of a high-frequency microphone preamplifier.

A simple FFT on the signals has shown an increase in amplitude with the increase of pressure difference (i.e. decrease of cavitation number), for almost every frequency bin analysed. Therefore, a Proper Orthogonal Decomposition has been applied to the signal in the frequency domain, to reduce the dimensionality of the problem. The sorted variances of the principal components have been compared between the analysed conditions.

After a rapid decrease for the first principal components, the variance trend reaches a plateau, with increasing values for decreasing cavitation numbers (i.e. increasing cavitation occurrence). The average values of the variances for each case analysed have been compared, showing an increasing trend as the pressure difference between the inlet and the outlet of the spool valve goes from 4 to 6 bar. Moreover, it has been possible to notice a particular trend inversion of the value analysed, specifically at smaller openings, that suggests a characteristic behaviour of cavitation intensity, which seems to reach a stable condition after its first inception. These results are in line with what was found in a previous analysis [21]. Furthermore, the cumulative relative variance has been compared with a fixed value, and the number of principal components necessary to describe the required amount of variability of the phenomenon has been compared for each case analysed. Even if it is possible to spot a similar trend to the previous comparison, this second analysis did not bring clear results, and this suggests that better methods [25] should be adopted to properly compute the minimum number of principal components that describes the phenomenon with a good approximation.

This analysis is part of an ongoing project that aims to find an effective and reliable cavitation detection method, based on acoustic signal elaboration, which, as said, does not require invasive changes in the system to be acquired. The aim is to develop an intelligent controller able to change the operating conditions of the analysed component on the fly, in order to reduce the cavitation occurrence and its negative effects. This procedure will be also extended to other hydraulic components that can be actively controlled, which usually encounter cavitation occurrence in specific conditions, such as pumps and actuators.

References

1. Vacca A, Franzoni G (2021) Hydraulic fluid power: fundamentals, applications, and circuit design. Wiley
2. Borghi M, Milani M, Paoluzzi R (2005) Influence of notch shape and number of notches on the metering characteristics of hydraulic spool valves. Int J Fluid Power 6:5–18. https://doi.org/10.1080/14399776.2005.10781216
3. Li S, Du J, Shi Z, Xu K, Shi W (2022) Characteristics analysis of the pilot-operated proportional directional valve by experimental and numerical investigation. Energies 15:9418. https://doi.org/10.3390/EN15249418
4. Yu R, Lu X (2022) Design of noise-reducing two-stage cage control valve and its fluid characteristics and cavitation study. J Chem 2022. https://doi.org/10.1155/2022/7322655
5. Guan W, Jianfei D, Linyuan K, Wenghui W, Qianfeng G, Xuejun Z (2022) Study on the influence of structural parameters on the flow and cavitation characteristics of tandem multi-stage pressure-reducing valves. Flow Meas Instrum 87:102–230. https://doi.org/10.1016/J.FLOWMEASINST.2022.102230
6. Lu L, Wang J, Li M, Ryu S (2022) Experimental and numerical analysis on vortex cavitation morphological characteristics in u-shape notch spool valve and the vortex cavitation coupled choked flow conditions. Int J Heat Mass Transf 189:122–707. https://doi.org/10.1016/J.IJHEATMASSTRANSFER.2022.122707
7. Amirante R, Distaso E, Tamburrano P (2016) Sliding spool design for reducing the actuation forces in direct operated proportional directional valves: experimental validation. Energy Convers Manag 119:399–410. https://doi.org/10.1016/j.enconman.2016.04.068
8. Oshima S, Leino T, Linjama M, Koskinen KT, Vilenius MJ (2001) Effect of cavitation in water hydraulic poppet valves. Int J Fluid Power 2:5–13. https://doi.org/10.1080/14399776.2001.10781115
9. Martin CS, Medlarz H, Wiggert DC, Brennen C (1981) Cavitation inception in spool valves. J Fluids Eng 103(4):564–575. https://doi.org/10.1115/1.3241768
10. He J, Li B, Liu X (2019) Investigation of flow characteristics in the u-shaped throttle valve. Adv Mech Eng 11(3):1687814019830,492. https://doi.org/10.1177/1687814019830492
11. Fu X, Lu L, Ruan XD, Zou J, Du XW (2008) Noise properties in spool valves with cavitating flow. In: Intelligent robotics and applications. Springer, Berlin, Heidelberg, pp 1241–1249
12. Fu X, Lu L, Zou J (2011) Noise characteristics in spool valve with v-notches. In: Proceedings of the 8th JFPS international symposium on fluid power, Okinawa, pp 540–545
13. Osterman A, Hočevar M, Širok B, Dular M (2009) Characterization of incipient cavitation in axial valve by hydrophone and visualization. Exp Therm Fluid Sci 33(4):620–629. https://doi.org/10.1016/j.expthermflusci.2008.12.008
14. Messinò D, Sette D, Wanderlingh F (2005) Statistical approach to ultrasonic cavitation. J Acoust Soc Am 35:1575. https://doi.org/10.1121/1.1918760
15. McKee KK, Forbes GL, Mazhar I, Entwistle R, Hodkiewicz M, Howard I (2015) A vibration cavitation sensitivity parameter based on spectral and statistical methods. Expert Syst Appl 42:67–78. https://doi.org/10.1016/J.ESWA.2014.07.029
16. Wu P, Wang X, Lin W, Bai L (2022) Acoustic characterization of cavitation intensity: a review. Ultrason Sonochemistry 82:105–878. https://doi.org/10.1016/J.ULTSONCH.2021.105878
17. Jablonská J, Mahdal M, Kozubková M (2017) Spectral analysis of pressure, noise and vibration velocity measurement in cavitation. Meas Sci Rev 17:250–256. https://doi.org/10.1515/MSR-2017-0030
18. Romagnuolo L, Frosina E, Senatore A, Cesaro U (2022) Experimental investigation on noise due to the cavitation phenomenon in proportional spool valves. In: 2022 IEEE international workshop on metrology for automotive (MetroAutomotive), pp 65–69. https://doi.org/10.1109/MetroAutomotive54295.2022.9855150
19. Ye Y, Yin CB, Li XD, Zhou J, Feng W, Yuan F (2014) Effects of groove shape of notch on the flow characteristics of spool valve. Energy Convers Manag 86:1091–1101. https://doi.org/10.1016/j.enconman.2014.06.081

20. Romagnuolo L, De Rosa R, Frosina E, Senatore A (2024) Study of a proportional spool valve noise by means of functional data analysis: cavitation and intensity detection. Mech Syst Signal Proc 209:111,100. https://doi.org/10.1016/j.ymssp.2023.111100
21. Romagnuolo L, Frosina E, Amoresano A, Quaremba G, Spirto M, Senatore A (2023) Instability measurement of cavitation conditions in a spool valve through the definition of a cavitation instability index. Flow Meas Instrum 91:102,366. https://doi.org/10.1016/j.flowmeasinst.2023.102366
22. Frosina E, Marinaro G, Amoresano A, Senatore A (2021) A numerical and experimental methodology to characterize the gaseous cavitation in spool valves with u-notches. J Fluids Eng 143(10). https://doi.org/10.1115/1.4050849
23. Lu L, Zou J, Fu X (2012) The acoustics of cavitation in spool valve with u-notches. Proc Inst Mech Eng Part G: J Aerosp Eng 226(5):540–549. https://doi.org/10.1177/0954410011413221
24. Chatterjee A (2000) An introduction to the proper orthogonal decomposition. Curr Sci 78:808–817. https://www.jstor.org/stable/24103957
25. Stoica P, Selen Y (2004) Model-order selection: a review of information criterion rules. IEEE Signal Proc Mag 21(4):36–47. https://doi.org/10.1109/MSP.2004.1311138

Open Access This chapter is licensed under the terms of the Creative Commons Attribution 4.0 International License (http://creativecommons.org/licenses/by/4.0/), which permits use, sharing, adaptation, distribution and reproduction in any medium or format, as long as you give appropriate credit to the original author(s) and the source, provide a link to the Creative Commons license and indicate if changes were made.

The images or other third party material in this chapter are included in the chapter's Creative Commons license, unless indicated otherwise in a credit line to the material. If material is not included in the chapter's Creative Commons license and your intended use is not permitted by statutory regulation or exceeds the permitted use, you will need to obtain permission directly from the copyright holder.

Evaluating the Performance of Semi-autonomous Kinematically Redundant Loader Crane Operation

Amy Rankka, Marcus Rösth, and Alessandro Dell'Amico

1 Introduction

This study is part of a research aiming at reducing the energy consumption of the operation of loader cranes by both looking into the hydraulic system design and how the crane is operated. Conventionally, a human operator has full control over how the crane is operated, but when features with various degrees of automization are introduced, possibilities of optimizing the movements of the crane with regard to energy efficiency from a control system arises. One such feature is Crane Tip Control, CTC. CTC can be seen as an easy-of-use feature where instead of controlling each cylinder of the boom system of the crane individually, the operator controls the speed and direction of the crane tip. The commercially available version of CTC [2] is optimized for lifting capacity rather than energy efficiency.

1.1 Problem Formulation

In this paper, the potential on an energy efficient CTC strategy is investigated by an dynamic programming optimization. The optimized strategy is compared to the commercially available, version of CTC (will hence be denoted as the original CTC), by deploying both strategies on a laboratory crane. The original CTC algorithm is able to run in real time on the embedded control system of the crane. This is not the case for the dynamic programming optimization which cannot find a solution in real time due to the computational load. Another difference is that the original

---

A. Rankka (✉) · A. Dell'Amico
Division of Fluid and Mechatronic Systems, Linköping University, Linköping, Sweden
e-mail: amy.rankka@liu.se

A. Rankka · M. Rösth
Hiab AB, Hudiksvall, Sweden

© The Author(s) 2025
L. Ericson and P. Krus (eds.), *Advancements in Fluid Power Technology: Sustainability, Electrification, and Digitalization*, Lecture Notes in Mechanical Engineering,
https://doi.org/10.1007/978-3-031-84505-5_15

CTC only has a horizon of one timestep ahead, while dynamic programming requires a trajectory of substantial length to provide a useful optimization. Even if it were possible to run in real time, the dynamic programming approach is thus not suitable for online operation. However, solutions from the optimization can state whether there is any gain in trying to develop online methods with an energy minimization objective and if so, serve as a benchmark.

The two CTC strategies are also compared to a previously developed drive cycle, see [8], where the cylinder movements of the crane are set by a human operator that was not restricted to following a straight line for the crane tip. This comparison gives an indication of how good, in terms of energy efficiency, a motion automated with CTC can be compared to manual control.

The dynamic programming algorithm makes use of an energy consumption model of the crane. To investigate the accuracy of this model is an important part of the results, as it indicates how close the solution is to the real-world optimum, and what can be expected from developing online solutions for CTC or other automated features based on the same model.

The application for this study is a loader crane with three hydraulic cylinders used for performing motions in the xy-plane, the first boom cylinder "$1b$", the second boom cylinder "$2b$" and the extension cylinder "ext". The crane also has a slewing cylinder for rotating motions that is not considered here. The cylinders are controlled by a load sensing hydraulic system connected to a variable displacement pump. The geometry of the crane is described in Sect. 2.2.

1.2 Prior Art

A solution to the redundant kinematics can be found by analytical optimization methods such as the pseudoinverse Jacobian method, see for example [4]. The problem with these methods is that they are only able to find local optima and does not consider that optimizing the pose of the crane arm in a small part of the working area could mean an unfavorable pose, or even a singularity, when moving out toward a new position. One approach to calculating a (near-) optimal solution for a complete trajectory is to search the possible action space by a dynamic programming algorithm. This requires the trajectory to be known in advance and is too heavy a computation to be able to run in real time on an embedded system but can provide a benchmark for future online solutions.

Using dynamic programming for finding optimal solutions to the redundant kinematics was tested already in [3], even though the objective was maximizing tip speed rather than minimizing energy consumption. When compared to a local optimizing algorithm based on the pseudoinverse Jacobian method the dynamic programming algorithm was able to find significantly faster solutions.

A well-presented usage of dynamic programming together with energy efficiency as an objective for finding a solution to the redundant kinematics can be found in [6]. Experimental results on a forestry crane are presented in [5]. For a load sensing system, the dynamic programming solution was found to give significantly better results than the pseudoinverse Jacobian method for all the tested trajectories. The implementation of the dynamic programming algorithm in this work has been inspired by these two papers. The underlying model of the energy consumption used as a cost function uses a model of the cylinder chamber pressures that approximates the backpressure with the compensator setting on the meter out valve multiplied by a constant. For the reference system of this study, the backpressure is heavily dependent on the crane state, and modeling approach that takes this into account is necessary to get a proper estimation of the energy consumption.

A recent publication in the area is [11] where an arbitrary trajectory for a forestry crane between two corner points in a three-dimensional space is found while trying to minimize the energy consumption with an A* search algorithm. For the boom system movements, however, a strategy of always using the extension cylinder as little as possible is implemented regardless of whether it is energy efficient or not.

2 Optimization with Dynamic Programming

Dynamic Programming is a computational method of solving a discretized optimization problem introduced by Richard Bellman [1]. Two key elements of dynamic programming are; the Principle of Optimality: *An optimal policy has the property that whatever the initial state and initial decision are, the remaining decisions must constitute an optimal policy with regard to the state resulting from the first decision,* and a functional equation introduced in a general form as:

$$f(x) = \max_{0<y<x} [g(y) + h(x-y) + f(ay + b(x-y))] \tag{1}$$

This form of functional equation has later been denoted as the Bellman equation, stating a relation between the value function of a system, $f(x)$, in two consecutive time steps. The optimal policy for controlling the system is the sequence of y-values that maximizes the value function of the initial step. What Bellman showed is that the overall optimization problem can be broken down into subproblems, and that solving these local problems one by one results in finding the global optimum for a given initial state. When applied to the energy minimization problem of this study the Bellman equation can be written as:

$$C_{t,e} = \min_{\Delta e}[stepcost(t, e, \Delta e) + C_{t+T_s, e+\Delta e}] \tag{2}$$

The value function is replaced by a cost function, $C(t, e)$, since the optimization objective is to minimize the energy consumption rather than maximizing a value. The cost function is defined for a set of time, t, and extension cylinder position, e, values. Δe is the increment in extension position and the step cost is defined as the energy required to move from the crane pose at t, e to the pose at $t + Ts, e + \Delta e$

and will be further presented in Sect. 2.1. The crane pose should at each time step t result in a crane tip position (x_t, y_t) according to a pre-defined trajectory.

2.1 Energy Model

The energy consumption to be minimized by the optimization is that of the variable displacement pump, the step cost is thus defined according to Eq. 3. The pump delivers pressure and flow to a load sensing hydraulic system and the pump pressure, p_{pump} is thus controlled to be the load sensing pressure plus a fixed pressure margin, Δp_{pump}, according to Eq. 4.

$$stepcost(t, e, \Delta e) = \begin{cases} p_{pump} q_{pump} T_s, & \text{if no speed or acc. limits} \\ inf, & \text{otherwise} \end{cases} \quad (3)$$

$$p_{pump} = \max[p_{1b,cyl} + \Delta p_{1b}, p_{2b,cyl} + \Delta p_{2b}, p_{ext,cyl} + \Delta p_{ext}] + \Delta p_{pump} \quad (4)$$

where $p_{1b,cyl}$, $p_{2b,cyl}$ and $p_{ext,cyl}$ are the active side cylinder pressures and Δp_{1b}, Δp_{2b} and Δp_{ext} are the pressure drops between the cylinders and the main valve. These pressure drops are calculated according to the loss models defined in [7]. The main part of these losses originates from the load holding valves.

The cylinder pressures are given by Gaussian Process models trained on data collected on the real crane. The method for creating the models is described in [10].

The pump flow, q_{pump}, is the sum of the flow to the three cylinders, according to Eq. 5.

$$q_{pump} = q_{1b} + q_{2b} + q_{ext} \quad (5)$$

The cylinder flows are calculated as the flow needed to move each cylinder according to the cylinder position increments imposed by the extension cylinder movement from e to $e + \Delta e$, and the crane tip movement from (x_t, y_t) to (x_{t+T_s}, y_{t+T_s}).

The algorithm takes into account that there are acceleration and speed limitations on each cylinder. The step cost for any set of $(t, e, \Delta e)$ violating these limitations is thus set to a very large number.

2.2 Geometry Model

The increment in crane tip position and extension cylinder position is known for each step cost calculation. The first boom and second boom cylinder positions for a state can be calculated by the inverse geometry Eqs. 6 to 13 below. To get the

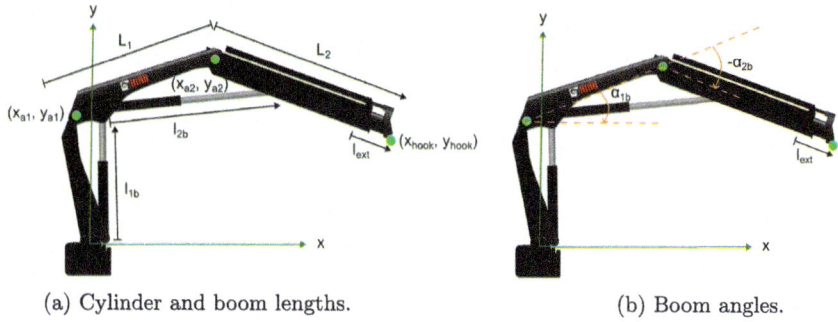

(a) Cylinder and boom lengths.   (b) Boom angles.

Fig. 1 Crane geometry

position increment, the positions are both calculated for e, (x_t, y_t) and for $e + \Delta e$, (x_{t+T_s}, y_{t+T_s}). $f_{linkage,1b}$ and $f_{linkage,2b}$ are geometric relations relating the crane boom angles of the first and second boom, respectively, to their cylinder lengths. For definition of angles, cylinder lengths and coordinates, see Fig. 1.

$$c = \sqrt{(x_{hook} - x_{a1})^2 + (y_{hook} - y_{a1})^2} \tag{6}$$

$$\alpha_1 = \cos^{-1}\left(\frac{L_1^2 + L_2^2 - c^2}{2L_1 L_2}\right) \tag{7}$$

$$\alpha_2 = \tan^{-1}\left(\frac{y_{hook} - y_{a1}}{x_{hook} - x_{a1}}\right) \tag{8}$$

$$\alpha_3 = \cos^{-1}\left(\frac{L_1^2 + c^2 - L_2^2}{2L_1 c}\right) \tag{9}$$

$$\alpha_{1b} = \alpha_1 + \alpha_2 \tag{10}$$

$$\alpha_{2b} = -(\pi - \alpha_3) \tag{11}$$

$$clen_{1b} = f_{linkage,1b}(\alpha_{1b}) \tag{12}$$

$$clen_{2b} = f_{linkage,2b}(\alpha_{2b}) \tag{13}$$

2.3 Dynamic Programming Algorithm

The minimization over the extension position represents searching for the best solution to positioning the redundant crane boom system at a given crane tip position. The crane tip position is represented by a trajectory in xy-coordinates discretized with a sample time $T_s = 0.5s$ and forms the input to the optimization. The output should be the optimal policy of control signals to each of the three cylinders at each time step together with the cost of the initial state, $C(t_0, e_0)$. The algorithm will calculate solutions for all possible initial extension positions, and to get the policy for a specific initial state, a forward pass through the grid is done after the optimization.

At every time step, the extension cylinder length, e, is discretized into 150 steps, giving a resolution of a little more than 1 cm. For a given extension position there can be only one set of first boom cylinder position, $clen_{1b}$, and second boom cylinder position, $clen_{2b}$ to fulfill the requirement of the hook position. If the solution exists, $clen_{1b}$ and $clen_{2b}$ can be calculated according to Eqs. 6 to 13.

A two-dimensional grid is created as $T \times E$, where $T = T_{end}/T_s$ is the number of time steps and $E = e_{max}/e_{resolution}$ is the number of discretized extension positions. Δe is discretized with the step E/T_s to avoid a drift in the extension position and $\Delta e \in [cvel_{ext,min} * T_s, cvel_{ext,max} * T_s]$, where $cvel_{ext,min}$ is the maximum speed in the negative direction and $cvel_{ext,max}$ is the maximum speed in the positive direction. It should be noted that the selected Δe will result in a control signal send the valve controlling the flow to the extension cylinder and that it is unnecessary to set E/T_s lower than the resolution of the valve command.

The iteration starts at the end of the trajectory, $t = T_{end}$, and proceeds backwards through time. At $t = T_{end}$, the cost of every valid extension position (where a solution for the first and second boom can be found) is set to 0 and to a very large number otherwise. At all other time steps, the cost, $C_{t,e}$, for each cell is calculated according to Eq. 2.

It should be noted that the algorithm will give an optimal solution to the discretized problem, for the real-world problem with continuous states it will only be a near-optimal solution. In [5], this is displayed in one of the test cases (vertical movement with the smallest possible initial extension length) where another optimization method yields a lower actual energy consumption.

Pseudocode for the algorithm can be found in the Appendix.

3 Tests on Crane

In [8], a drive cycle was developed for benchmarking the energy efficiency of different hydraulic systems. In this study, the hydraulic system is fixed but instead, the movements of the crane are optimized for energy efficiency. In its most complete form, the drive cycle includes speed references for each cylinder that are defined by a human operator, and as such, it is part of the comparison that is done in Sect. 3.2 and denoted "Non-CTC drive cycle". Before the cylinder speed references are added to the drive cycle, it consists of corner points that define the start, one intermediate and the end position of each lift. Crane tip trajectories can then be defined as straight lines between these corner points and different CTC strategies can be evaluated on these. The speed profile of each trajectory is set to complete the motion in a time given by the drive cycle while allowing for acceleration and deceleration of the movements in each direction.

For this paper, only the trajectories without any external load in the hook apart from the grapple tool are included. This is considered sufficient to verify the methodology. These trajectories, 12 in total, are given as input to the dynamic programming optimization and the reference speed and position trajectory outputs for each cylinder are saved. The speed trajectories are then upsampled to the sample time of the real-time system of the crane.

The crane tip trajectories from the drive cycle are also given as input to the original CTC algorithm. This algorithm is normally used online but to provide the same conditions as for the dynamic programming algorithm, the cylinder speed strategies for the whole trajectories are calculated offline assuming an ideal integration of the cylinder positions. These cylinder speed trajectories together with reference position trajectories are saved for later deployment on the crane. The algorithm could not find a valid solution to one of the trajectories which then had to be removed from the test set.

3.1 Experiment Setup

The crane used in the experiments is the same medium sized crane with a load sensing hydraulic system used as reference in [8] and [7]. The cylinder speed trajectories are sent to a feed-forward model representing the spool-curves of the hydraulic valve developed by [9] and commands to the spools of each cylinder are then sent to the embedded control system of the crane by a real-time Python script.

To ensure a decent reference following that will allow for a fair comparison between the two different strategies, feedback control of each cylinder position is implemented. The initial state of each trajectory is set the same for both algorithms.

The 11 trajectories that both algorithms have found valid solutions for are operated on the crane with a small external load, representing the grapple, attached to the crane tip. The crane movements, together with the load sensing pressure are measured and recorded for each trajectory and for each CTC strategy. The recordings from the dynamic programming CTC strategy are denoted as "DP-CTC", and the recordings from the original CTC strategy are denoted as "Org-CTC".

3.2 Results

The average pump energy per trajectory is 15.4 Wh for the optimized strategy and 18.9 Wh for the reference strategy, the reduction is 18.5%. The energy consumption for each individual trajectory is displaced in Fig. 2 together with the reference consumption for the dynamic programming strategy and the drive cycle with cylinder speed references defined by a human operator (non-CTC drive cycle). The average energy for the non-CTC drive cycle is 17.9 Wh but there is a large variation over the trajectories of its energy performance compared to the CTC strategies.

Notably, the improvement in energy consumption for the optimized CTC compared to the original does not originate from lower pressure levels, the average pressure level is quite similar for the two strategies, and sometimes even higher for the

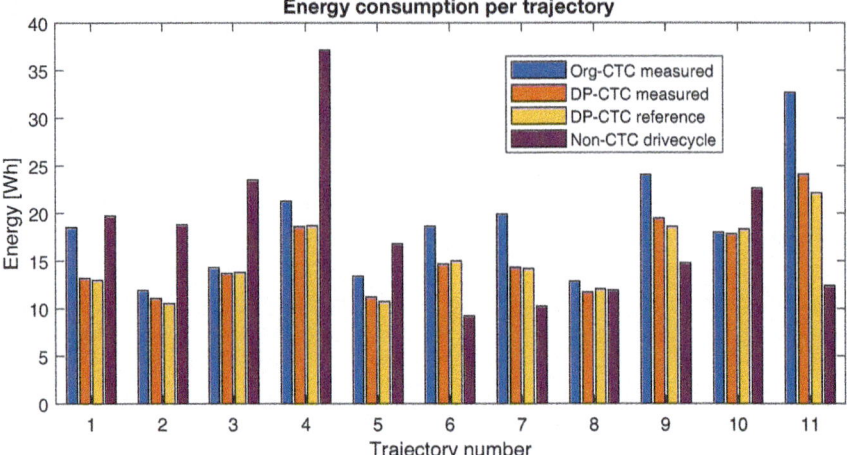

Fig. 2 Comparison of energy consumption for the two CTC strategies, the reference DP consumption, and the non-CTC drive cycle, for the 11 valid trajectories. The measured values for DP-CTC are close to their reference values for all trajectories. The energy reduction for DP-CTC compared to Org-CTC and Non-CTC varies over the trajectories

dynamic programming strategy, which can be seen in four example trajectories in Fig. 3. However, part of the improvement comes from reduced losses from simultaneous operation of several cylinders. An estimation of the simultaneous losses based on the measured cylinder pressures and speeds, and models for pressure losses in load holding valves and hoses gives that they are 21% lower for the optimized CTC. The improvement also comes from lower oil consumption, that is a smaller traveled cylinder area distance, which is displayed in Fig. 4.

It should be noted that even though a feedback control is active, the reference following of the crane tip trajectory is not perfect, but considered to be good enough for making a fair comparison between the strategies. An example of the reference following for four trajectories is displayed in Fig. 5.

Each trajectory represents a lift, the crane tip is lifted in the first part of the trajectory and lowered in the last part. Given that the system does not regenerate any energy and that the pressure required for lowering the first and second boom is independent of the load, a small difference in boom system lever between a reference and an actual trajectory in the end of the motion has a very low impact on the energy consumption.

3.2.1 Model Fit

The total amount of energy consumed during crane operation with the dynamic programming strategy is 102% of the estimated consumption from the algorithm. The pump pressure mostly follows the reference well, which can be seen in Fig. 3.

Pump pressure comparison

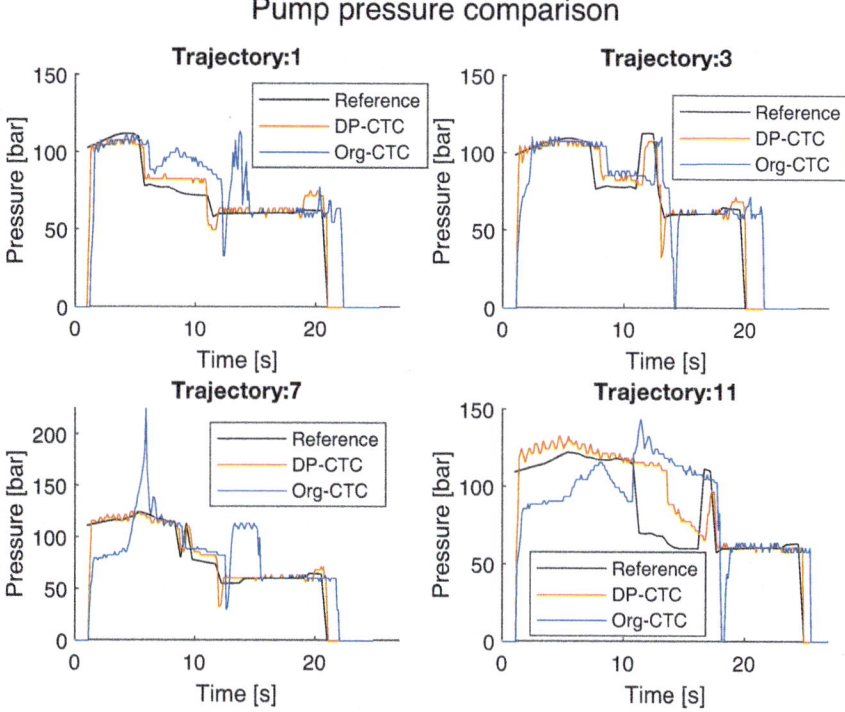

Fig. 3 Comparison of measured pump pressure for the two CTC strategies, and with estimated reference pressure for the DP-CTC, for four of the tested trajectories. The measured pressure for DP-CTC is rather close to the reference for all trajectories. Although there are variations between the DP-CTC and Org-CTC pressures, their average levels are similar

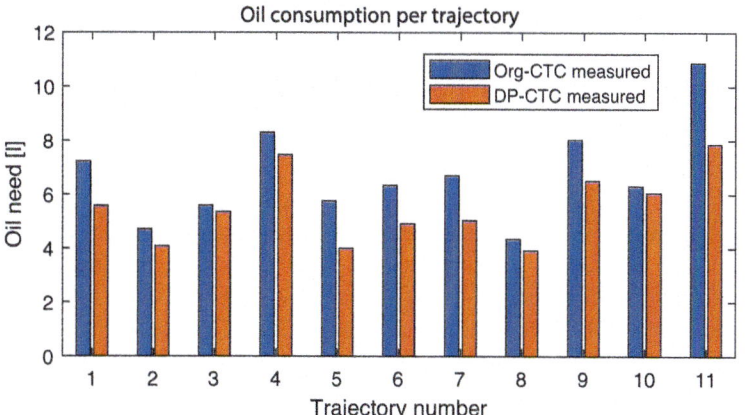

Fig. 4 Comparison of oil need for the two CTC strategies for the 11 valid trajectories. The oil need is noticeably higher for Org-CTC than for DP-CTC for all trajectories

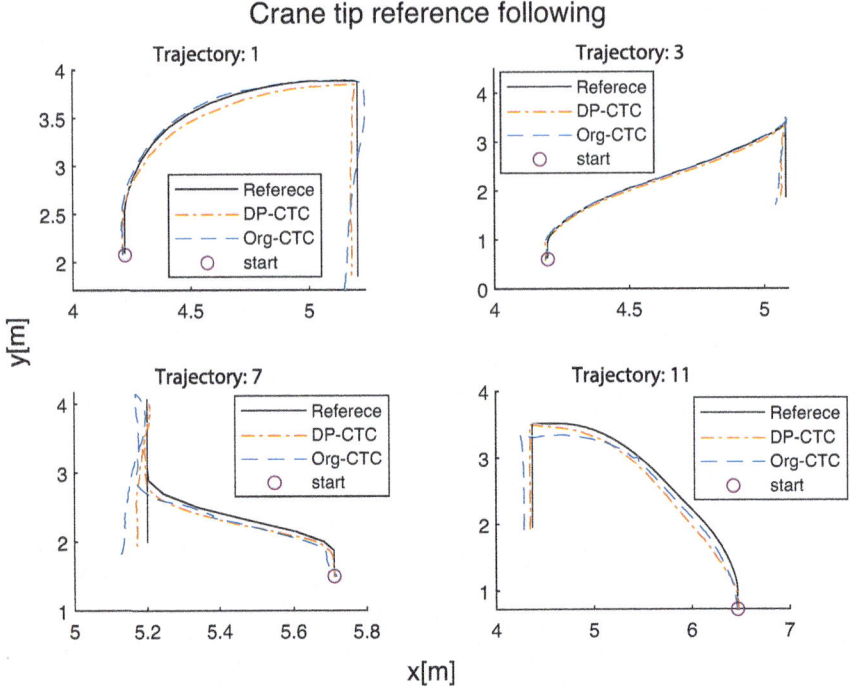

Fig. 5 Measured crane tip trajectories for the two CTC strategies compared to the reference trajectories

The energy for each individual trajectory can be compared in Fig. 2. The mean value of the relative error is -1.47% and the standard deviation is 3.83%.

4 Discussion

The dynamic programming optimization was able to find a more energy efficient strategy for moving the different crane cylinders than the original. The improvement was both due to a lower oil consumption and due to less co-operation of cylinders with different pressure levels. It should be noted that the external load of the crane was light, and therefore the absolute difference in pressure levels of the different cylinders was rather small. With a higher load, there will be a bigger difference between for example cylinders working against and with gravity and consequently the losses from simultaneous operation could have a bigger impact on the total consumption, which further highlights the importance of reducing them.

When comparing the measurement results from operation of the optimized CTC with the estimated energy consumption from the algorithm, there are two causes for the small difference. One is the model error and the other is the difference in actual crane tip trajectory. When comparing measurement results from the two different CTC strategies, the difference in crane tip trajectory following is considered to be small in comparison to the difference in energy consumption. Modeling errors can be overlooked in this comparison, the important result is that the optimized strategy performs significantly better in practice.

5 Conclusion

The dynamic programming algorithm found better, in terms of energy efficiency, CTC strategies for operating the three cylinders of the boom system than the original algorithm for all the tested trajectories. This was expected, given that the original strategy does not explicitly consider energy consumption at all. The point of interest is that the total energy reduction is as high as 17.5%, which gives an incentive for developing, and provides a benchmark for, online CTC solutions with an energy minimization objective.

Compared to a human operator, an automated CTC strategy for moving the crane between two points needs to be optimized for energy efficiency to perform better. On some trajectories, even the optimized CTC is much worse than the human operator, which indicates that the most energy efficient way of moving the crane tip between two points is not always to follow a straight line. If a completely automated motion is considered, strategies with a relaxed condition for the crane tip position should be considered.

The models used to estimate the energy consumption in the dynamic programming algorithm were found to have a good accuracy with a total difference of only 2% between the estimated and measured energy consumption. This shows that the models can be used as part of a simulation model of the crane for the future development of online CTC solutions.

Acknowledgements The authors would like to thank Hiab AB for their support during the work with this paper. The work in this paper was sponsored by the Swedish Energy Agency and is part of STEALTH II - Sustainable, Electrified and Automated Load Handling, App. no 44427-3.

6 Dynamic Programming Optimization Pseudocode

The pseudocode below for the dynamic programming optimization is presented in Algorithm 1. The matrix variables $costMap$, $cvelMap_{1b}$, $cvelMap_{2b}$ and $cvelMap_{ext}$ are all of size TxE and represent the output from the algorithm. $costMap$ states the optimal cost for starting at each extension position at each time step (including both the initial time step and all intermediate time steps). $cvelMap_{1b}$, $cvelMap_{2b}$ and $cvelMap_{ext}$ states the optimal cylinder speed commands at each extension position at each time step. $\mathcal{E}$ is the set of valid extension cylinder lengths, $\mathcal{T}$ is the set of time steps and $\mathcal{U}$ is the set of valid extension cylinder velocities.

The function $invGeometry()$ is defined by Eqs. 6 to 13. The equivalents of the functions $estPumpPressure()$ and $estCylPressures()$ are explained in Sect. 2.1.

Algorithm 1 Dynamic Programming Optimization

Require: x_{hook}, y_{hook}
  Initialize: $costMap \leftarrow inf$
  Initialize: $uMap_{1b}, uMap_{2b}, uMap_{ext}, eNextMap \leftarrow 0$
  for all $e \in \mathcal{E}$ **do**
    $validEndPos = checkEndPos(e, x_{hook}[T_{end}], y_{hook}[T_{end}])$
    if $validEndPos$ **then**
      $costMap[T_{end}, e] = 0$
    else
      $costMap[T_{end}, e] = inf$
    end if
  end for
  for all $t \in \mathcal{T}$ **do**
    $t = 1$
    for all $e \in \mathcal{E}$ **do**   ▷ $clen_{ext} = e$
      $clen_{1b}, clen_{2b} = invGeometry(e, x_{hook}[t], y_{hook}[t])$
      for all $u \in \mathcal{U}$ **do**
        $cvel_{ext} = u$
        $\Delta e = cvel_{ext} * T_s$
        $e_{next} = e + \Delta e$
        $clen_{1b,next}, clen_{2b,next} = invGeometry(e_{next}, x_{hook}[t+1], y_{hook}[t+1],)$
        $cvel_{1b} = \frac{clen_{1b,next} - clen_{1b}}{T_s}$
        $cvel_{2b} = \frac{clen_{2b,next} - clen_{2b}}{T_s}$
        $q_{ext} = calculateFlow(cvel_{ext})$
        $q_{1b} = calculateFlow(cvel_{1b})$
        $q_{2b} = calculateFlow(cvel_{2b})$
        $p_{cyl,1b}, p_{cyl,2b}, p_{cyl,ext} = estCylPressures(clen_{1b}, clen_{2b}, e, M_{load}, q_{1b}, q_{2b}, q_{ext})$
        $q_{pump} = q_{1b} + q_{2b} + q_{ext}$
        $validSolution = checkSolution()$
        if $validSolution$ **then**
          $p_{pump} = estPumpPressure(cvel_{ib}, cvel_{ob}, cvel_{ext}, p_{cyl,1b}, p_{cyl,2b}, p_{cyl,ext})$
          $step\_cost = q_{pump} * p_{pump} * T_s$
        else
          $step\_cost = inf$
        end if
        $totCost = step\_cost + totCostMap[e_{next}, t+1]$
        if $totCost < totCostMap[e, t]$ **then**   ▷ Save state if the lowest cost so far
          $costMap[t, e] = totCost$
          $cvelMap_{ext}[t, e] = cvel_{ext}$
          $cvelMap_{1b}[t, e] = cvel_{ib}$
          $cvelMap_{2b}[t, e] = cvel_{ob}$
        end if
      end for
    end for
  end for

References

1. Bellman R (1957) Dynamic programming. Princeton University Press
2. Hiab Official Webpage (2020) Ctc-crane tip control taking ease of use to a new level, July 2020
3. Löfgren B (2004) Kinematic control of redundant Knuckle Booms. Licentiate thesis, Jan 2004
4. Nilsson R (2009) Inverse kinematics. Master's thesis
5. Nurmi J, Mattila J (2017) Global energy-optimal redundancy resolution of hydraulic manipulators: experimental results for a forestry manipulator. Energies 10
6. Nurmi J, Mattila J (2017) Global energy-optimised redundancy resolution in hydraulic manipulators using dynamic programming. Autom Const 73:120–134
7. Rankka A, Dell'Amico A (2021) Loss analysis and concept comparison for electrically driven hydraulic loader crane
8. Rankka A, Dell'Amico A, Krus P (2019) Drive cycle generation for a hydraulic loader crane
9. Taheri A, Gustafsson P, Rosth M, Ghabcheloo R, Pajarinen J (2022) Nonlinear model learning for compensation and feedforward control of real-world hydraulic actuators using gaussian processes. IEEE Robot Autom Lett 7:9525–9532
10. Taheri A, Pettersson R, Gustafsson P, Pajarinen J, Ghabcheloo R (2023) Towards energy efficient control for commercial heavy-duty mobile cranes: modeling hydraulic pressures using machine learning, May 2023
11. Zhidchenko V, Komarov T, Handroos H (2023) Towards energy-efficient semi-autonomous operation of hydraulic mobile cranes, Oct 2023

Open Access This chapter is licensed under the terms of the Creative Commons Attribution 4.0 International License (http://creativecommons.org/licenses/by/4.0/), which permits use, sharing, adaptation, distribution and reproduction in any medium or format, as long as you give appropriate credit to the original author(s) and the source, provide a link to the Creative Commons license and indicate if changes were made.

The images or other third party material in this chapter are included in the chapter's Creative Commons license, unless indicated otherwise in a credit line to the material. If material is not included in the chapter's Creative Commons license and your intended use is not permitted by statutory regulation or exceeds the permitted use, you will need to obtain permission directly from the copyright holder.

Design of Hydraulic Power Take-Offs for Wave-Powered Reverse Osmosis Desalination: Meeting Constraints on Pressure Variation

Jeremy Simmons and James Van de Ven

1 Introduction

Researchers and government agencies have recently turned their attention toward renewable energy, such as solar, wind, and ocean wave energy, as direct sources of power for desalination of seawater [1, 12, 16, 18]. There is specific interest in reverse osmosis (RO) [13, 14]. This is a membrane-based process used to separate water from dissolved solids using high pressure as a driving force that is three to six times more energy efficient than thermal desalination processes. As a way to reduce cost and improve the efficiency of a wave-powered RO process, the reverse osmosis process can be integrated into the hydraulic circuit of a wave energy converter's (WEC) power take-off (PTO) (the subsystem responsible for loading the WEC and converting power) [3, 10, 18, 20, 29]. This avoids losses in the conversion of power to and from mechanical and electrical power when pressurizing feedwater.

An example architecture for a wave-powered reverse osmosis system is shown in Fig. 1. This system is capable of co-generating fresh water and electricity. This system includes (1) a seawater compatible pump that is driven directly by the oscillating WEC, (2) an RO membrane module receiving pressurized seawater from the WEC-driven pump, (3) an energy recovery unit (ERU) that recovers available power from the waste brine stream of the RO module and drives additional seawater to the RO feed inlet, (4) a charge pump that drives an intake flow and provides an elevated pressure at the WEC-driven pump inlet (to avoid cavitation), and (5) a hydraulic motor and electric generator that produce electrical power for onsite electrical power consumption and to regulate the mean pressure at the RO feed inlet. Additional components for this system that are not shown are pressure relief valves (PRVs) used to limit peak pressures.

J. Simmons · J. Van de Ven (✉)
University of Minnesota, Minneapolis, MN, USA
e-mail: vandeven@umn.edu

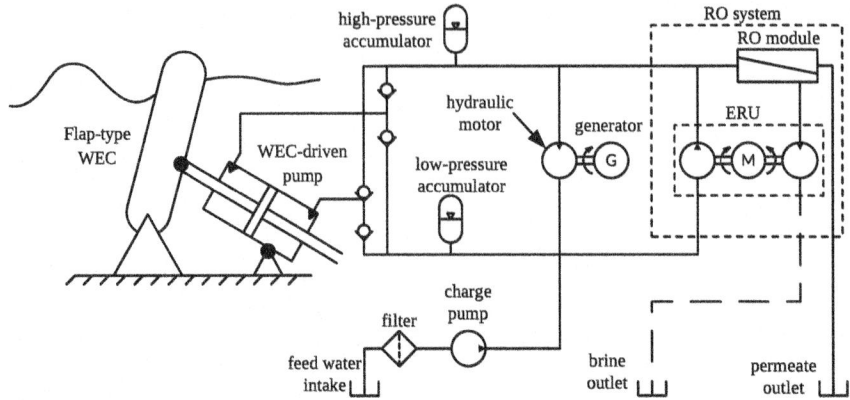

Fig. 1 A wave-powered reverse osmosis (RO) system comprised of a flap-type wave energy converter (WEC) and hydraulic power take-off operating with seawater as the working fluid to produce fresh water (Figure adapted from [23, 24])

Conversion of wave energy is a considerable technical challenge. The wave energy conversion process is a characteristically variable process. It involves absorbing power under the force of slow, irregular waves with a high degree of variation. For example, the ratio between the peak and mean power absorbed by a WEC has been estimated to be anywhere between 7 and 58 to 1, depending on the system design and load control scheme [15, 26]. In contrast, the conventional RO process is characteristically steady, with changes to operation being made very slowly and only as feedwater temperature changes or the membrane performance degrades. In fact, several sources recommend that start-up and shut-down of RO systems should be performed relatively slowly, with a rate of change in pressure less than 70 kPa per second, to avoid mechanical damage to the membrane and membrane housing [8, 17, 27].

The recommended limit to the rate of change in feed pressure is of particular interest in this work because it is a dominant constraint in the selection of accumulator volumes. This constraint is specified for start-up and shut-down without an extension to normal operation being made explicit by the literature. However, in normal operation of conventional RO systems, the pressure is held nearly constant, so it is understandable that an explicit statement would not be made. Furthermore, the value of the constraint seems not to be substantiated in any literature. The lack of clarity on this issue suggests that the constraint may be conservative, based on experience, and/or that the underlying mechanisms of failure motivating the constraint are not well understood. However, both [17, 27] claim that this constraint relates to mechanical damage to the membrane and membrane housing and suggest compaction of the membrane and cracking of the housing to be modes of failure.

Early work on wave-powered RO systems has dismissed this constraint. In the work by Folley et al., the constraint was acknowledged but neglected for lack of substantiation in literature and based on an argument that more robust RO system components would be produced if a wave-powered RO industry created the demand [11].

Recently, work by Sitterly et al. [25] and Das et al. [7] have acknowledged a potential for the pressure variations in wave-powered RO systems to harm RO membranes. Both works studied the effect of pressure variation using an experimental system with pressure being controlled. Sitterly et al. controlled pressures to match numerical results of a wave-powered RO system model presented in [29]. As part of their analysis, the authors compared the performance of the RO module before and after the membrane elements were subjected to the pressure variation experiments. They reported a decrease in membrane permeability of 7.4% and a decrease in water flux of 18.4%. It is important to note that this change came after a standard break-in procedure where membrane compaction is expected. The authors hypothesized that the pressure variation drove additional compaction that would not have occurred otherwise. The conclusion was that the pressure variation did not have significant effects on the membrane performance and integrity. However, from a system-level perspective, having membrane productivity decline 18.4% is substantial.

Das et al. [7] controlled pressure in two ways: first as sinusoidal variations between 35 and 65 bar and then as rectified sinusoidal functions with pressure varying between 0 and 70 bar. Membrane integrity tests showed no significant change after the simple sinusoidal pressure tests. However, after the more extreme rectified sinusoidal tests, the membrane integrity tests showed a two to four times increase in the salinity of the permeate suggesting significant degradation of the membranes' integrity.

This paper addresses the design impacts associated with meeting the pressure rate-of-change constraint on several alternative PTO architectures. Design performance is evaluated by two design metrics: the total volume of accumulators deployed and the amount of power lost during operation. The baseline architecture is also evaluated without the rate-of-change constraint enforced. Each architecture is modeled mathematically and numerically simulated to estimate its performance. Grid studies are used to find optimal values for the design parameters relevant to each architecture.

The proposed architectures are described in the following section, Sect. 2. The methods of this study are described in Sect. 3 and include the mathematical models, the method of solving the models, and the procedure for optimizing and comparing the design performance of each architecture. The results of the study are presented in Sect. 4 and are followed by a discussion and suggestions for future work in Sect. 5. Conclusions drawn from the study are presented in Sect. 6.

2 Proposed Power Take-Off Architectures

The baseline architecture for this design study, shown in Fig. 1, includes a first-order low-pass filter in the form of a single high-pressure accumulator. A second-order low-pass filter is easily constructed with the addition of a resistive element and a second accumulator as shown in Fig. 2a. The resistive element could be a passive element

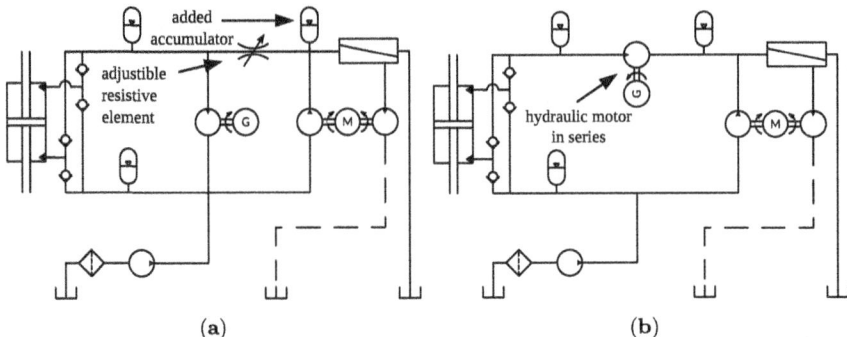

Fig. 2 Proposed PTO architecture schematics: (**a**) a parallel-type power take-off architecture with a second-order low-pass filter (**b**) and a series-type power take-off architecture (figure adapted from [23])

or an actively controlled valve. The passive element accomplishes the formation of a second-order low-pass filter while an actively controlled valve provides the opportunity of controlling the flow to the RO feed inlet and thereby rate of change in pressure. Both approaches reduce the total accumulator volume required to meet the pressure rate-of-change constraint.

A third alternative is the series-type PTO architecture proposed in [23] and illustrated in Fig. 2b. In this architecture, the hydraulic motor of the parallel-type architecture is placed in series with the WEC-driven pump and RO module. This way the hydraulic motor, which can operate as either a motor or a pump, has direct control of the flow reaching the RO feed inlet from the WEC-driven pump.

3 Methods

The goals of this work are to (1) understand how the selection of architecture influences the required accumulator volume and energy losses in the system when the rate-of-change constraint is enforced and (2) understand how volume requirements compare when the constraint is not enforced. The metric assumed for energy losses is the average power losses associated with managing the pressure variation; specifically, the losses from the activation of pressure relief valves at the outlet of the WEC-driven pump and the feed inlet of the RO module, the power losses from the hydraulic motor and electric generator, and throttling losses of the resistive element.

Numerical models of the system are used to simulate the system and perform a grid search to identify optimal performing designs for each PTO architecture. The design variables considered in the grid search (where applicable) are the total high-pressure accumulator volume, the proportion of volume placed at the RO feed inlet (versus the outlet of the WEC-driven pump), and the resistance of the resistive element.

To identify the cost of the constraint to the design performance, the optimal design performance for the baseline architecture is evaluated with and without enforcing the constraint on pressure rate of change.

The hypothetical system being considered is one installed in the nearshore environment at Humboldt Bay, CA and which is being driven by the Oyster 1, a bottom hinged flap-type WEC (like the one illustrated in Fig. 1) designed and tested by the former Irish company Aquamarine Power [2]. This selection of site and WEC design builds on prior work on wave-powered RO [23, 28]. The displacement of the WEC-driven pump and RO membrane area are selected based on the design study presented in [23]. The parameters are selected from the parallel-type and series-type PTO architecture having a fixed pump displacement and fixed RO membrane area and are based on having the same RO membrane area and annual rate of freshwater production.

A single sea condition is evaluated which is selected from the sea states specified for the Wave-to-Water Prize competition [5]. The selected sea state from that set, having a significant wave height of 2.64 meters and a peak period of 9.86 s, contributes the most to the annual available wave energy and when taken as a corner condition (maximum sea height and wave period), accounts for approximately half of the available wave energy annually (this is based on data from a near shore buoy in Humboldt Bay, CA for the rate of occurrence of sea states [28] with a weighting by available wave power).

The following subsections present the mathematical models used to simulate the WEC and proposed PTO architectures, the control methods, the method of solving the numerical problem of simulating the system, and the procedures used to carry out the design study.

3.1 Modeling

The WEC/PTO systems are modeled as being a dynamic system, described by a set of ordinary differential equations. The WEC and PTO subsystem models are coupled. The velocity of the WEC is an input to the PTO subsystem model and the reaction torque of the WEC-driven pump is an input to the WEC subsystem model. The inputs to the WEC subsystem model are the surface elevation of the ocean waves around the WEC and excitation force of the waves on the WEC.

Wave Energy Converter Model The model of the WEC and the parameters used in the study are identical to those developed by the authors in [23]. The state variables describing the WEC subsystem are the position and velocity of the WEC, and a set of states describing the forces on the WEC due to wave refraction.

The forces on the WEC include the wave excitation force, radiation damping, a hydrostatic restoring force, and weight of the device. The wave elevation and excitation force are constructed from the Pierson Moskowitz spectrum [9] using the superposition of 1000 sinusoidal signals having frequency dependent amplitudes and phase. The phase of each frequency component is additionally modified by a uniformly distributed random value in the range $[-\pi, \pi]$. Frequency components are distributed based on an equal energy method such that each interval of the wave elevation power spectrum has an equal area. The frequency dependent coefficients for the amplitude and phase were obtained using the open source boundary element method solver Nemoh [6], as was an impulse response function for the radiation damping. The methods presented in [21] were used to identify a 3rd order system model approximating the impulse response function.

Power Take-Off Model Each PTO architecture shares several components that are modeled identically. The differences between the PTO models are in the parameter values of the model (specifically the WEC-driven pump displacement and the hydraulic motor displacement), the flow connections, and the control algorithms applied.

A schematic showing the modeled variables and node connections for these PTO architectures is given in Fig. 3. This schematic accounts for each architecture with the optional configuration displayed with dashed-line boxes. For the baseline architecture, the added resistive element and additional accumulator associated with the pressure p_f are excluded. For the parallel-type architectures, the hydraulic motor discharges to the low-pressure branch. For the series-type PTO architecture, it discharges to the RO feed inlet. The low-pressure branch, which includes the charge pump and low-pressure accumulator, were included in the model but are not described here because they have little effect on the behavior of the high-pressure branch. The pressure of the low-pressure branch in every simulation was approximately constant at 0.44 MPa.

The state variables for these PTO models are the pressures at nodes having compressible volumes and the integral of the error for a proportional-integral pressure regulating controller determining the shaft speed of the electric generator.

The derivatives of pressure states are a function of the capacitance of the node, the net flow into the node, and changes to the fluid volume:

$$C_i(p_i) \frac{dp_i}{dt} = \sum q_{in} - \sum q_{out} - \frac{dV_i}{dt} \quad (1)$$

where p_i is the pressure of node i, $C_i(\cdot)$ is the nonlinear function describing the capacitance of the node as a function of pressure, q_{in} and q_{out} are flowrates into and out from the node, respectively, and V_i is the volume of the node.

The pumping chambers of the WEC-driven pump are modeled as volumes of compressible fluid. The working fluid is assumed to be a mixture of seawater and entrained gas with the properties given in Table 1. The seawater is assumed to be linearly compressible while the entrained gas is modeled as an ideal gas compressed isothermally. Therefore, the capacitance of these volumes is described by

$$C_{f,i}(p_i) = \frac{V_i \left(1 + \beta \alpha \frac{p_0}{p_i^2}\right)}{\beta} \quad (2)$$

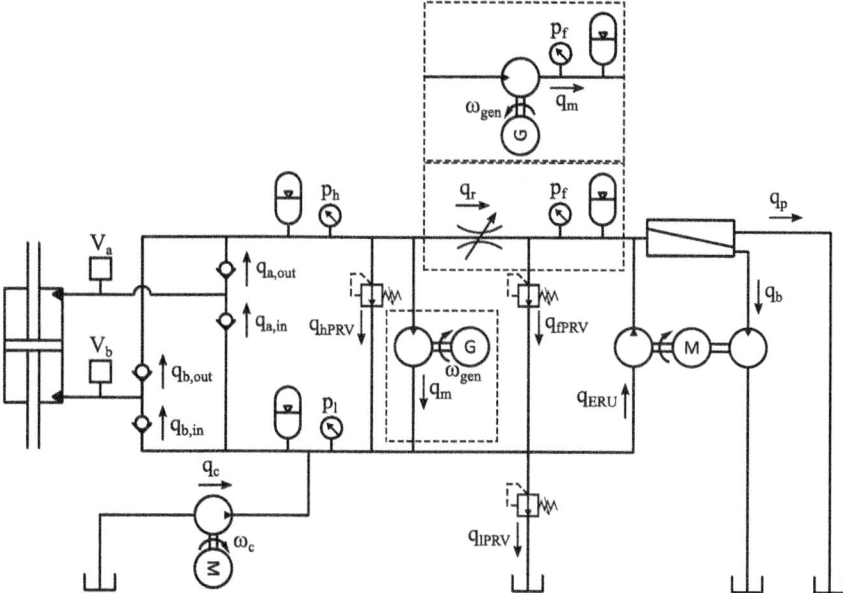

Fig. 3 Modeling schematic with optional configurations (i.e., with resistive element and additional accumulator and parallel-type or series-type architecture) are indicated with dashed line boxes

Table 1 Working fluid parameters

Parameter	Value	Units
Density, ρ	1025	kg/m^3
Viscosity, dynamic, μ	$9.4 \cdot 10^{-4}$	Pa·s
Bulk modulus, β	$2.2 \cdot 10^9$	Pa
Entrained air fraction (at 1 atm), α	0.0001	–
Atmospheric pressure	101300	Pa

where β is the bulk modulus of the liquid phase (i.e., seawater) and α is the volume fraction of entrained air at the reference pressure, p_o. The volume of each pumping chamber is a function of the WEC position and is assumed to change proportionally with the position of the WEC. The maximum extent of the WEC-driven pump is assumed to be at $\pm \pi/2$ radians from the vertical position of the flap-type WEC. An additional dead volume, equal to 10% of the total swept volume, is added to each pumping chamber to account for non-swept volume such as porting and piping.

The accumulators are modeled as being ideal gas volumes being compressed isothermally:

$$C_{g,i}(p_i) = V_{c,i} \frac{p_{c,i}}{p_i^2} \qquad (3)$$

where p_c and $V_{c,i}$ are reference values for pressure and volume of the gas at node i.

The flow through the resistive element of the pressure filter and the check valves of the WEC-driven is modeled as flow through an orifice such that

$$q_i(\Delta p_i) = k_{v,i} \frac{\Delta p_i}{|\Delta p_i|} \sqrt{|\Delta p_i|} \qquad (4)$$

where $k_{v,i}$ is a flow coefficient for valve i, Δp_i is the difference between the upstream and downstream pressure. There is no flow through the check valves when the pressure differential is below the cracking pressure and there is a pressure margin over which the area of the valve increases with the pressure differential until it is fully open. For the marginal regime, the flow coefficient for the check valves is assumed to vary linearly with the pressure difference between zero and the maximum value. Parameters for the check valves are given in Table 2.

The pressure relief valves are modeled with a flow coefficient varying linearly with poppet position (assuming a force balance between static pressure acting on the poppet faces and a linear spring, as was assumed in [23]) giving the flow rate

$$q_i = \frac{1}{C_{prv,i}} \left(p_i^{\frac{3}{2}} - p_{cr,i} p_i^{\frac{1}{2}} \right) \qquad (5)$$

where, for valve i, $p_{cr,i}$ is the cracking pressure and $C_{prv,i}$ is determined by a desired peak flowrate and peak pressure. The parameters used for the pressure relief valves are given in Table 3.

Table 2 WEC-driven pump check valve parameters

Parameter	Value	Units
Cracking pressure	$1 \cdot 10^5$	Pa
Margin to fully open	$1 \cdot 10^5$	Pa
Flow coefficient, inlet	15.18	L/Pa$^{1/2}$
Flow coefficient, outlet	10.11	L/Pa$^{1/2}$

Table 3 Pressure relief valve parameters

Parameter	Value	Units
PRV at WEC-driven pump outlet		
Cracking pressure, $p_{cr,hPRV}$	$20 \cdot 10^6$	Pa
Coefficient, $C_{prv,hPRV}$	$2.2361 \cdot 10^9$	Pa$^{3/2} \cdot$s/m^3
PRV at RO feed inlet		
Cracking pressure, $p_{cr,fPRV}$	$8.3 \cdot 10^6$	Pa
Coefficient, $C_{prv,fPRV}$	$1.4405 \cdot 10^9$	Pa$^{3/2} \cdot$s/m^3

Mechanical and flow losses of the hydraulic motor are modeled using the McCandlish-Dory model [19] with constant coefficients and fixed displacement. The flow rate through the pump/motor is modeled as

$$q_m = D_m \omega_m \left(1 - \lambda \left(C_s \frac{|\Delta p|}{\mu |\omega_m|} + \frac{\Delta p_i}{\beta}(V_r + 1)\right)\right) \qquad (6)$$

where D_m is the volumetric displacement per radian, ω_m is the shaft speed in radians per second, μ is the dynamic viscosity of the working fluid. The torque of the motor and generator are modeled as

$$T_m = D_m \Delta p \left(1 + \lambda \left(C_v \frac{\mu |\omega_m|}{|\Delta p|} + C_f\right)\right) \qquad (7)$$

The coefficients of Eqs. 6 and 7 are given in Table 4.

The WEC-driven pump and electric generator are assumed to have a fixed efficiency with values in Table 4. The WEC-driven pump displacement is derived from results in [23]. The displacement of the hydraulic motor was chosen for the parallel-type architecture based on a grid study considering the performance of the pressure regulation and the power losses of the motor, generator, and pressure relief valves; this study is presented in [22]. For the series-type architecture, it was chosen to provide the flowrate required to achieve steady operation of the RO module at 8 MPa.

Table 4 WEC-driven pump, hydraulic motor and generator parameters

Parameter	Parallel-type	Series-type	Units
WEC-driven pump			
Displacement	0.23	0.163	m³/rad
Efficiency, mechanical		0.9	–
Hydraulic motor and generator			
Generator efficiency		0.9	–
Maximum speed		1750	rpm
Motor displacement,	1000	2300	cc/rev
Laminar flow loss coefficient, C_s		3.0554·10⁻¹⁰	–
Volume ratio, V_r		1.103	–
Viscous torque loss coefficient, C_v		7.1755·10⁵	–
Coulomb torque loss coefficient, C_f		0.0259	–

Table 5 Reverse osmosis module and energy recovery unit (ERU) parameters

Parameter	Value	Units
Membrane area, S_{ro}	3,700	m^3
Permeability coefficient, A_{perm}	$2.57 \cdot 10^{-12}$	m^3/(N·s)
Recovery ratio	0.25	–
ERU Volumetric efficiency	0.95	–
ERU Mechanical efficiency	0.95	–

The rate of permeate production by the RO module is modeled as

$$q_p = S_{ro} A_{perm} \left(p_f - p_{osm} - p_p \right) \tag{8}$$

where p_{osm} is the osmotic pressure of the seawater solution, p_p is the pressure at the permeate outlet, S_{ro} is the total installed membrane surface area, and A_{perm} is a coefficient describing the permeability of the membranes. The seawater is assumed to have an osmotic pressure of 2.275 MPa. The permeate outlet pressure is assumed to be atmospheric pressure. The permeability coefficient is taken from [29], which is based on results of the WAVE design tool offered by FilmTec for a particular configuration of RO membrane elements [4].

The ERU is assumed to maintain a constant recovery ratio in the RO process of 25% (i.e., the ratio of permeate production to feedwater intake), which is achieved by controlling the ERU shaft speed. Mechanical and volumetric losses from the ERU's hydraulic motor and pump are modeled with constant efficiency. The electric motor makes up for the difference in torque between the hydraulic motor and pump. The hydraulic motor and pump are assumed to have the same volumetric displacement. Parameters used for the RO module and ERU are given in Table 5.

Control There are two control schemes implemented for the PTO subsystem. First, the hydraulic motor and generator are used to regulate the RO feed pressure using a feed-back control loop. Second, the flow coefficient for the active resistive element is used to limit the rate of change in feed pressure. In this case, the flow coefficient is prescribed by feed-forward control.

Proportional-integral control is used for pressure regulation. The error is the difference between the pressure at the RO feed inlet and the nominal set point. The control signal is a nominal shaft speed for the generator. For the parallel-type PTO architectures, only the proportional term is used. The proportional and integral terms are both used for the series-type PTO architecture. The value for the proportional gain in both cases is $5 \cdot 10^{-4}$ rad/s/Pa. The integral gain is $5 \cdot 10^{-6}$ rad/s^2/Pa.

The feed-forward control specifying the flow coefficient of the resistive element is based on a model for the pressure node capacitance and the flowrates associated with the RO module. In practice, the model would have to estimate these values and may include an observer that is informed by sensor measurements. For this study, these values are calculated by reversing the calculations given above. The feed-forward control identifies an ideal flow coefficient based on a positive rate of change in feed pressure equal to the prescribed limit. The flow coefficient command is bound by an upper bound attributed to the size of the valve. The ideal flow coefficient is

$$k_{v,ideal} = \frac{p_h - p_f}{|p_h - p_f|} \frac{C_f(p_f) + q_f - q_{ERU,f}}{\sqrt{p_h - p_f}} \tag{9}$$

3.2 Numerical Solution

These models are solved using the Euler method with a time step of $5 \cdot 10^{-6}$ seconds. This time step was selected based on convergence of a mass and energy balance. Similarly, the length of time simulated is based on a convergence study on the power captured by the WEC and metrics of pressure variation (i.e., mean, minimum, maximum, and variance). The simulations are 2250 s long with only the last 2000 seconds contributing to the performance calculations.

3.3 Design Studies

The PTO architectures are compared based on power losses and the total accumulator volume. The Pareto front is approximated using a multi-variable grid study. The pressure rate-of-change constraint is enforced through eligibility to the Pareto optimal set.

The variables of each grid study are specified in Table 6. Total volume refers to the total volume for the high-pressure accumulators. The distribution of volume between the accumulator upstream and downstream is indicated by the proportion of the total volume placed at the RO feed inlet. Along with the range of values, the number of grid points and the distribution scheme for the grid points are specified. Two grid

Table 6 Grid study parameters: bounds, number of grid points and spacing scheme

PTO architecture	Total volume	Portion at RO feed	Flow coefficient
	(L)		(L/kPa$^{1/2}$)
Baseline	500–30,000 (80/log)	–	–
Passive element, with constraint	5,000–15,000(10/log)	0.2–0.35(10/equal)	0.1–15(40/log)
Passive element, without constraint	1,000–15,000(80/log)	0.05–0.9(10/equal)	0.1–30(20/log)
Active element	5,000–15,000(10/log)	0.01–0.99(40/equal)	40(1/–)
Series-type	1,000–15,000(20/log)	0.01–0.95(40/log)	–

point distribution schemes are used; a constant spacing between grid points and a log-scale spacing.

4 Results

In this section, results are presented from the grid study for the baseline case, a comparison of the Pareto optimal performance results for each PTO architecture, and time-series results for the proposed PTO architecture with representative selections of design parameters.

Grid study results for the influence of accumulator volume on the peak rate of change in feed pressure and mean power loss for the baseline PTO architecture are given in Fig. 4. The mean power losses include the pressure relief valves and the hydraulic motor and generator and are normalized to the mean power capture by the WEC. The lowest total high-pressure accumulator volume meeting the constraint on the pressure rate of change is 17,600 liters. Above 3,000 liters of accumulator volume, power losses are near constant at about 2.8%. The pressure relief valves are not used in this range, only the hydraulic motor and generator contribute to these losses. Below 3,000 liters, the pressure relief valve at the RO feed inlet is activated and increases the losses up to 5.5% with 500 liters of accumulator volume.

The Pareto optimal performance for all the PTO architectures is given in Fig. 5. This figure shows the power losses as a function of the total accumulator volume. Results for the baseline architecture are given without regard for the pressure rate-of-change constraint (indicated by "no constraint" in the legend). However, the minimum volume needed to meet the constraint is indicated with a vertical dash-dotted line.

These data show a clear ranking in the performance of the architectures. First, the parallel-type architecture with the additional resistive-capacitive network performs better with an active element than a passive element. Second, for accumulator volumes less than 8,000 liters, the series-type architecture outperforms the parallel-type architectures. However, above about 8,000 liters, the power losses are greater for the series type architecture; this is due to the losses of the hydraulic motor being greater in the series configuration, at 3.1%, than in the parallel configuration at 2.8%. Third, when the pressure rate-of-change constraint is not enforced, the baseline outperforms all cases where the constraint is enforced.

To compare required accumulator volume between the architectures, a target limit value of 5% loss is considered, which is just less than double the power loss of the baseline architecture with 17,600 liters. At 5% combined power loss, the parallel-type architecture requires 9,090 liters with a passive resistive element and 7,920 liters with an active element. The series-type architecture requires 4,370 liters. Respectively, these values account for reductions in the required volume from the baseline of 48, 55, and 75%. When the pressure rate-of-change constraint is not observed, the baseline architecture requires 670 liters to achieve 5% combined power loss. This is 96% less volume than is required to meet the rate-of-change constraint.

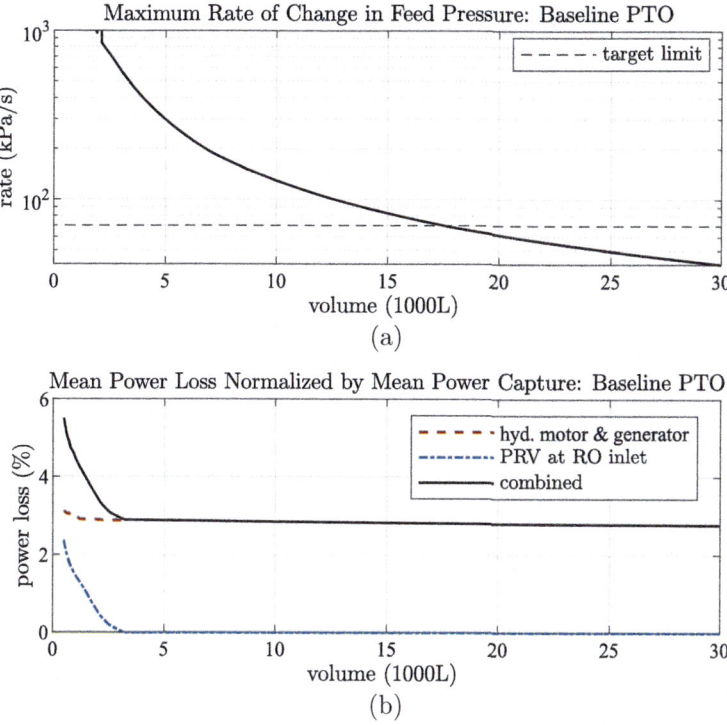

Fig. 4 Grid study results for the baseline PTO architecture: rate of change in feed pressure (**a**) and power loss (**b**)

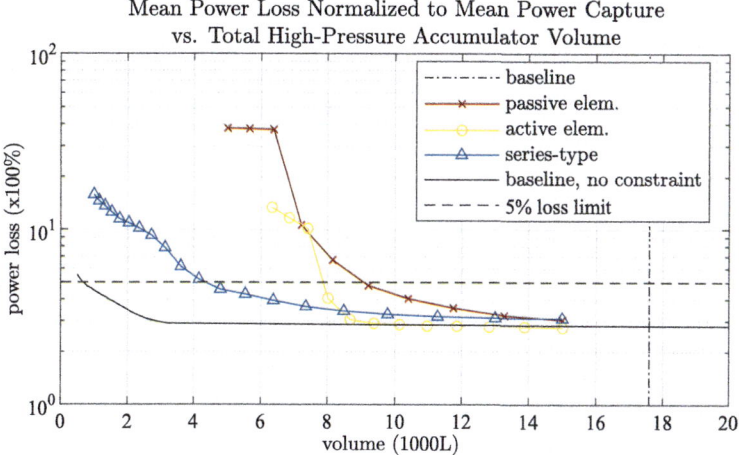

Fig. 5 Comparison of the design performance of each PTO architecture

Permeate production has been left out of this analysis since, by design, the permeate rate should be consistent for all cases. This is confirmed by the results for each design meeting the 5-% combined loss. The baseline produced $1870\,m^3$/day when meeting the constraint and $1890\,m^3$/day without. The parallel-type produced $1860\,m^3$/day with the passive pressure filter and $1900\,m^3$/day with the active pressure filter. The series-type produced $1850\,m^3$/day. All values are within 3% difference.

Time-series results related to pressure variation illustrate the behavior of these systems. These are presented for the designs meeting the 5-% combined loss. First, results for the parallel-type architecture with a passive resistive element and 9,090 liters of total volume are given in Fig. 6. The other design parameter values for this case are a flow coefficient of $4.06\,L/s/kPa^{1/2}$ and 28.3% of the accumulator volume at the RO feed inlet. These results show (1) pressure at the two accumulator banks, (2) the behavior of the hydraulic motor responding to the difference between the nominal RO feed pressure and the actual pressure, and (3) the rate of change in feed pressure compared to the target limit of 70 kPa/s. Two observations are notable. First, there is only a small difference in pressure between the two accumulators (i.e., about 0.14 MPa on average). Second, the rate of change in pressure approaches the peak of 70 kPa/s for only a small fraction of time within the 2000 second simulation. For comparison, the 97th and 99th-percentile values of the rate-of-change magnitude are indicated and are 38 and 48 kPa/s, respectively.

Results for the parallel-type architecture with an active resistive element and 7,920 liters of total volume are given in Fig. 7. This design has 94.7% of the accumulator volume at the RO feed inlet. These results include (1) the pressure at the two accumulators, (2) the rate of change in feed pressure, and (3) the flow coefficient of the resistive element compared to the value determined from the feed-forward control law. A notable observation is that several instances are shown where the pressure upstream of the resistive element peaks to extreme values, yet these values are still well below the pressure relief valve setting of 20MPa. Another is that the rate of change in feed pressure and the valve coefficient command signal have higher frequency content that is not seen with the passive element. Finally, the feed-forward command signal for the flow coefficient has a mean value ($543\,L/s/kPa^{1/2}$) that is an order of magnitude greater than the maximum value (saturation limit) of $40\,L/s/kPa^{1/2}$ that was chosen.

Finally, time-series results for the series-type architecture having 4370 liters of total accumulator volume are given in Fig. 8. This design has 1.12% of the accumulator volume placed at the RO feed inlet. These results include the pressure at the RO feed inlet and the pressure upstream of the hydraulic motor. The rate of change is not presented because it is near zero, being under direct control by the hydraulic motor. The RO feed pressure is controlled well and is nearly constant while the upstream pressure varies significantly. It is notable that the upstream pressure falls below the RO feed pressure for several periods within the simulation. In these cases, the hydraulic motor is in a pumping mode.

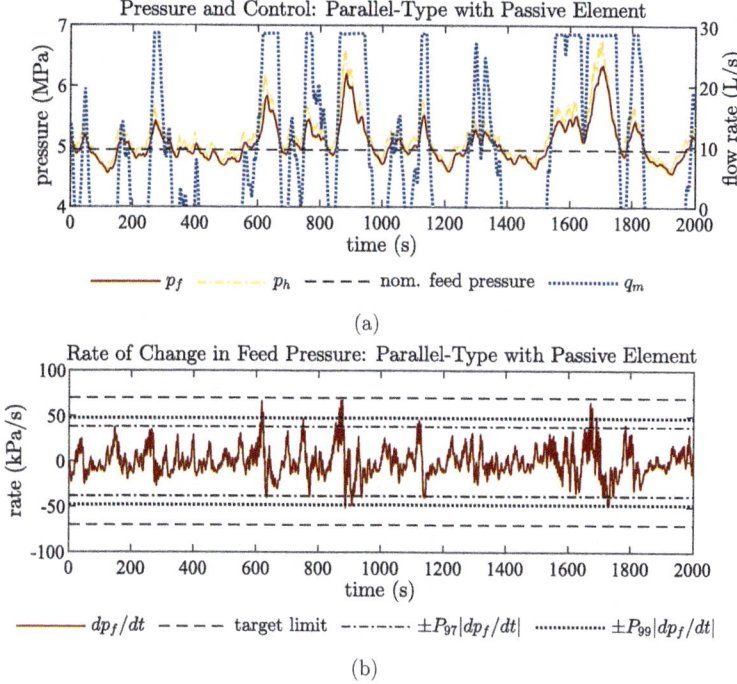

Fig. 6 Time-series results related to pressure variation for the parallel-type architecture with a passive resistive element: pressure and hydraulic motor flow rates (**a**) and rate of change in RO feed pressure (**b**)

5 Discussion

The results of this study suggest the series-type architecture is a superior design to the parallel-type architecture with respect to controlling the pressure variation at the RO module. This is in addition to this architecture using a 29% smaller WEC-driven pump, based on the results of the study presented in [23]. However, if the parallel-type architecture is chosen, adding a resistive-capacitive network to form a second-order low-pass filter can reduce the total accumulator volume by about half.

The results of this study also show that there is a significant cost associated with enforcing the 70 kPa/s constraint on the RO feed pressure rate of change. Treating the constraint as a hard limit in this study showed that an order of magnitude more accumulator volume is required than would be deployed otherwise. Yet the need for this constraint appears to be not well understood. Future work is needed to clarify what should be the constraints placed on the variation in pressure for wave-powered RO systems. Clarity on this issue will contribute significantly to our ability to design a robust and cost-effective system.

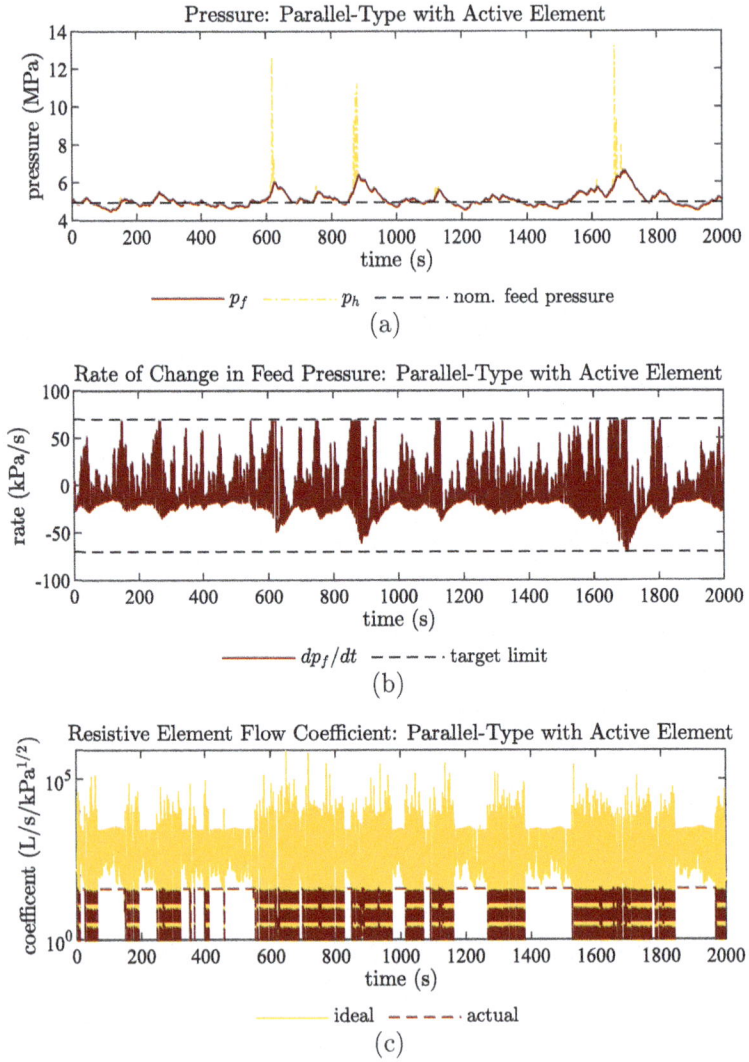

Fig. 7 Time-series results related to pressure variation for the parallel-type architecture with an active resistive element: pressure (**a**), rate of change in RO feed pressure (**b**), and flow coefficient of the resistive element (c)

A specific recommendation would be to clarify whether the rate of change in pressure contributes to mechanical failure or just degraded performance, whether that mechanism is accumulative, and to what degree. It was noted that the portion of time where peak rate of change is observed for the parallel-architecture is relatively small and that the 97th and 99th-percentile values for the rate of change are significantly lower than the peak. This highlights the fact that taking the limit of 70kPa/s as a

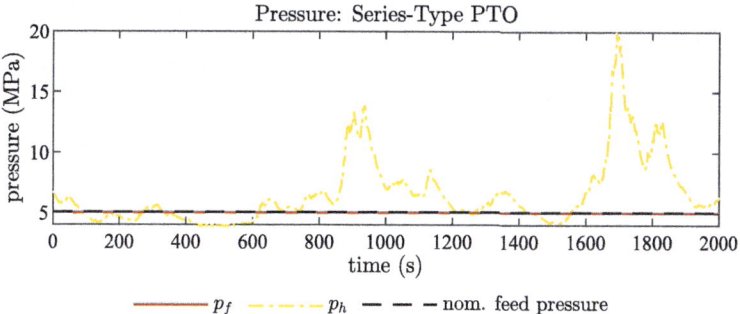

Fig. 8 Time-series results for pressure for the series-type PTO architecture

hard limit may be overly conservative. If a mechanism of failure or performance degradation is cumulative, a probabilistic distribution for the rate of change may be a more appropriate feature to consider in the design of these systems. If so, the system may tolerate a greater degree of pressure variation and require significantly less accumulator volume than observed in this study.

A trend in how accumulator volume is distributed in the system is illustrated by the study. Although the data are not given for all Pareto optimal results, the designs selected for comparison are representative of the trends between each architecture. Due to the direct control of flow at the RO feed inlet with the series-type architecture, very little accumulator volume, (i.e., 1.12%) is placed at the RO feed inlet. In contrast, 28.3% of the volume is placed at the RO feed inlet in the parallel-type architecture with the passive resistive element and 94.7% with the active resistive element. The active resistive element case is at the other extreme from the series-type PTO. This may be driven by the cases where the rate of change is negative, as in Fig. 7b at about 1700 s. In these cases, the resistive element is unable to force an increase in flow. This is where the active resistive element is limited; it is capable of retarding flow but not driving flow.

There may be room to further reduce the accumulator volume required by these systems. This study has assumed an accumulator charge pressure of 4 MPa regardless of the use case. The capacitance of an accumulator is greatly affected by charge pressure with it being more advantageous to have the charge pressure close to the operating pressure. Attention to this margin may yield a reduction in the required volume. This is especially notable for the series-type architecture where the upstream pressure is expected to be greater than the RO feed pressure. However, there likely needs to be a significant margin to accommodate the high level of flow and pressure variation in these systems. Another opportunity for reduced accumulator volumes is with higher working pressure limits since the pressure relief valves drive the increased power losses at low accumulator volumes.

A limitation of this study is that the design of this system was only examined for a single sea condition. The sea condition that was selected was argued to be a reasonable corner condition, but there is a motivation to extend that corner condition to higher energy sea states. This would require greater accumulator volumes than found in this study because of the higher peak power input to the system. Generally, the accumulator volume requirement is dominated by relatively infrequent sea conditions. A trade-off between production and component size requirements needs to be negotiated. This adds further motivation for clarifying and justifying the constraint. If the constraint were to be relaxed, this trade-off could be shifted toward higher production and lower accumulator volume requirements.

6 Conclusions

This study used numerical models to simulate the dynamics of a wave-powered RO system with a variety of power take-off architectures. A constraint found within the RO industry for the operation of an RO system is that the rate of change in feed pressure should not exceed 70 kPa/s. However, it is not clear what the mechanism of failure is that motivates the constraint. This study highlights the significance of this constraint in the design of a wave-powered RO system and the influence the power take-off architecture has on meeting this constraint. The results of this study showed that the constraint on the rate of change in feed pressure requires an order of magnitude greater high-pressure accumulator volume for the baseline system than if the constraint was not enforced. Furthermore, improvements over the baseline architecture were demonstrated for three proposed architectures. The study found that the addition of a resistive-capacitive network in the hydraulic circuit reduced the required high-pressure accumulator volume by 48% when the resistive element was passive and 55% when the resistive element was actively controlled through feed-forward control. The study also found that a series-type architecture, where a hydraulic motor is placed in series with the RO module, provides a reduction of about 75% in required accumulator volume while also requiring a 29% smaller WEC-driven pump.

7 Data Availability

Custom software and data used in this work are available at: https://github.com/novaTehnika/2023-DynPTOModelDesignStudies/releases/tag/publications2024 (accessed 29 April 2024).

Acknowledgements This material is based upon work supported by the U.S. National Science Foundation under award No. CMMI-2206018. Any opinions, findings and conclusions or recommendations expressed in this material are those of the author(s) and do not necessarily reflect the views of the U.S. National Science Foundation.

References

1. American-Made Challenges (2020). https://americanmadechallenges.org/. Accessed 06 Jul 2020
2. Oyster 1 (2010). http://www.aquamarinepower.com/technologies/oyster-1/. Accessed 29 Nov 2010
3. Resolute Marine Energy (2019). http://www.resolutemarine.com/. Accessed 15 Apr 2019
4. WAVE Water Treatment Design Software (2023). https://www.dupont.com/water/resources/design-software.html. Accessed 21 Aug 2023
5. Wave Energy Prize Experimental Sea State Selection (2017) In: International conference on offshore mechanics and arctic engineering, vol 10. Ocean Renewable Energy. https://doi.org/10.1115/OMAE2017-62675
6. Babarit A, Delhommeau G (2015) Theoretical and numerical aspects of the open source BEM solver nemoh. In: 11th European wave and tidal energy conference (EWTEC2015)
7. Das TK, Folley M, Lamont-Kane P, Frost C (2024) Performance of a SWRO membrane under variable flow conditions arising from wave powered desalination. Desalination 571:117,069
8. Dupont: Filmtec™ Reverse Osmosis Membranes Technical Manual. Dupont Water Solutions, Edina, MN, USA (2023). Form No 45-D01504-en, Rev 16
9. Falnes J, Kurniawan A (2020) Ocean waves and oscillating systems: linear interactions including wave-energy extraction, vol 8. Cambridge University Press
10. Folley M, Suarez BP, Whittaker T (2008) An autonomous wave-powered desalination system. Desalination 220(1–3):412–421. https://doi.org/10.1016/j.desal.2007.01.044
11. Folley M, Whittaker T (2009) The cost of water from an autonomous wave-powered desalination plant. Renew Energy 34(1):75–81. https://doi.org/10.1016/j.renene.2008.03.009
12. García Rodríguez L (2003) Renewable energy applications in desalination: state of the art. Solar Energy 75(5):381–393. https://doi.org/10.1016/j.solener.2003.08.005
13. Ghaffour N, Missimer TM, Amy GL (2013) Technical review and evaluation of the economics of water desalination: current and future challenges for better water supply sustainability. Desalination 309:197–207. https://doi.org/10.1016/j.desal.2012.10.015
14. Greenlee LF, Lawler DF, Freeman BD, Marrot B, Moulin P (2009) Reverse osmosis desalination: water sources, technology, and today's challenges. Water Res 43(9):2317–2348. https://doi.org/10.1016/j.watres.2009.03.010
15. Hansen RH, Andersen TO, Pedersen HC (2011) Model based design of efficient power take-off systems for wave energy converters. In: Proceedings of the 12th scandinavian international conference on fluid power, Tampere, Finland, pp 18–20
16. Kalogirou SA (2005) Seawater desalination using renewable energy sources. Prog Energy Combust Sci 31(3):242–281. https://doi.org/10.1016/j.pecs.2005.03.001
17. Kucera J (2015) Reverse osmosis: industrial processes and applications. Scrivener Publishing, Salem, MA
18. Leijon J, Boström C (2018) Freshwater production from the motion of ocean waves—a review. Desalination 435:161–171. https://doi.org/10.1016/j.desal.2017.10.049
19. McCandlish D, Dorey RE (1984) The mathematical modelling of hydrostatic pumps and motors. Proceedings of the Institution of mechanical engineers, Part B: management and engineering manufacture, vol 198(3), pp 165–174. https://doi.org/10.1243/PIME_PROC_1984_198_062_02
20. Nolan G, Ringwood J (2006) Control of a heaving buoy wave energy converter for potable water production. In: 2006 IET Irish signals and systems conference, pp 421–426
21. Perez T, Fossen TI (2009) A matlab toolbox for parametric identification of radiation-force models of ships and offshore structures. Model Identif Control 30(1), 1–15. https://doi.org/10.4173/mic.2009.1.1
22. Simmons JW (2024) Modeling and design of hydraulic power take-offs for ocean wave powered reverse osmosis desalination. PhD thesis, University of Minnesota

23. Simmons JW, Van de Ven JD (2023) A comparison of power take-off architectures for wave-powered reverse osmosis desalination of seawater with co-production of electricity. Energies 16(21):7381
24. Simmons JW, Van de Ven JD (2023) Limits on the range and rate of change in power take-off load in ocean wave energy conversion: a study using model predictive control. Energies 16(16):5909
25. Sitterley KA, Cath TJ, Jenne DS, Yu YH, Cath TY (2022) Performance of reverse osmosis membrane with large feed pressure fluctuations from a wave-driven desalination system. Desalination 527:115–546. https://doi.org/10.1016/j.desal.2022.115546
26. Tedeschi E, Carraro M, Molinas M, Mattavelli P (2011) Effect of control strategies and power take-off efficiency on the power capture from sea waves. IEEE Trans Energy Convers 26(4):1088–1098. https://doi.org/10.1109/TEC.2011.2164798
27. Wilf M, Awerbuch L (2007) The guidebook to membrane desalination technology: reverse osmosis, nanofiltration and hybrid systems: process, design, applications and economics. Balaban Desalination Publications, L'Aquila, Italy
28. Yu YH, Jenne D (2017) Analysis of a wave-powered, reverse-osmosis system and its economic availability in the united states. In: International conference on offshore mechanics and arctic engineering, vol 57786. American Society of Mechanical Engineers, p V010T09A032
29. Yu YH, Jenne D (2018) Numerical modeling and dynamic analysis of a wave-powered reverse-osmosis system. J Mar Sci Eng 6(4):132. https://doi.org/10.3390/jmse6040132

Open Access This chapter is licensed under the terms of the Creative Commons Attribution 4.0 International License (http://creativecommons.org/licenses/by/4.0/), which permits use, sharing, adaptation, distribution and reproduction in any medium or format, as long as you give appropriate credit to the original author(s) and the source, provide a link to the Creative Commons license and indicate if changes were made.

The images or other third party material in this chapter are included in the chapter's Creative Commons license, unless indicated otherwise in a credit line to the material. If material is not included in the chapter's Creative Commons license and your intended use is not permitted by statutory regulation or exceeds the permitted use, you will need to obtain permission directly from the copyright holder.

Experimental Bladder Accumulator-Based Passive Resonator

Ville Närvänen

1 Introduction

Hydraulic cylinders are commonly used in industrial rotor systems for adjustable support purposes. One example is a paper winder, which utilizes hydraulic cylinders for paper roll support and rider roll height adjustment. Due to natural instabilities in the rolling process, vibration is generated and can be anywhere in 10–75 Hz range depending on the specific machine type [1].

Gas-charged pressure accumulators are typical components utilized widely in several hydraulic systems, bladder, diaphragm, and piston types being the most common. Along with operating as hydraulic energy storages, accumulators are often used near hydraulic pumps to attenuate pressure pulsations because they can store and release hydraulic fluid in response to pressure oscillation, functioning as attenuating gas spring-dampers. Pressure accumulators are most effective in damping low-frequency (<100 Hz) pulsations [2].

If an accumulator is in direct connection with a hydraulic cylinder chamber via an open fluid line, a hydraulic equivalent similar to a mass–spring system with a damper is created. Such system will be referred as hydraulic resonator in this paper. Resonator has a direct effect on the dynamic stiffness of the hydraulic cylinder, depending on oscillation frequency, fluid line length, accumulator throat dimensions, accumulator volume, and accumulator type. By modifying the aforementioned attributes, the hydraulic cylinder's dynamic stiffness in response to specific low-frequency vibration can be increased. Although hydraulic systems utilizing accumulator-based cylinder vibration dampers exist, experimental research is lacking.

In this research, a simple passive resonator consisting of bladder accumulator and fluid line is constructed and its effect on hydraulic cylinders dynamic stiffness

V. Närvänen (✉)
Aalto University, Espoo, Finland
e-mail: ville.narvanen@aalto.fi

characteristics is simulated and experimentally tested. Resonator is tested in experimental setup, which consists of load cylinder and test cylinder attached in series. Load cylinder generates an average load of 50 kN and sinusoidal oscillation. Configuration is tested in 15–30 Hz frequency range with two oscillation amplitudes. Simulations are built in MATLAB–Simulink environment.

2 Materials and Methods

2.1 Simulation

Simple linear model of resonator system was constructed in MATLAB–Simulink environment to define resonator's mechanical parameters.

Due to the dynamics of the test system, fluid compressibility becomes relevant factor. All volumes are considered as hydraulic capacitances, which are defined as

$$C = \frac{V_f}{B} \quad (1)$$

where B is the bulk modulus of hydraulic fluid and V_f static fluid volume.

Piston movement due to vibration generates pressure change in fluid volume. Pressure change is estimated with the following equation:

$$\Delta p = \frac{1}{C} \Delta V_c \quad (2)$$

where ΔV_c is the relative volume change to static volume.

Due to choke from cylinder chamber to resonator pipe, fluid acceleration has significant inertia. This inertia in relation to pipe volume is considered hydraulic inductance [2]:

$$I_{hydr} = \frac{\rho L_p}{A_p} \quad (3)$$

where ρ is the fluid density and A_p/L_p are the resonator pipe area and length, respectively.

Kajaste et al. [4] presented transfer function describing pressure accumulator model with linearized friction:

$$\frac{Q_{acc}}{P_{acc}} = \frac{C_{acc} * s}{\frac{s^2}{\omega_{acc}^2} + \frac{2c_m}{\omega_{acc}} * s + 1} \quad (4)$$

where Q_{acc} is the flow output of the accumulator, P_{acc} is the pressure input to the accumulator, C_{acc} is the accumulator capacitance, ω_{acc} is the accumulator specific angular velocity, and c_m is the damping coefficient. Accumulator specific angular velocity is specified in Eq. 5:

$$\omega_{acc} = \frac{1}{\sqrt{\rho * C_{acc} * \frac{L_{th}}{A_{th}}}} \qquad (5)$$

where ρ is the hydraulic oil density, L_{th} is the length of accumulator throat, and A_{th} is the area of accumulator throat.

To keep the specific angular velocity of the resonator system as linear as possible in response to pressure pulsations, accumulator volume should be sufficiently large to minimize capacitance change. Based on Eqs. (4) & (5), a nominal volume of 4 L was selected, as accumulators of that size are readily available. A resonator pipe diameter of 15 mm was selected.

Simulink model of test system was built (Fig. 1).

Model accepts oscillation amplitude as input and estimates pressure force in cylinder piston. Using Simulink Model linearized plugin, frequency response of the system was mapped in relation to resonator pipe length and diameter and hydraulic cylinder piston position. Frequency where dynamic stiffness is peaked to its maximum value is hereinafter referred to as optimal frequency.

The following parameters were also considered in simulation but left out for brevity: Flow losses in resonator pipe, free air in bulk modulus and pressure losses over throttle points [5, 6].

2.2 Experimental Setup

The experimental setup consists of two 80/125 hydraulic cylinders attached in a series. Both cylinders are operated with Parker DFplus directional valves. The left cylinder, here on "load cylinder", generates an average load of 50 kN and sinusoidal oscillation. The right cylinder, here on "test cylinder", is connected to the load cylinder via a steel sled (Fig. 2).

Hydraulic passive resonator consists of a 2-meter long, 15 mm diameter pipe which connects the bottom of test cylinder and bladder accumulator of 4 L nominal volume (Fig. 3). Pipe dimensions and accumulator volume were selected based on simulations.

Components/sensors and their locations are specified in Fig. 4 and Table 1.

Ball valve (9) separates resonator from test cylinder. Resonator can be "activated" by opening the valve and allowing flow to resonator pipe.

One test cycle consists of single test cylinder piston thrust outwards at constant speed of 2 mm/s, while load cylinder generates average load of 50 kN and sinusoidal oscillation. Piston is moved from 100 mm position to 600 mm, length indicating piston side chamber length. A 0.8 mm throttle is installed to cylinder inlet, restricting inlet flow back into the system during oscillation.

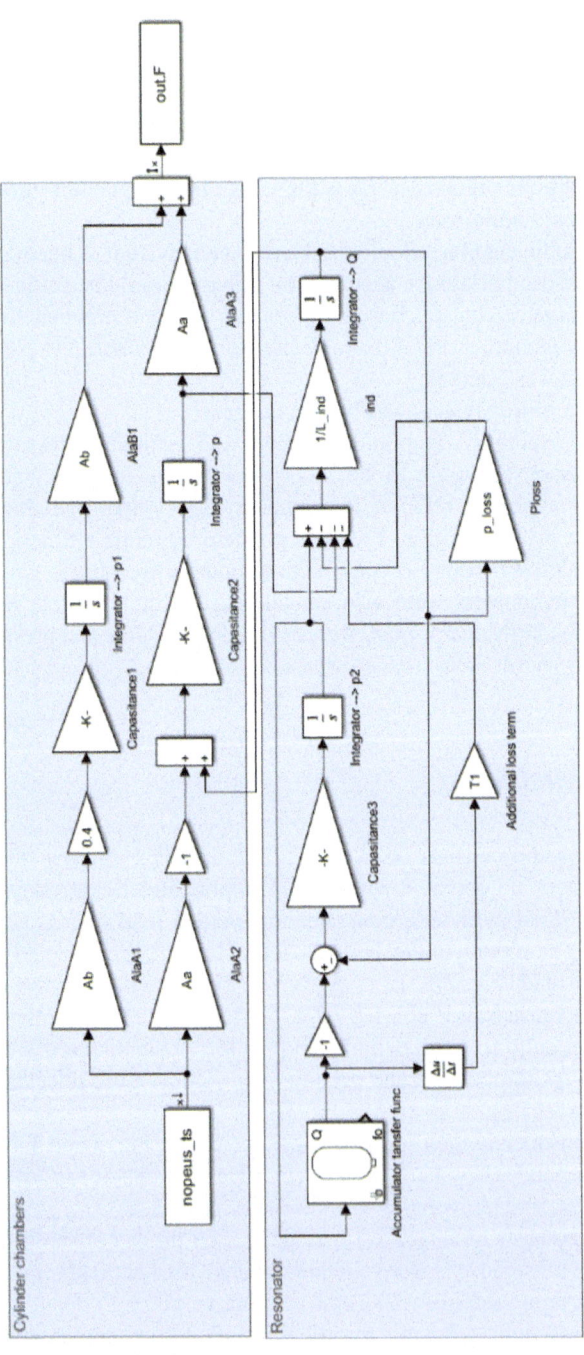

Fig. 1 Simulink model of experimental system

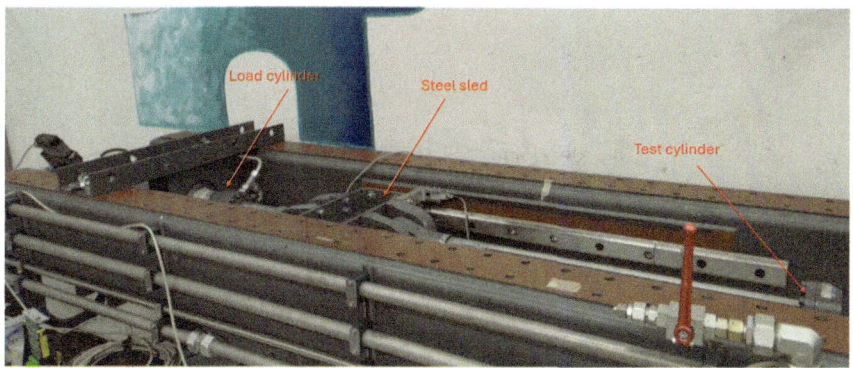

Fig. 2 Picture of the experimental system

Fig. 3 Passive resonator

Oscillation is generated by applying sinusoidal voltage output via OWON AG 1012F oscilloscope directly to control signal of the proportional valve (8.1). Two voltage amplitudes are used, 10Vpp and 15Vpp.

Piston oscillation amplitude is calculated from accelerometer (4.1–2) data by integrating Fourier transforms. Two accelerometers are used to negate any possible cylinder movement from data due to structure yield. Piston force oscillation is calculated from chamber pressures (5.1–2).

Larita oil 32 was used. The average oil temperature was kept at 48 ± 1 °C.

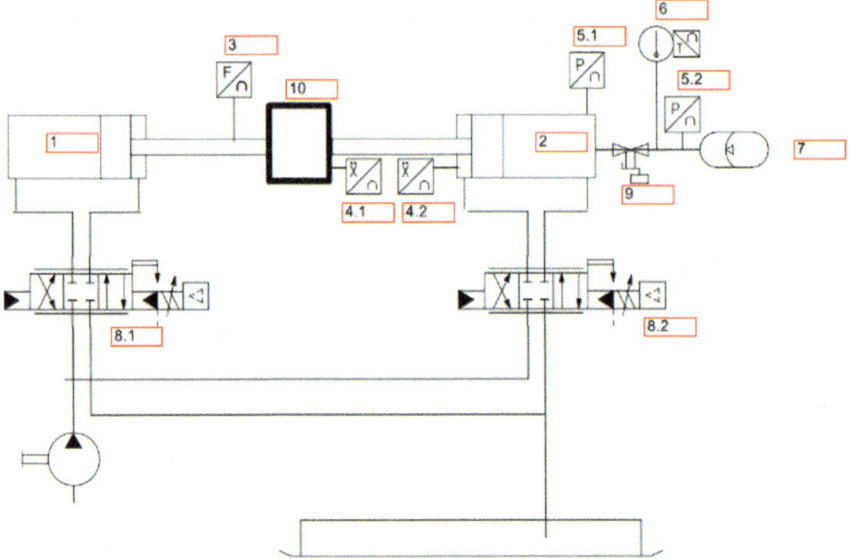

Fig. 4 Experimental setup

Table 1 Component list

Number	Component
1–2	Hydraulic cylinder 80/125/800
3	Load cell
4.1–2	Accelerometer
5.1–2	Pressure sensor
6	Temperature sensor
7	Pressure accumulator
8.1–2	Proportional Directional valve
9	Ball valve
10	Steel sled

3 Results

3.1 Simulation Results

Referring to Eqs. (2), (3), and (4), resonator has three variables that must be considered when system dimensions are designed, resonator pipe length, resonator pipe area, and accumulator volume. Resonator component dimensions are selected based on the desired attenuation frequency.

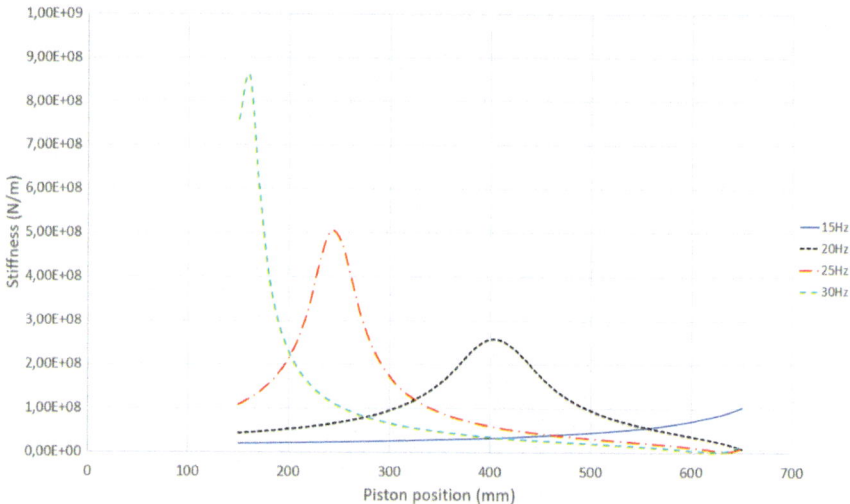

Fig. 5 Single simulated run of dynamic stiffness in relation to piston position. Resonator pipe length was set to 2 m. Depending on oscillation frequency, stiffness forms in "peaks" depending on optimal cylinder chamber volume

Simulation output plots the expected dynamic stiffness behavior of the test cylinder in response to static oscillation and cylinder volume change. Figure 5 demonstrates output of single simulation run.

Several simulations were run for several resonator pipe lengths and peaks stiffness values were extracted. From collected data, surface plot seen in Fig. 6 was constructed. Plot describes the relationship between piston position, resonator pipe length and optimal frequency.

Optimal frequency is highly depended on piston position due to change in cylinder chamber volume. To achieve 15–30 Hz attenuation range within the piston movement range, a 2-meter-long resonator pipe is required. Expected optimal frequencies with the aforementioned pipe length relative to piston position are plotted in Fig. 7:

Optimal frequency in relation to piston position on non-linear, demonstrating deeper curve in the beginning of the motion (low chamber volume) and smoothing out towards the end of the piston movement.

3.2 Experimental Results

Test cylinder baseline stiffnesses were measured for comparison by closing ball valve between test cylinder and resonator.

Measurement results with activated resonator by opening ball valve between test cylinder and resonator are presented in Fig. 9.

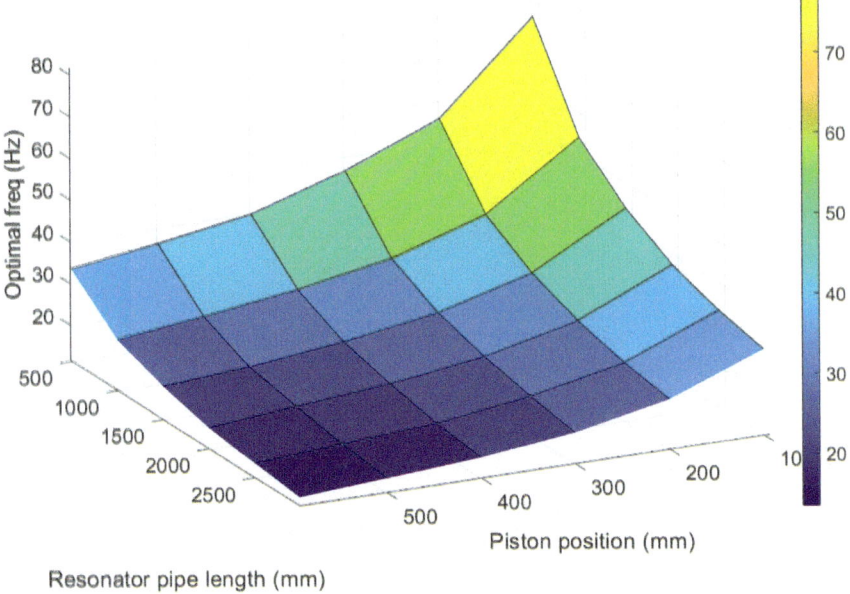

Fig. 6 Optimal frequency in relation to resonator pipe length and piston position. Resonator pipe diameter 15 mm

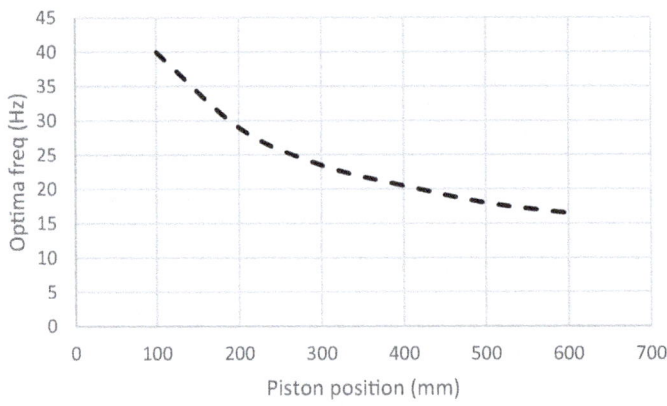

Fig. 7 Expected optimal frequencies of system with 2 m long and 15 mm diameter resonator pipe

As predicted in simulation, optimal frequency is depended on piston position, which can be seen as stiffness peaks for specific frequencies in Fig. 6.

In Tables 2 and 3, maximum relative stiffness increase from baseline are presented.

Although peak stiffness can be 5–6 times that of baseline, this only applies to narrow volume range of the cylinder chamber in specific frequency. Outside of

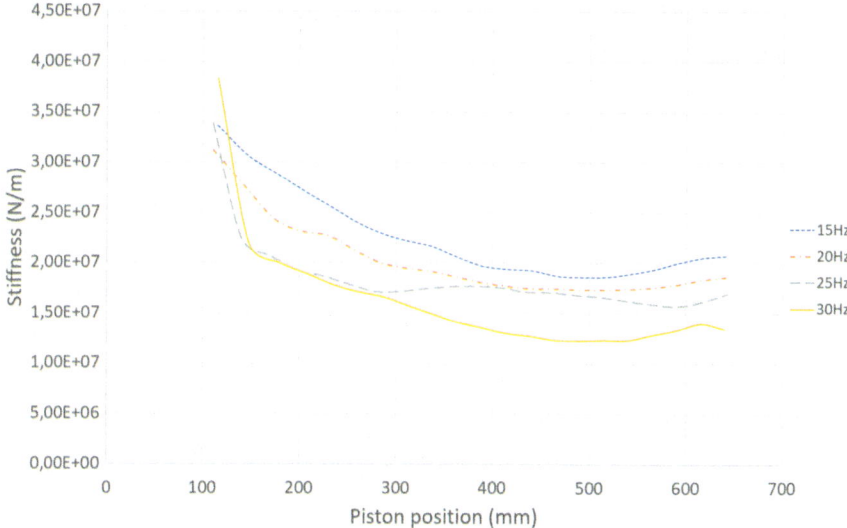

Fig. 8 Test cylinder baseline dynamic stiffnesses without resonator. Oscillation frequency affected the stiffness response of the cylinder. Three oscillation amplitudes were tested and had little to no effect on measured stiffness. Decreasing trend in dynamic stiffness is due to increased chamber volume naturally lowering test cylinder stiffness [5]

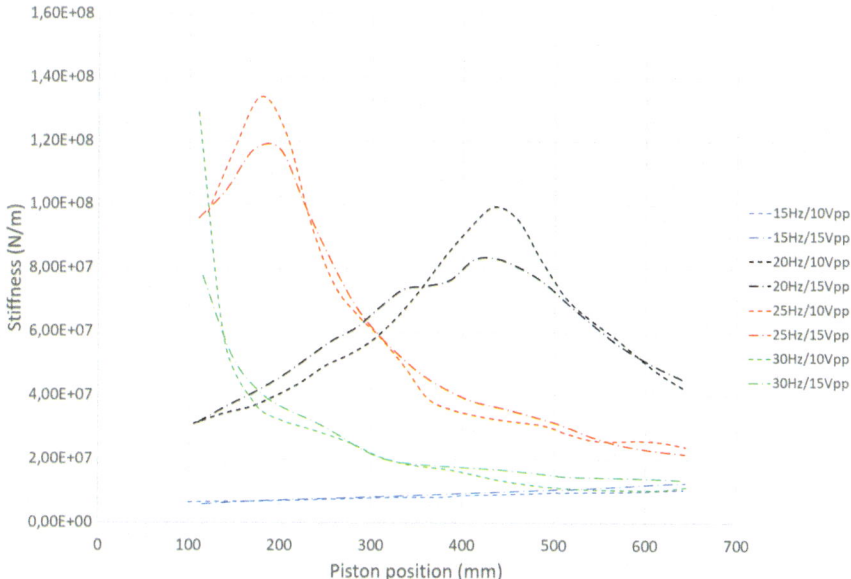

Fig. 9 Test cylinder dynamic stiffness with activated resonator. "Vpp" refers to oscilloscope amplitude setting

Table 2 Relative stiffness increase, Vpp 10

Oscilloscope setting: 10 Vpp				
Frequency (Hz)	Piston position (mm)	Resonator OFF (N/m)	Resonator ON (N/m)	Maximum difference to baseline (%)
15	200	$2.74*10^7$	$0.69*10^7$	−75
20	440	$1.75*10^7$	$9.98*10^7$	+470
25	190	$2.43*10^7$	$13.30*10^7$	+447
30	105	$3.84*10^7$	$12.91*10^7$	+236

Table 3 Relative stiffness increase, Vpp 15

Oscilloscope setting: 15 Vpp				
Frequency (Hz)	Piston position (mm)	Resonator OFF (N/m)	Resonator ON (N/m)	Maximum difference to baseline (%)
15	200	$2.74*10^7$	$0.70*10^7$	−74
20	440	$1.75*10^7$	$8.29*10^7$	+374
25	190	$2.43*10^7$	$11.74*10^7$	+383
30	105	$3.84*10^7$	$7.77*10^7$	+102

optimal frequency range, cylinder stiffness can be lower than baseline, as can be seen at 15 Hz oscillation frequency.

A 30 Hz oscillation frequency is barely within optimal range, elevating cylinder stiffness at the very beginning of piston motion. Towards the end of piston stroke, stiffness falls below that of baseline.

Interestingly, oscillation amplitude affects the stiffness, higher amplitude resulting in lower measured stiffness.

4 Discussion

Simulated results showed stiffness around three times of measured from experimental system (Fig. 5, 9). Also, grouping of stiffness peaks was tighter in relation to piston position than measured. For example, piston position change between 20 and 25 Hz peaks was 170 mm in simulation and 240 mm in measured results. Consideration of more energy loss parameter must be done to compensate for too high simulated stiffness results.

Experimental results demonstrated hydraulic resonator effect on hydraulic cylinders dynamic stiffness. Peak stiffness up to six times that of baseline was achieved at optimal frequencies. However outside optimal frequencies resonator can lower cylinder stiffness under that of baseline.

Oscillation amplitude had a visible effect on measured stiffness when resonator was activated. Similar effect was not apparent with baseline results when resonator was deactivated. More in-depth research on non-linear behaviors of the resonator system is conducted in the future.

Due to the cylinder volume being directly tied to optimal frequency, this type of passive resonator is most effective on static holding and frequency situation. It can also potentially be used to "skip" problematic frequencies during piston movement by controlled activation via valve.

Secondary, more specific possible use case is to consider rotor systems with an "evolving" frequency range during operation, where frequency follows downward curvature. One very specific example is paper winder, where paper roll harmonics evolve in similar way to cylinder optimal frequency as demonstrated in Fig. 5 [7].

For more generalized solutions, several accumulators could be utilized in tandem to widen the effective frequency range.

References

1. Virtanen T (2006) Fault diagnostics and vibration control of paper winders. Helsinki University of Technology
2. Ijas M (2007) Damping of low frequency pressure oscillation
3. Johnston DN (2002) Prediction of hydraulic inductance. In: ASME International mechanical engineering congress and exposition, Vol 36312, pp 85–92
4. Kajaste J (1999) Of the capability of component models to predict the response of a fluid power system with a long pipeline and an accumulator. Acta polytechnica scandinavica, Mechanical engineering series 139:50–60
5. Feng H, Du Q, Huang Y, Chi Y (2017) Modelling study on stiffness characteristics of hydraulic cylinder under multi-factors. J Mech Eng 63:447–456
6. Funk JE, Wood DJ, Chao SP (1972) The transient response of orifices and very short lines
7. Jorkama M, Von hertzen R (1997) Two-drum winder run simulation model

Open Access This chapter is licensed under the terms of the Creative Commons Attribution 4.0 International License (http://creativecommons.org/licenses/by/4.0/), which permits use, sharing, adaptation, distribution and reproduction in any medium or format, as long as you give appropriate credit to the original author(s) and the source, provide a link to the Creative Commons license and indicate if changes were made.

The images or other third party material in this chapter are included in the chapter's Creative Commons license, unless indicated otherwise in a credit line to the material. If material is not included in the chapter's Creative Commons license and your intended use is not permitted by statutory regulation or exceeds the permitted use, you will need to obtain permission directly from the copyright holder.

Investigation of the Heat Conduction in Axial Piston Pumps by Direct Measurement and Simulation

Roman Ivantysyn, Ahmed Shorbagy, Amey Vedpathak, and Jürgen Weber

1 Introduction

In the realm of hydraulic systems, axial piston pumps stand out for their robustness, efficiency, and versatility, finding applications in fields ranging from mobile machinery to aerospace. These systems' performance is closely linked to their thermal behavior, as temperature variations significantly affect both oil properties and component materials. This sensitivity underscores the pivotal role of heat convection coefficients as a boundary condition in the simulation models of these pumps. For example, if the heat convection coefficients are set too high, then the components will dissipate more heat to the environment resulting in lower part temperatures. Lower part temperatures result in different thermal deformations, which influence the resulting gap heights. In [5] it was shown that varying the heat convection coefficients boundary condition can change the part temperature by more than 30 K, causing vastly different gap heights and resulting in false temperature hot spots zones on the valve plate.

Traditionally, the determination of these coefficients has leaned heavily on computational fluid dynamics (CFD) tools, which typically were heavily simplified, for example by ignoring the tilted swash plate. Other approaches include analytical estimations, which are often derived from abstract geometries like spheres or pipes rather than from direct measurements within the pumps themselves.

The significance of this research lies in measuring these heat convection coefficients directly on the cylinder block surface for different operating conditions in an otherwise unmodified axial piston pump while simultaneously monitoring its

---

R. Ivantysyn (✉) · A. Shorbagy · A. Vedpathak · J. Weber
Institute of Mechatronic Engineering, Technische Universität Dresden, Dresden, Germany
e-mail: roman.ivantysyn@tu-dresden.de

J. Weber
e-mail: fluidtronik@mailbox.tu-dresden.de

© The Author(s) 2025
L. Ericson and P. Krus (eds.), *Advancements in Fluid Power Technology: Sustainability, Electrification, and Digitalization*, Lecture Notes in Mechanical Engineering,
https://doi.org/10.1007/978-3-031-84505-5_18

thermal behavior. By measuring these coefficients directly within the pump environment and comparing these measurements against sophisticated CFD simulations, this research endeavors to bridge the divide between theoretical models and their practical applicability.

2 Literature

2.1 Heat Convection Simulation and Measurement

The investigation of heat convection coefficients within piston pumps marks a significant advancement in hydraulic system simulations. Prior studies have extensively analyzed heat transfer phenomena in simplified geometries, such as concentric and eccentric cylinders with inner rotating components [1], providing foundational insights into the mechanics of convective heat transfer. Research in smoothed particle hydrodynamics (SPH) has further expanded the understanding of boundary conditions essential for accurate simulations [2], emphasizing the method's adaptability to complex boundaries [12].

Further contributing to this field, a study documented by SAE International presents an innovative experimental approach to determine the heat transfer coefficient (HTC) within piston cooling galleries. Utilizing an inverse heat transfer method based on surface temperature measurements with an infrared camera, the study offers a novel methodology for spatially resolved HTC determination inside piston cooling galleries. This approach not only advances the understanding of the piston thermal management but also opens new avenues for optimizing piston cooling strategies to enhance durability while minimizing parasitic losses and overall heat rejection [3].

In the area of pumps and hydraulic systems, the selection of fundamental research on the area of heat transfer within pumps has been limited. However, as high-fidelity gap simulations require accounting for both the pressure and temperature influence on the fluid and the solid bodies, they require heat convections as their boundary conditions. There have been multiple approaches in the past to estimate these. Most researchers have chosen to modify the geometry and use simple CFD models to estimate the flow conditions around these parts [6–23]. Others have based their approach on geometrical similarities, which can be found in literature through Nusselt and Prandtl relations [20, 11]. Both approaches have not been directly validated, rather they were used to parametrize the thermal component models, and if the measured temperatures match the predicted, they were assumed to be correct. In the past, the author has used a mixture of these approaches. CFD simulation results were scaled to new dimensions based on geometrical similarities. Nusselt and Prandtl numbers were used when shapes were simple such as the piston, pipes, or the cylinder block. The author conducted several temperature measurements inside axial piston pumps, such as the valve plate temperature, and validated the model using these measurements [5, 11–7]. However, it was discovered that the chosen approach did not always yield adequate results, where simulated part temperatures were either too high or too low, but the overall trend matched. After adjusting the convection coefficients, the temperatures matched the measurements [10]. While this approach works on a case-by-case basis, it lacked a fundamental character that can be applied to pumps in general. This issue is meant to be addressed with this paper including novel measurement of the actual heat convection coefficients.

3 Research Goal and Approach

The core objective of this research is to measure and simulate heat convection coefficients in piston pumps, aiming to bridge the gap between theoretical models and their practical applicability. This involves a detailed investigation into how various simulation tools can accurately reflect the complex thermal behavior observed in real-world piston pump operations using a simplified geometry where either analytical results or measurements are published. After the appropriate simulation tool was chosen, the full pump including the entire rotating kit in its tilted geometry was simulated using different boundary conditions. Adding complexity step-by-step in order to improve the results fidelity. A significant part of this goal is to determine the level of detail required in simulations to ensure they provide reliable predictions that can inform effective pump design and optimization strategies.

These results will then be compared to the direct measurement of the heat convection coefficients inside the working pump. These measurements will serve as the benchmark for the simulation, and a discussion highlighting the necessary steps to sufficiently reach accurate results will be presented. To show what level of accuracy is necessary, different heat convection boundary conditions were simulated in the pump modeling software Caspar FSTI, including the influence on thermal convergence and power loss.

Lastly, the measured heat convection coefficients will be analyzed and their dependence on shaft speed, system pressure, and stroke angle will be given. This research approach, combining empirical measurement with advanced simulation techniques, underscores the commitment to not only advance theoretical understanding but also ensure practical relevance and applicability in improving piston pump designs and operations.

4 Simulation of Heat Convection Coefficients in Piston Pumps

4.1 Overview

The simulation segment of the paper is dedicated to a comprehensive analysis of heat convection coefficients within piston pumps. It leverages two tools: 1. An advanced computational fluid dynamics (CFD) tool to model the thermal behavior and heat transfer mechanisms critical for the pump's efficiency and reliability and 2. A specialized numerical tool that captures the power losses and temperatures inside the lubricating gap of the pump, called Caspar FSTI. To choose the correct CFD tool a preliminary study was conducted with simplified geometries, three tools were considered: Two finite volume (FVM) CFD models, Simerics MP + , Ansys CFX, and the particle-based Dive Solutions model, which was included for its unique approach to handling complex fluid dynamics problems and complex geometries, allowing for easy set-up.

4.2 Simplified Test Cases and Tool Comparison

The simplified geometrical models were chosen to resemble the basic elements of the piston pump. The first case consists of two flat planes sliding alongside each other, for which an analytical solution exists. The second geometry consists of two cylinders with a small fluid gap in between each other, where one of them was rotating. This flow geometry is comparable to the case flow between the rotating cylinder block and the housing, which is a Taylor–Couette Flow. An Ansys solution was found in literature, which was remodeled and used as a comparative benchmark [13]. These models served to assess each tool's accuracy in predicting heat convection coefficients under controlled conditions. Although the results of the SPH simulation and both FVM simulations were in good agreement with each other, the results of the SPH simulation for the complete pump housing were off by an order of magnitude likely due to boundary layer limitations. Hence, SPH will not be discussed for the pump CFD simulation in this paper but can be found in [21]. While Ansys allows for very custom solutions, setup can be cumbersome, especially when requiring moving meshes as are required in an asymmetric pump housing. Simerics MP + delivered good results and proved to be a good compromise between easy setup and reliable results.

4.3 Pump Simulation

With the preliminary assessment complete, Simerics MP + was selected for detailed simulation due to its robust handling of the specific challenges presented by rotating and inclined piston pump geometries and its efficient simulation of heat transfer processes. This section describes the results for the pump geometry within Simerics MP + and how some crucial boundary conditions were determined. The pump model included all the important features, such as slipper hold-down mechanism, pistons, slippers, and the swash plate geometry, were retained whereas non-essential details such as the setting mechanism were removed from the negative volume. The focus of the 3D CFD simulations was to determine the heat transfer coefficients on the surface of the cylinder block and the other components in the rotating kit and to determine how speed, pressure, and pump displacement affect it.

An overview of the performed pre-studies and a flow chart of the order of succession for the pump geometry investigation are given in Fig. 1. Key parameters such as fluid properties, operational conditions (e.g., temperature, pressure), and geometrical details of the pump components were initially unknown and were determined using gap simulations performed with Caspar FSTI, which is a specialized numerical tool that considers the fluid in the narrow lubricating gaps only, which does not require the calculation of turbulences in the flow and exhibits a constant pressure across the height of the gap. These simplifications result in a faster solving of the Navier–Stokes equation than a traditional CFD tool, allowing for additional considerations of micro-motion and the effect of dynamic part deformation on the fluid film height. In turn, the gap simulations require temperature and heat convection as boundary conditions on the solid parts that define the gap. As these were unknown at the beginning of this study as well, they were estimated based on previous pump models such as published in [17]. The gap simulations delivered the power loss and therefore heat flow around the gap and the expected solid body temperatures of the cylinder block. The fluid temperatures such as the leakage temperature and the resulting case wall temperatures, which are also an input for the Simerics CFD model, were determined using a thermal steady-state model, similar as published by [15]. After the cylinder block measurements were concluded, the simulation was rerun using the measured boundary conditions such as actual leakage and housing temperature, which barely changed the result of the heat convection (compare Sim. A with B) in Fig. 3). The gap simulations however showed a higher dependency on the thermal boundary conditions, rendering the need for these complex CFD models to provide accurate heat convection coefficients. The goal of the exercise was to demonstrate that the heat convection coefficients can be predicted using CFD models in a sufficient accuracy, capturing the transient behaviors as the pump operates under various conditions. The heat convection are important boundary conditions for the gap simulation, changing the result significantly if false values are used. Both the CFD results as well as the gap simulation were validated using novel measurements using temperature fields and a direct measurement of the surface heat convection.

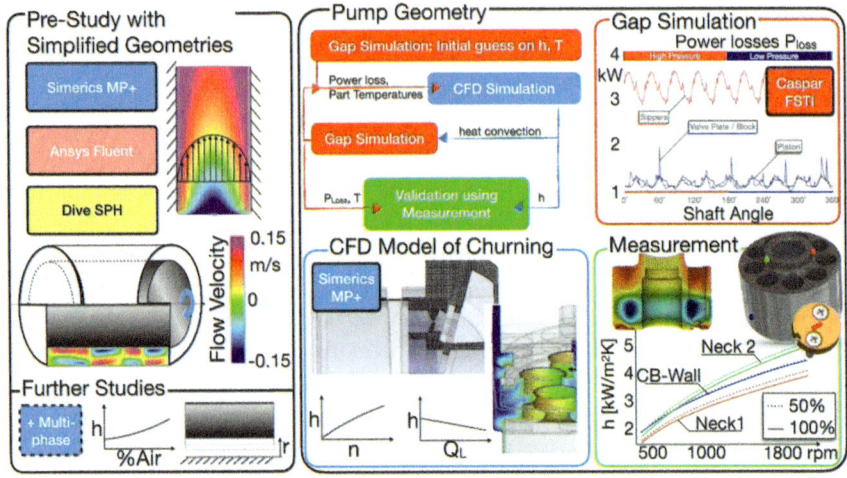

Fig. 1 Overview of the pre-studies performed and the flow chart for the pump simulations

Next some key assumptions for the CFD model will be discussed. The initial CFD model, which was set up before the experimental results were captured, assumed that the heat transfer occurred from the cylinder block to the case drain oil. Hence a set value of heat flux was given on the surface of the cylinder block. This value was determined from the hydromechanical losses from previous pump measurements. It was assumed that all the power loss due to hydromechanical reasons is converted into heat energy. This is equivalent to the stirrer work in thermodynamics and hence increases the internal energy of the control volume, i.e., oil in the pump casing.

After the block temperature measurements were conducted, it was discovered, that most of the cylinder block's surface absorbs heat from the leakage oil rather than dissipating it. The heating of the oil is predominantly in the fluid gaps and successively the leakage flow. Hence, for the improved simulation model, a specified negative value of heat flux was prescribed on the surface of the cylinder block which absorbs the heat from the casing oil. This value was taken as 10% of the total hydromechanical losses occurring in the system, which roughly corresponds to the surface area compared to all other parts in the case. This value was determined by experiment, varying from 5 to 20% as can be seen in Fig. 2 above. As cylinder block temperature measurements were available (red line), it was possible to calibrate this value based on the temperature prediction. The left graph shows the study at 500 rpm and 100 bar. For this operating point, the best match was close to 7%. Ten percent was chosen in order to be on the conservative side, especially at low displacements and low speeds the factor could change. The right bar graph shows the resulting surface temperatures assuming the 10% factor across all speeds for full displacement. With higher speed the temperature difference between measurement and simulation decreases, which could have a number of reasons, one of them could be the single-phase assumption. It is also possible that the 10% factor decreases with higher speed. In order not to overfit the data, the 10% was kept constant for all simulations.

The resulting heat transfer coefficients predicted from Simulations A (constant heat source) and B (10% heat absorption with leakage flow) are shown in Fig. 3. The red line shows the measured values, which were captured with a custom-built sensor, which are described more detail in the next chapter. Both simulations match the measurement in terms of magnitude for pump speeds between 1000 and 1500 rpm, although there are discrepancies in the trend when regarding the entire speed range. At higher speeds, both models underestimate the convection coefficient, whereas

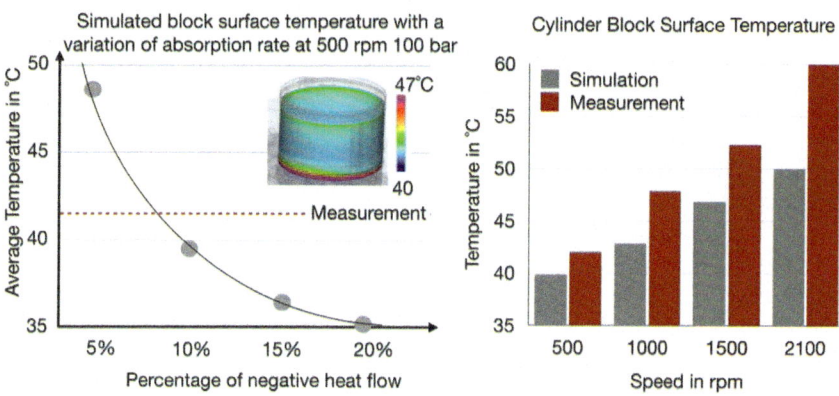

Fig. 2 Determination of heat absorption rate using measured cylinder block surface temperature

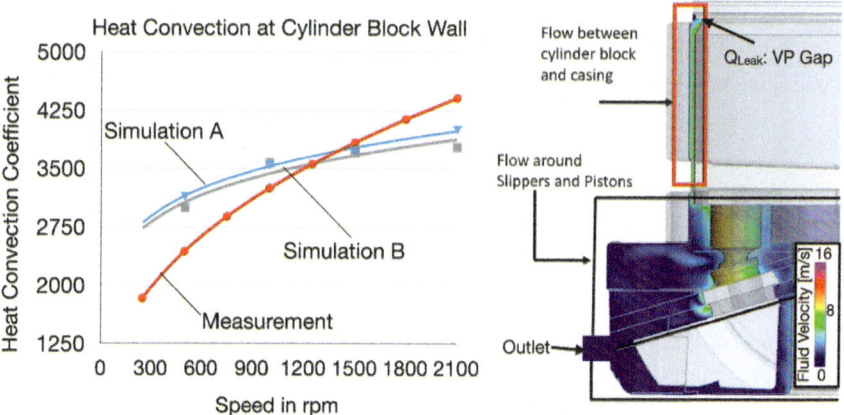

Fig. 3 Simulated heat convection coefficient using two approaches: Simulation A (blue): Cylinder block as a constant heat source with no leakage. Simulation B (gray): 10% Heat absorption rate with 1.0 l/min leakage (100 bar) from the valve plate (VP) gap. Measurement of 100 bar and full displacement is shown in red. Simulation setup and velocity field at 1000 rpm are shown on the right

at lower speeds, they overestimate it. By comparing the simulated and measured temperature trends (Fig. 2 right) for Simulation A, a direct correlation between heat convection and temperature can be derived.

In Simulation B, the heat flow is reversed (from the fluid to the block surface, rather than the other way around) and an attempt was made to enhance flow circulation around the cylinder block by introducing gap leakage. This leakage redirected the hot oil from the valve plate gap towards the leakage port at the opposite end of the pump, as shown on the right in Fig. 3. Adding these features did not change the trend of the simulated heat convection. When increasing the leakage flow from 1 l/min (measured at 100 bar) to 10 l/min (highest measured value at 300 bar), the convection coefficient result changed by less than 1%, indicating that the pressure has little influence on the heat transfer between block and fluid. This is confirmed by measurement as will be shown in the next section (compare Fig. 8).

It can be concluded, that the CFD model using Simerics MP + gave sufficiently well results, matching measurements in most of the speed regimes as well as correctly predicting the leakage independence. However, the trend with speed is not matched with the current model assumptions. Both simulation models (A & B) regarded single-phase flow only, which means that no air was considered in the fluid. From experience the housing of pumps usually contains some amount of air and the churning of the oil in the case creates a foamy mixture. Therefore, multiphase flow will be implemented in the future, to further improve the model. Studies with the simplified Taylor–Couette cylinder model have confirmed that a multiphase approach significantly increased the heat convection rate at higher speeds. Using this simplified model, a study of the influence of the fluid gap between rotating block surface and housing was conducted. The result was that the smaller the gap the more influence the multi-phase approach had. The tested pump has a gap of 5 mm, while many other pumps have larger gaps. When the gap was increased to 25 mm the effect of accounting for both phases started to diminish. This means that this pump could require more simulation effort than other pumps, due to its small fluid gap.

5 Measurement

5.1 Test Rig Setup

To validate the simulation model, a unique test rig was developed. The setup is shown in Fig. 4 below. The goal of the setup was to measure the temperature field and heat convection coefficients on the outer surface of the cylinder block in its original housing, without modifying the functionality of the pump. To accomplish this the cylinder block of a 160 cc pump was equipped with 26 temperature sensors and 4 heat convection coefficient sensors. The temperature measurements were published in [8].

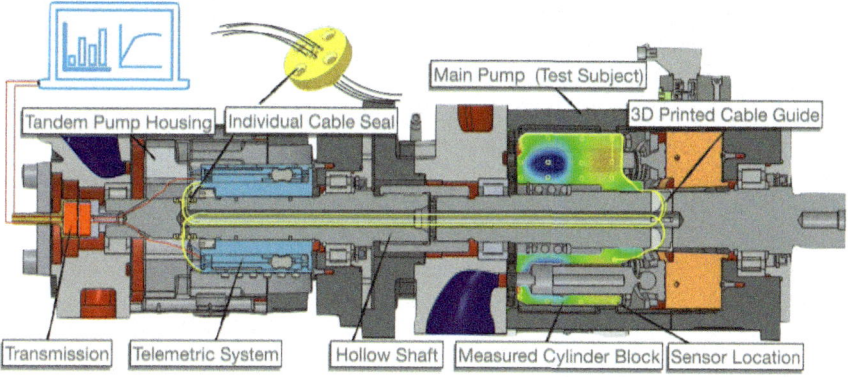

Fig. 4 Measurement set-up

The setup consists of a tandem pump arrangement, which was used to house the custom telemetric system. The sensor locations and temperature field are illustrated, with the exact sensor locations shown in Fig. 4. The sensor cables are guided from the cylinder block through hollow shafts using 3D-printed parts to guide the cables during pump assembly. To seal the case properly, the cables were individually guided through a cable seal, that was then sintered shut. The thermocouples were of type K with metal mantel, which allowed them to be sintered into the cylinder block. The telemetric system converts the analog temperature signals to digital bits that can be transmitted wirelessly to the acquisition system. The power is also transmitted through this brushless system. This setup allows for very reliable and accurate measurements, not only of temperature but also of other signals. In other publications, simultaneous pressure and temperature measurements are shown, that were recorded using this setup [9].

5.2 Measuring the Heat Convection Coefficient

To measure the heat convection coefficient a custom sensor and telemetric system was developed. Uffrecht et al. firstly introduced such a system including the miniature sensor, which is published in [19–4]. The sensor consists of a thermistor that can simultaneously be used as a heating element, all while being insulated from the surrounding material with a heat-isolating material. The equation for the heat transfer is given in Eq. 1 below. With a known current, the heat loss Q is given. Using an insulated material, the convection to the material is being kept to a minimum, reducing the heat loss to a convection through the surface of the sensor, hence the area A is known. To determine the heat convection coefficient h it is necessary to measure the temperature ΔT, which is the body surface temperature T and the surrounding Temperature T_{inf}. The sensor accomplishes this by acting as both the heat source and the temperature sensor itself. When powered only a short time the temperature can be measured, and by continuously powering the sensor and its surrounding fluid it will heat up. To further increase its accuracy, different currents are used at specific times, changing the heat flux and body temperature. This allows to calculate change in temperature ΔT with a change in power level ΔP.

$$Q = h \cdot A \cdot (T - T_{\text{inf}}) \tag{1}$$

Figure 5 shows a depiction of the sensor and the installation location on the cylinder block neck. On the right side, the heat flow paths are shown, and the temperature rise of the sensor with different electric currents. As can be seen, the electric power level is very low (<0.12W), while the temperature rise is significant ($\Delta T > 15$ K), which confirms good thermal insulation from the surrounding metal parts. To give some perspective, to achieve similar temperature rises in the cylinder block, the heat flow from the gaps is close to 1 kW of power loss. The sensor was programmed to cycle between three current levels, with several minutes in between to guarantee thermal equilibrium. To guarantee reliable results it was important to first reach thermal equilibrium in the entire pump. This process required monitoring the temperature of the cylinder block until the temperature change was less than 0.1 K in 2 min. The transient temperatures of a high-speed medium-pressure operating condition are shown in Fig. 6 on the right. The left side shows the sensor locations superimposed on the simulation result of Caspar FSTI, which was used for these simulations [8]. For this particular measurement, the operating point was the first of the day, essentially illustrating how long it takes to achieve thermal steady state from a cold state. The first temperature rise is due to the acceleration from 0 to 2100 rpm, while the second (which also increases the pressure to about 100 bar and the dropped back to 50 bar. The second rise is from increasing the pressure to 200 bar. The temperature increases quite rapidly but then takes about 10–15 min before being close to the steady state levels. The remaining time was used to adjust the pump cooling to keep a constant 35°C inlet temperature (T_{LP}). As can be seen, the cylinder block temperatures range between 48°C and 61°C, where sensors 7 and 8 are a close representation of the temperature inside the displacement chamber. Only after all sensors showed no change in temperature, the heat convection measurements were performed. The simulation results match the measurement quite well; however, this is after the boundary conditions of inlet, outlet, and leakage temperature as well as the heat convection coefficient of the cylinder block wall were updated from the measurements.

Initially, the boundary conditions were estimated using the preliminary CFD simulation, which underpredicted the heat convection at 2100 rpm, yielding lower block core temperatures as can be seen in Fig. 7. This is due to the fact that the cylinder block has three heat sinks: The oil in the displacement chamber, the gap losses and the leakage oil surrounding it. When the heat convection coefficient to the oil is too low, then less heat is absorbed by the surrounding fluid, increasing the block temperature in the gap and decreasing it at the surface. At this operating condition, the cylinder block/valve plate gap has the highest losses, essentially heating the block

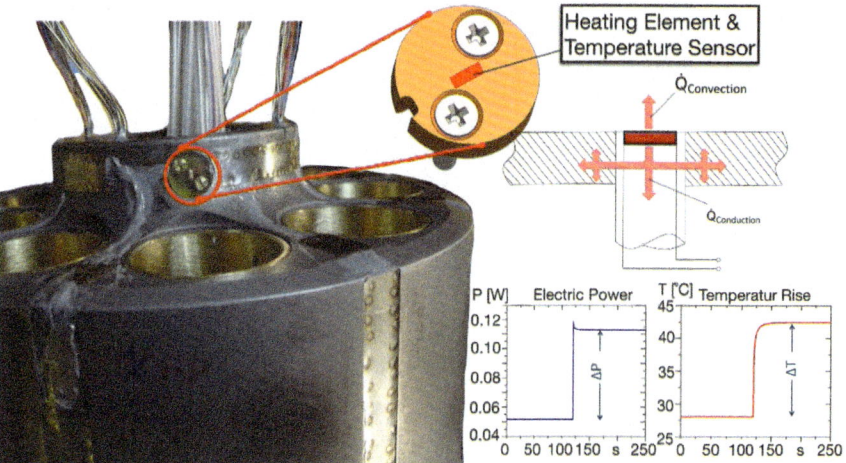

Fig. 5 Heat convection sensor mounting location on the neck, including glue protrusion (left). Right: Heat flow paths and temperature rise with electrical power inside the sensor

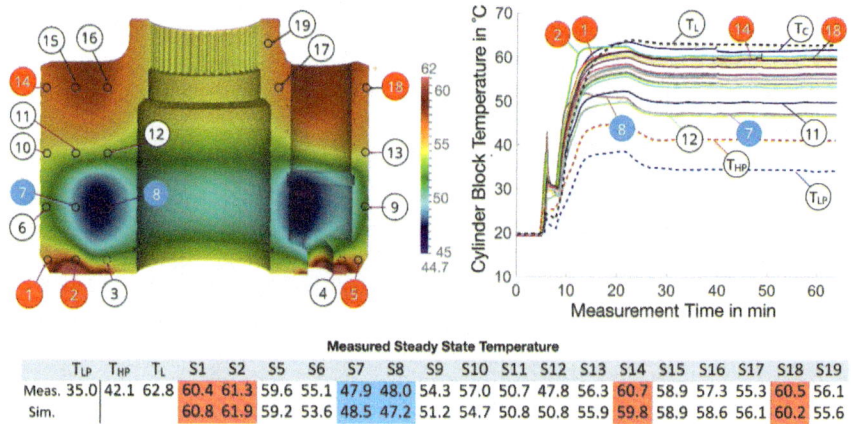

	T_{LP}	T_{HP}	T_L	S1	S2	S5	S6	S7	S8	S9	S10	S11	S12	S13	S14	S15	S16	S17	S18	S19
Meas.	35.0	42.1	62.8	60.4	61.3	59.6	55.1	47.9	48.0	54.3	57.0	50.7	47.8	56.3	60.7	58.9	57.3	55.3	60.5	56.1
Sim.				60.8	61.9	59.2	53.6	48.5	47.2	51.2	54.7	50.8	50.8	55.9	59.8	58.9	58.6	56.1	60.2	55.6

Fig. 6 Temperature field and transient temperatures at 2100 rpm, 200 bar, and 100% displacement. Left: simulated temperature field. Right: transient temperatures. Table: comparison between the measured and simulated steady-state temperatures

from the bottom up. At the same time, the displacement chamber removes a large portion of this heat. With a lower leakage temperature and lower heat convection coefficients (left figure), the upper part of the block gets only heated from the piston/bushing losses. When inputting the measured leakage temperatures, the cylinder block reaches a higher core temperature, while remaining at similar heat levels at the bottom gap.

The change in boundary conditions has a significant effect on the simulated power losses, decreasing the losses in the cylinder block/valve plate gap by more than 500 W

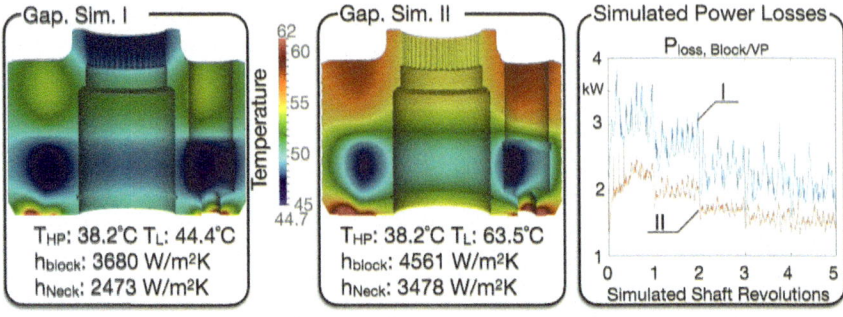

Fig. 7 Changes in simulated temperature field with different boundary conditions for 2100 rpm, 200 bar, and 100% displacement using gap simulation software Caspar FSTI at 35°C inlet temperature. Left: Sim I—Lower leakage temperature and lower heat convection coefficients. Middle: Sim II—Boundary conditions according to the measurement. Right: Simulated power loss at the cylinder block/valve plate gap

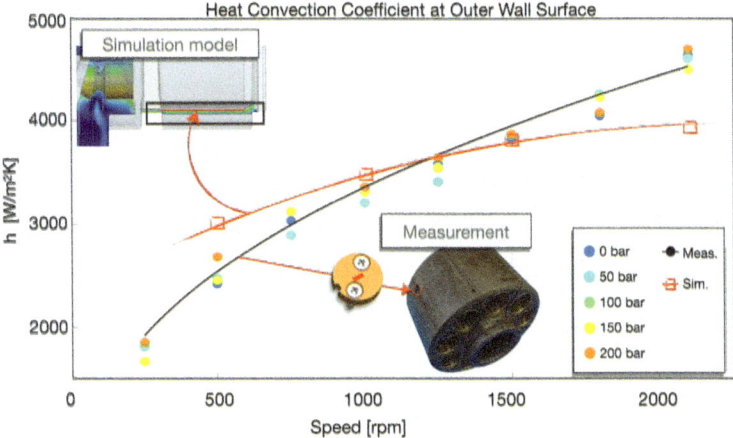

Fig. 8 Measured heat convection coefficient at full pump displacement at various pressures in comparison to simulation results

for the given operating condition. The block also reaches thermal stability much faster, which is indicated by the stabilizing power loss versus shaft revolutions. Caspar FSTI calculates the component temperature levels after every completed revolution, essentially the block temperature increases with each revolution. The gap losses decrease with each simulated revolution, which can be explained by the decrease in viscosity in the gap. As this gap is mainly dominated by viscous friction at this operating point, these losses decrease with temperature. Once the temperature and power losses do not change anymore, the simulation is completed.

The previous gap simulation result illustrates the influence that the boundary conditions have on the gap performance. It can be concluded that correct fluid temperatures surrounding the pump and heat convection levels have a significant effect on the simulation accuracy. To predict the correct heat convection coefficients, it is essential to develop a CFD model that yields accurate results, that reflect the reality. This model was described in Chap. 3. To validate and parameterize this CFD model the measurements of this convection coefficient are an essential step. Figure 8 shows the resulting levels for the heat convection coefficient of the cylinder block outer wall surface at various pressure and speed levels. The superimposed red line shows the simulation results from the CFD program. Both the measurement and simulation predict a significant rise of the heat convection with shaft speed, whereas the pump pressure has nearly no impact on the convection level, albeit small differences are visible but no clear trend.

The simulation tends to overpredict at low speed while underpredicting at high speed. However, the overall levels are quite comparable and are sufficiently accurate to yield good gap simulation results.

Figure 8 shows the measurement result of the outer wall convection sensor and its strong speed dependence and low-pressure dependence. The sensors at the cylinder block neck showed a similar trend, where the pressure level had nearly no impact on the heat convection. This trend is confirmed by the leakage study performed in the CFD model, suggesting that leakage flow and therefore pressure have little to no effect on the heat convection in pumps.

Figure 9 illustrates the impact the rotational radius and the displacement angle have on the heat convection coefficient. The full lines are the measurements at full pump displacement, while the broken line represents 50% displacement. The colors show the measurement results of the outer wall sensor (blue) and Neck Sensors 1(green) and 2 (red) for the first measurement. The measurements were repeated with a completely new block setup, now with two sensors at the neck and two sensors at the inner surface of the block. The neck sensors of the repeated measurement results are shown in black, both for each individual sensor and the average. It can be seen that these sensors deviate from the average the higher the speed, suggesting strong turbulences. The shown curves represent the average of all pressures, as indicated in Fig. 8. When comparing the different sensor locations at the same displacement (full line 100%), it becomes obvious that sensor Neck 2 from the first measurement (red) exhibits different trends. Before the repeated measurements were conducted it was assumed that Neck 1 was the wrong curve, Neck 2 exhibited the expected trend: Smaller rotational radius should yield lower flow velocities, which in turn should result in lower heat convection. However, after both neck sensors in the repeated measurement run confirmed the Neck 1 trend, it should be clear that Neck 2 was the abnormality. This probably was caused by the way this sensor was mounted, where glue that was used for fixating the wires was protruding from the wall surface. This small protrusion caused the heat convection to change by more than 20%, in this case decreasing it. In the repeated measurements it was made sure that no surface protrusion was altering the flow across these sensors, see pictures in Fig. 9. The conclusion that can be drawn from these results is that the proximity to the

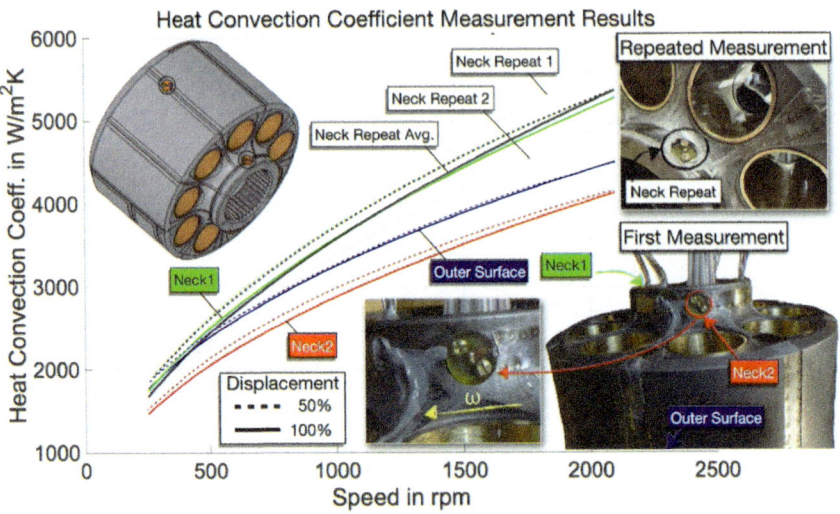

Fig. 9 Heat convection coefficient with speed at different locations

oscillation pistons is causing turbulences that increase the heat convection drastically. Interestingly the small protrusion just next to sensor Neck 2 caused a change in the flow pattern, decreasing its heat convection. When reflecting how the heat convection can impact power output of the pump (see Fig. 7) this finding could start a new investigation on surface topology of moving pump parts to artificially influence their heat convection.

When regarding the influence due to displacement the following observations can be made: The outer wall surface of the block shows no significant impact of the swash plate angle, especially at high speed, as both blue lines are nearly on top of each other. The heat convection around the neck region increases its heat convection with a decreasing swash plate angle. This trend matches expectations and can be explained as follows: [22] demonstrated both in theory and practice that the piston area that is exposed to the churning oil increases the lower the swash plate angle is, increasing the churning losses of the pump. This also causes higher turbulences around the piston at lower displacements, increasing the heat convection coefficient as well. All sensors on the neck exhibit this predicted behavior, confirming the theory.

The trend with displacement at the neck is shown in Fig. 10 for a larger displacement range. The trend is linear, decreasing with a slope of $-47\ W/m^2K$ per swash plate angle β ($\beta_{max} = 15°$) for this pump at this region. It should be noted that the slope decreases with speed, making the effect more prominent with lower speed.

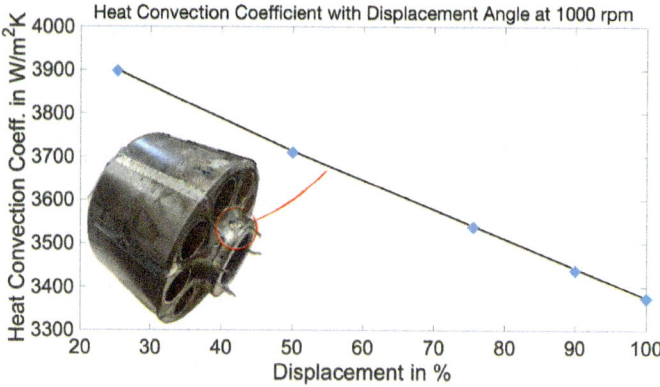

Fig. 10 Heat convection coefficient with pump displacement at the cylinder block neck region

6 Conclusion and Outlook

This paper presents novel findings on heat convection of the rotating cylinder block utilizing an innovative test rig to measure the temperature field and the heat convection coefficients within a fully functional pump across various operational regimes. The measurements were used to parametrize and validate a numerical gap simulation built with Caspar FSTI and a CFD pump model of the churning oil surrounding the rotating kit crafted in Simerics MP + . This CFD software package was chosen for its efficient setup, particularly when handling moving meshes, compared to the more time-consuming setup processes of Ansys CFX and Dive SPH. Although the Ansys and Dive models exhibited promising trends in simpler geometries, their limitations in complex setups requiring a detailed resolution of the boundary layer influenced their non-selection for this study.

The measurements were performed on a 160 cc open-circuit pump equipped with a custom-built system consisting of a high bandwidth telemetric system with 26 temperature sensors, and a series of heat convection sensor on the block surface. Findings underscored a strong dependence of heat convection on shaft speed, attributed to increased surface velocity and influenced by factors such as turbulence, which are not strictly linear. There was also a linear dependence on the swash plate angle near the pistons, confirming assumptions that heat convection increases with lower displacement angles. The dependence on pressure and leakage flow was observed to be negligible.

Simulations demonstrated how initial temperature assumptions and power losses from gap simulations are sufficient to predict the heat convection in the pump. The added complexity of adding a measured leakage flow and carefully updating boundary conditions to measurement did not change the results significantly. Overall, the model successfully predicted the general trend of heat convection, though it tended to overpredict at lower speeds and underpredict at higher speeds.

The CFD model incorporated the rotating piston and slippers in an angled swash plate. It was shown how accounting for these complexities is significant as they have a profound effect on the heat convection on the upper region of the cylinder block. The outer wall of the block is not as affected by the pistons, which was confirmed by the measurements, as only small changes were recorded with different displacements.

Considering free air in the churning oil can improve results, especially at high speeds; however, this necessitates more simulation power and requires a precise prediction at what speed the air dissolves and how much is in the solution. Initially, studies with a simplified geometry have shown that adding 10% free air can increase the heat convection by more than 50% for small fluid gaps.

Repeated measurements at the neck region validated initial sensor readings that were originally thought to be faulty. These findings highlight that turbulences around the piston cause a deviation from the theoretical models at the upper neck region, which also cause strong fluctuations in the readings, particularly at high speeds. The study also demonstrated that topological changes, like a small protrusion in flow paths, can alter heat convection significantly. Gap simulations were used to demonstrate that these changes in heat convection coefficient can influence the gap and result in significant changes in power losses, primarily due to a change in part temperature. Future simulations should extend these investigations to other surfaces like slippers and pistons to comprehensively understand heat convection dynamics across different pump components.

References

1. Abed WM, Al-Damook A, Khalil WH (2021) Convective heat transfer in an annulus of concentric and eccentric cylinders with an inner rotating cylinder. Int J Heat Technol 39(1):61–72
2. Adami S, Hu X, Adams N (2012) A generalized wall boundary condition for smoothed particle hydrodynamics. J Comput Phys 231(21). https://doi.org/10.1016/j.jcp.2012.05.005a
3. Bing X, Al E (2015) Modeling and analysis of the churning losses characteristics of swash plate axial piston pump. International Conference on Fluid Power and Mechatronics
4. Eschmann G, Kuntze A, Uffrecht W, Odenbach S (2014) Measurement if heat transfer coefficients un gaseous flow - first test of a recent sensor concept for stationary and oscillating flow. ASME Turbo Expo 2014:1–11
5. Ivantysyn R, Shorbagy A, Weber J (2017) An approach to visualize lifetime limiting factors in the cylinder block / Valve plate gap in axial piston pumps, in FPMC2017–4327, pp 1–12
6. Ivantysyn R, Weber J (2018) Investigation of the thermal behaviour in the lubricating gap of an axial piston pump with respect to lifetime, in 11. IFK 2018
7. Ivantysyn R, Weber J (2016) Transparent Pump–An approach to visualize lifetime limiting factors in axial piston pumps. In: ASME 2016 9th FPNI Ph.D Symposium on Fluid Power, Florianapolis, Brazil
8. Ivantysyn R, Weber J, Kunze A, Uffrecht W (2024) Thermal analysis of the cylinder block of an axial piston pump–The key to monitoring efficiecny, in IFK2024, pp 1–21
9. Ivantysyn U, Kuntze W (2024) Efficiency forecasting in axial piston pumps: leveraging high speed fluid temperature correlations for condition monitoring. In: JFPS. Hiroshima, S. 1–15.
10. Kozsar E (2020) Experimentelle Untersuchung des thermischen Einlaufverhaltens von Gleitschuhen eines Axialkolbenmotors

11. Michel S (2021) Elektrisch-hydrostatische Kompaktantriebe mit Differentialzylinder für die industrielle Anwendung
12. Di Monaco A, Manenti S, Gallati M, Sibilla S, Agate G, Guandalini R (2011) SPH Modeling of solid boundaries through a semi-analytic approach. Eng Appl Comput Fluid Mech 5(1):1–15. https://doi.org/10.1080/19942060.2011.11015348
13. Neale A, Al E (2006) CFD calculation of convective heat transfer coefficients and validation–Part I: Laminar flow in Annex 41–Kyoto
14. Pelosi M (2012) An investigation on the fluid-structure interaction of piston/cylinder interface, Purdue University
15. Pohl E, Weber J (2014) Efficient model-based thermal simulation method demonstrated on a 24-ton wheel loader. Ifk 2024:1–12
16. Schenk A (2014) Predicting lubrication performance between the slipper and swashplate in axial piston hydraulic machines, Purdue University
17. Shorbagy A, Ivantysyn R, Berthold F, Weber J (2022) Holistic analysis of the tribological interfaces of an axial piston pump–Focusing on pumps efficiency, in IFK2022, Aachen
18. Uffrecht W, Heinschke B, Günther A, Caspary V, Odenbach S (2015) Measurement of heat transfer coefficients at up to 25,500g - a sensor test at a rotating free disk with complex telemetric instrumentation. Int J Therm Sci 96:331–344. https://doi.org/10.1016/j.ijthermalsci.2015.03.006
19. Uffrecht W, Günther A (2012) Electro-thermal measurment of heat transfer coefficients, in ASME Turbo Expo 2012
20. VDI-Wärmeatlas (2006), vol 10. Accessed on 01 Aug 2014
21. Vedthpathak A (2023) Determination of heat convection coefficients in the churning oil in axial piston pumps by means of SPH and CFD simulation and validation by measurement
22. Xu B, Li Y, Zhang J, Chao Q (2015) Modeling and analysis of the churning losses characteristics of swash plate axial piston pump. In: 2015 International conference on fluid power and mechatronics (FPM), pp 22–26. https://doi.org/10.1109/FPM.2015.7337078
23. Zecchi M (2014) A novel fluid structure interaction and thermal model to predict the cylinder block/valve plate interface performance in swash plate type axial piston machines, Purdue University, West Lafayette, IN, 2013. Accessed on 20 Aug 2014

Open Access This chapter is licensed under the terms of the Creative Commons Attribution 4.0 International License (http://creativecommons.org/licenses/by/4.0/), which permits use, sharing, adaptation, distribution and reproduction in any medium or format, as long as you give appropriate credit to the original author(s) and the source, provide a link to the Creative Commons license and indicate if changes were made.

The images or other third party material in this chapter are included in the chapter's Creative Commons license, unless indicated otherwise in a credit line to the material. If material is not included in the chapter's Creative Commons license and your intended use is not permitted by statutory regulation or exceeds the permitted use, you will need to obtain permission directly from the copyright holder.

Numerical Analysis of a High-Power Piezoelectric Pump using Computational Fluid Dynamics (CFD) Simulations

Francesco Sciatti, Vincenzo Di Domenico, Paolo Tamburrano, Nathan Sell, Andrew R. Plummer, Elia Distaso, Giovanni Caramia, and Riccardo Amirante

1 Introduction

Piezoelectric materials, characterized by reduced moving parts, wide operating frequency, fast response, lightweight nature, and stable performance, are increasingly capturing attention in the field of fluid power [1–9]. The advent of Micro-Electro-Mechanical Systems (MEMS) has significantly fueled this interest, as these smart materials, harnessing the inverse piezoelectric effect and converting electrical energy

---

F. Sciatti (✉) · V. Di Domenico · P. Tamburrano · E. Distaso · G. Caramia · R. Amirante
Department of Mechanics, Mathematics and Management (DMMM), Polytechnic University of Bari, Bari, Italy
e-mail: francesco.sciatti@poliba.it

V. Di Domenico
e-mail: v.didomenico@phd.poliba.it

P. Tamburrano
e-mail: paolo.tamburrano@poliba.it

E. Distaso
e-mail: elia.distaso@poliba.it

G. Caramia
e-mail: giovanni.caramia@poliba.it

R. Amirante
e-mail: riccardo.amirante@poliba.it

N. Sell · A. R. Plummer
Centre for Power Transmission and Motion Control (PTMC), University of Bath, Bath, UK
e-mail: nps22@bath.ac.uk

A. R. Plummer
e-mail: arp23@bath.ac.uk

into mechanical energy, are actively under investigation as potential key components in the development of advancing piezoelectric pumps [10, 11].

Piezoelectric pumps, commonly contracted to piezopumps, present promising applications in several fields such as biomedicine, robotics, aerospace, electronics, chemistry, and automotive, owing to their advantages in accurate precision flow control, compact structure, absence of magnetic influence, and low noise [12–15]. Within piezoelectric pumps a classification can be made between "valved piezopumps" and "valveless piezopumps," depending on whether the pump incorporates movable valves [16]. Valved piezopumps can achieve higher flow rates and back pressures but may reduce the survival rate of possible transported living organisms [17]. In contrast, valveless piezopumps, with a simpler structure, decrease the probability of crushing damage to the possible transported substance, despite the smaller flow rate delivered (typically in the order of mL/s) [17]. Consequently, the latter finds commercial applications, especially in medical dispensing [18].

In the field of fluid power, where higher flow rates and back pressures are crucial, noteworthy valved piezopumps under research are piezohydraulic pumps, also recognized as piezostack pumps [19]. These pumps operate as reciprocating positive displacement pumps, utilizing a piezostack actuator as the driving power source for fluid delivery. This characteristic makes them well-suited for "high-power" applications ranging from 1 to 100 W [20]. The term "high-power" is significant because piezopumps with power outputs greater than 20W are uncommon, with only one example cited in literature [21].

Composed of stacked piezoelectric elements enclosed between electrodes, piezostack actuators feature layer thicknesses ranging from 25 to 100 μm and total lengths up to around 200 mm, offering free strokes of approximately 300 μm [1, 22]. The force applied by the piezoelectric material is determined by its face area, and commercially available stacks with diameters exceeding 50 mm can generate maximum forces (or blocking forces) of up to 70 kN [22]. It has been proven that selecting the correct preload value enables a longer lifetime for these piezoelectric actuators, with optimum preload ranging from 20 to 50 percent of the blocking force [23, 24]. However, a notable limitation in using piezostack actuators is their low stroke, significantly smaller compared to their length. To address this, mechanical amplification methods can be employed to increase displacement, enabling substantial deformation (up to 1 mm [25]).

To meet the previously specified power requirements, piezohydraulic pumps utilize hydraulic motion accumulation instead of mechanical methods. This system consists of a piezostack actuator driving a piston at high frequency, coupled with a pair of check valves regulating the flow at the inlet and outlet of the pump chamber. Figure 1 shows the operating cycle of piezohyraulic pumps.

As shown in Fig. 1, the working process of a conventional piezohydraulic pump consists of four stages: compression, delivery, expansion, and intake. During the compression stage, both inlet and outlet check valves remain closed. Upon the application of a voltage signal to the stack, it expands, inducing compression of the fluid within the pump chamber and consequently increasing the chamber pressure. Once the chamber pressure equals that of the outlet environment, the delivery stage begins.

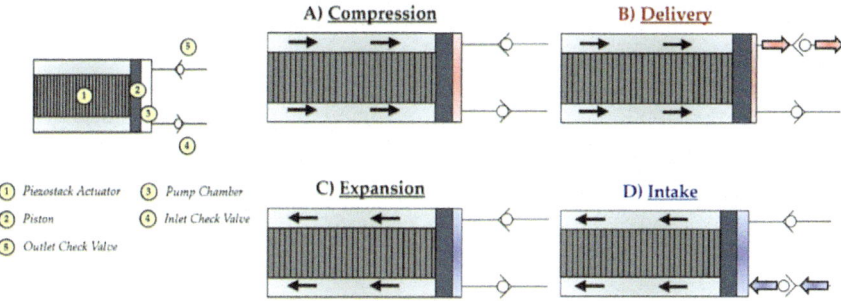

Fig. 1 Operating cycle of piezohyraulic pumps

Here, the outlet check valve opens, enabling the flow of fluid from the pump chamber to the outlet environment. Following this, there is the expansion stage; upon removal of the voltage signal from the stack, it contracts, leading to a reduction in pressure within the chamber toward the level of the inlet environment. The final stage, the intake stage, occurs when these pressures reach equilibrium. At this point, the inlet check valve opens, allowing fluid to re-enter the pump chamber.

Leveraging the broad operating frequency of piezoelectric materials, and thus, supplying the piezostack actuator with a high-frequency sinusoidal voltage signal (exceeding 1 kHz) allows for a substantial flow despite the stack's limited displacement. Indeed, increasing the operating frequency turns out to be as a simple and convenient approach to enhance the performance of piezoelectric pumps without the need to increase their size or input power [26].

The correct choice of valves is also crucial in this context. While some studies argue for active valves to boost operating frequency and overall pump performance [27, 28], others propose maintaining passive valves and transitioning from ball-type check valves to reed valves [29]. The latter, utilizing flexible materials, reduces the moving mass and, thus, enhances bandwidth in the region of 1 kHz [30, 31].

Nevertheless, the high driving frequency exposes piezohydraulic pumps to severe cavitation risks [32], potentially resulting in component damage, efficiency reductions, vibrations, and noise [33]. Cavitation, involving the formation and implosive collapse of vapor bubbles under high-pressure conditions [34], has captured the attention of researchers due to its impact on the performance of piezoelectric pumps operating with "high" flow rates. In their investigation of cavitation in piezoelectric pumps, Zhang et al. pointed out that during liquid suction, the expanding pump chamber reduces pressure, causing dissolved gas to escape and triggering cavitation [32]. He et al. examined the adverse effects of cavitation on piezoelectric pumps, discovering that it reduces the bulk modulus of the working fluid, severely decreasing efficiency [35].

Given the substantial influence of cavitation on the performance of piezoelectric pumps, the aim of this paper is to initiate a numerical investigation using CFD software to assess the potential for cavitation initiation in a specific piezohydraulic pump developed at the University of Bath. Specifically, the study focuses on simulating two diverse oil flow scenarios through the piezohydraulic pump with a fixed inlet pressure and pump chamber pressure and varying inlet reed valve opening.

To begin, the architecture and operating principle of the piezohydraulic pump under investigation are presented. Then, the pump 3D CAD model, the simplified 2D domain considered, and its corresponding computational grid using unstructured tetrahedral meshes are illustrated. Finally, the results of the simulations are examined in order to evaluate potential situations that may lead to cavitation.

2 Piezohydraulic Pump Architecture

The piezohydraulic pump developed at the University of Bath [20] is shown in a detailed longitudinal section view in Fig. 2a, with a comprehensive bill of materials provided in Table 1. Additionally, a photograph of the fully assembled pump and its components is presented in Figs. 2b and c, respectively.

Observing Fig. 2, it is evident that the oil enters the pump through the inlet end cap (1) and passes through eight orifices of the inlet plate (2), entering the pump

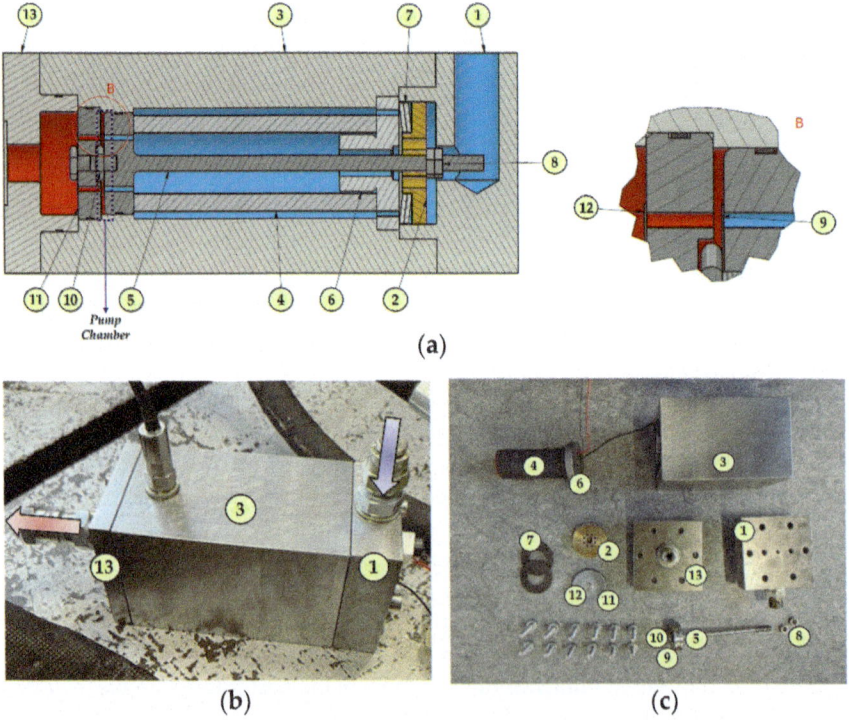

Fig. 2 Piezohydraulic pump architecture realized at the University of Bath [20]: **a** Longitudinal section view; **b** Photograph of assembled pump prototype; **c** Photograph of pump components

Table 1 Bill of materials of the piezohydraulic pump—Refer to Fig. 2

Part number	Quantity	Description
1	1	Inlet end cap
2	1	Inlet plate
3	1	Pump body
4	1	Ring stack
5	1	Piston
6	1	Piston clamp
7	2	Belleville washer
8	2	Bolt
9	1	Inlet reed valve
10	2	Valve screw
11	1	Valve plate
12	1	Outlet reed valve
13	1	Outlet end cap

body (3) and the hole of the ring stack (4). Positioned on the left end side of the stack is the piston (5), which, with its rod passing through the stack hole, clamps and preloads the stack using the piston clamp (6), Belleville washers (7), and bolts (8). Featuring twelve orifices, the piston is equipped with an inlet reed valve (9) fixed onto its left face using a valve screw (10). A similar arrangement is found on the valve plate (11), where twelve orifices are covered by an outlet reed valve (12) positioned onto the left face of the valve plate using another valve screw. The extension and retraction of the stack govern the opening and closing of the inlet and outlet reed valves, allowing oil to either flow within the pump chamber (intake) or exit (delivery) based on internal pressure. The pump chamber, designed with a height of only 1 mm, aims to increase the maximum displacement of both the inlet and outlet reed valves. Finally, the pressurized oil, coming from the outlet reed valve, exits the pump through the outlet end cap (13).

As shown in Fig. 2, this pump utilizes passive disc-style reed valves to achieve pumping frequencies exceeding 1 kHz and to regulate the flow of oil into and out of the pump chamber. The driving force behind this mechanism is the ring stack actuator manufactured by PI Ceramic, specifically the model PICA P025.50H, with a maximum height of 66 mm. This actuator is capable of providing the highest maximum free stroke value of 80 μm. Additionally, it offers a maximum blocking force of 9600 N and a stiffness of 120 N/μm. For detailed specifications, please refer to Table 2 [36].

This pump prototype was experimentally tested within a simple loading system, shown in Fig. 3. In this setup, the oil pumped from the piezohydraulic pump traveled through a pipe with a length of 3 m and a diameter of 6 mm. At the end of this pipe, a variable area orifice, functioning as a loading valve, was connected. This allowed for the adjustment of the orifice area, enabling a range of different load pressures to be generated. The oil then entered a small hydraulic volume equivalent to the

Table 2 PICA P025.50H Ring stack: Specifications [36]

Parameter	Value	Unit
Cross-sectional area	2.89	cm^2
Outer diameter	25	mm
Inner diameter	16	mm
Length	66	mm
Maximum voltage	1000	V
Free stroke	80	μm
Blocking force	9.6	kN
Capacitance	1.2	nF
Natural Frequency	17	kHz

output chamber of the pump, measuring 0.01 L. The pump's inlet was maintained at a constant pressure of 20 bar throughout the experiment. This measure aimed to prevent cavitation at the inlet and ensure the fluid's stiffness remained consistent. However, the experimental results revealed that at an operating frequency of 1250 Hz and a mean load pressure difference of 15 bar, the minimum chamber pressure dropped approximately to 0 bar [20]. This value was significantly lower than the inlet pressure of 20 bar, suggesting the potential occurrence of cavitation due to the high-pressure drop across the inlet reed valve. Under these conditions, the piezohydraulic pump demonstrated a capacity of delivering 1.05 L/min and approximately 30 W of hydraulic power [20].

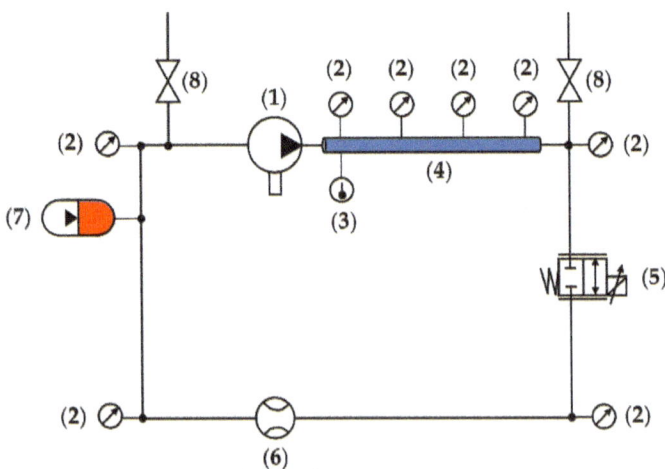

Fig. 3 Hydraulic circuit scheme of the test rig: (1) Piezohydraulic pump; (2) Pressure sensor; (3) Temperature sensor; (4) Pipe; (5) Loading valve; (6) Flow sensor; (7) Accumulator; (8) Needle valve

3 Fluid Dynamic Analysis

3.1 3D–2D Model

A commercial computer-aided design (CAD) software, specifically Autodesk Inventor Professional, was used to create the 3D model of the piezohydraulic pump shown in Fig. 4a. Since the aim of this simulation is to provide an understanding of the operating conditions of the pump that might lead to cavitation during the intake stage, the 3D model of the pump was simplified into a 2D model through three approximations. Firstly, a simplified 3D domain was delineated by marking two cutting planes (in yellow) on the 3D model (refer to Fig. 4a), constituting the first approximation. Figure 4b illustrates the distribution of oil within the pump between these two planes, specifically highlighting the oil within the inner part of the piezostack, the twelve piston orifices, and a portion of the pump chamber, considering a specific opening of the inlet reed valve. Notably, the oil between the piezostack and the pump body was omitted due to its passive role in the pump's operation. The oil contained in the simplified 3D domain shown in Fig. 4b exhibits axial symmetry, leading to the second approximation: examining only a slice of this simplified 3D domain, containing only one piston orifice (highlighted in red in Fig. 4b) to reduce computational costs. Furthermore, since the focus is on understanding velocity and pressure fields during the intake stage across the inlet reed valve, a third approximation was considered. Only the mid-plane section of the selected slice (highlighted in Fig. 4c) was analyzed to investigate the potential occurrence of cavitation phenomenon. This approach resulted in a simplified 2D model of the piezohydraulic pump for the purpose of analysis.

3.2 Computational Grid

The 2D model system resulting from the previous considerations was discretized using mesh generation software for CFD, specifically Cadence Fidelity Pointwise. Unstructured tetrahedral meshes were employed in this initial analysis. The grid was designed for two simulated openings of the inlet reed valve: the maximum possible opening ($v_{opn_{max}} = 0.7$ mm) and the minimum opening ($v_{opn_{min}} = 0.1$ mm). Mesh setup involved using a small number of points, leading to large interval sizes ranging from 0.1 mm to 0.9 mm for edges not defining restricted sections of the pump. Conversely, a large number of points, resulting in very small interval sizes ranging from 0.0025 mm to 0.01 mm, were used for meshing the edges of the piston orifice and the inlet reed valve. This approach aimed to identify pressure and velocity gradients and flow swirls in detail. The resulting total number of cells was approximately 56,874 for the inlet reed valve opening at $v_{opn_{max}} = 0.7$ mm and 54,031 for $v_{opn_{min}} = 0.1$ mm.

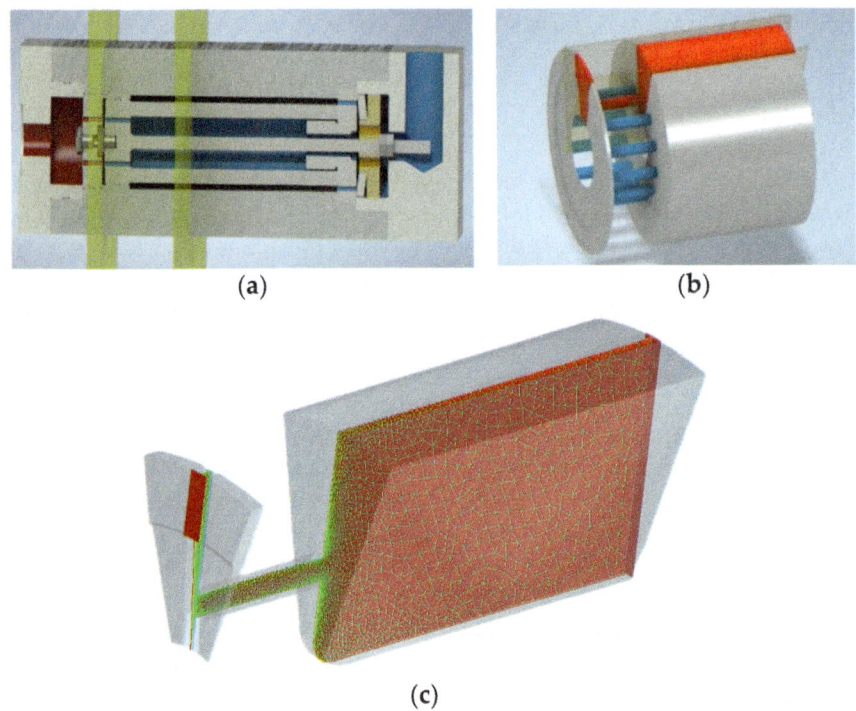

Fig. 4 a Longitudinal section view of the piezohydraulic pump assembly, showing two cutting planes used to simplify the 3D domain; **b** Oil distribution within the simplified 3D domain of the piezohydraulic pump, highlighting a slice of the total oil volume due to the symmetry of the geometry; **c** Simplified 2D model displaying the mid-plane section of the selected slice

Figure 5 provides a view of the computational grid at $v_{opn_{max}} = 0.7$ mm and a zoomed view of the region between the piston orifice and the inlet reed valve, highlighting the inlet, outlet, and walls considered in the analysis.

After discretizing all edges, the entire flow domain was meshed. The final step before exporting the mesh involved defining the boundary conditions. With reference to Fig. 5, the following boundary conditions were defined:

- Pressure Inlet: representing the constant pressure of the oil at the pump inlet (p_{in});
- Pressure Outlet: representing the pressure value set for the oil in the pump chamber during the intake stage (p_{cham});
- Walls: representing all remaining lines in this initial simplified analysis, highlighted in light blue in Fig. 5;
- Oil zone: representing all the region between inlet and outlet and surrounded by the walls in Fig. 5.

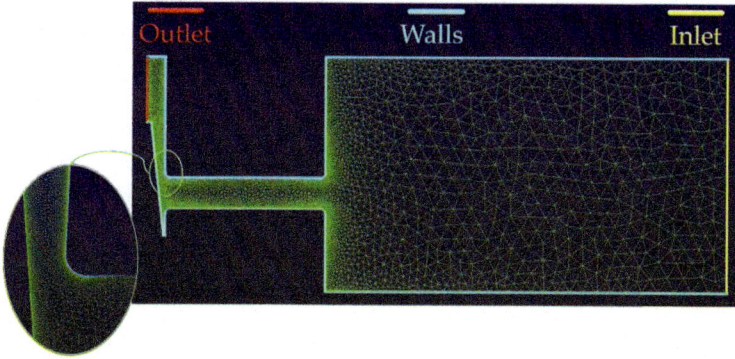

Fig. 5 View of the computational grid at $v_{opn_{max}} = 0.7$ mm and zoomed view of the region between the piston orifice and the inlet reed valve, highlighting the inlet, outlet, and walls considered in the analysis

3.3 Governing Equations and Solution Strategy

The steady-state Reynolds-averaged Navier–Stokes (RANS) equations were solved using Ansys Fluent, a CFD commercial software for fluid simulation. For solving, the semi-implicit method for pressure-linked equations (SIMPLE) pressure-based segregated algorithm was selected. Among the available segregated algorithms in Ansys Fluent (including SIMPLE, SIMPLEC, and PISO), the SIMPLE algorithm was chosen for this problem based on its suitability for the physics involved.

To predict cavitation, three main methods can be employed, namely the multiphase flow model, the homogeneous equilibrium model, and the interface tracking model [37–39]. In this study, since the main focus is on severe variation of density, the multiphase flow model is chosen. Specifically, the Schnerr and Sauer equations were used for simulating cavitation in a multiphase flow, as well documented in [40].

Vapor was treated as incompressible. Specifically, in the numerical model, vapor phase exists only when pressure values are below the vapor pressure, enabling vapor density to be assumed constant.

The liquid oil was set with a density of 866 kg/m^3 and a dynamic viscosity of 0.027 kg/(m·s), while the vapor oil was set with a density of 4 kg/m^3 and a dynamic viscosity of $3 \cdot 10^{-6}$ kg/(m·s).

In CFD programs, there are several turbulent models available, including the standard $k - \varepsilon$, RNG $k - \varepsilon$, realizable $k - \varepsilon$, and SST $k - \omega$ models [41]. The choice of turbulent model significantly affects the accuracy of simulation results. Studies have shown that the SST $k - \omega$ model provides reliable results with an error margin of around ± 5% [42]. Therefore, in this study, the SST $k - \omega$ method was employed to predict turbulence.

The turbulence equations for k and ω, as well as the momentum equations, were solved initially using a first-order upwind discretization and then a second-order upwind discretization to stabilize the simulation. For pressure interpolation, the

Table 3 Scaled residuals after 5,000 iterations (opening inlet reed valve condition $v_{opn_{min}} = 0.1$ mm)

Equation	Scaled residual
Continuity	$9.01\ e^{-07}$
x-Momentum	$2.23\ e^{-07}$
y-Momentum	$2.69\ e^{-07}$
k	$2.16\ e^{-08}$
ω	$5.77\ e^{-08}$
Vapor	0

PRESTO! scheme, specifically designed for flows with steep pressure gradients, was selected [43]. Additionally, a first-order upwind scheme was employed for the discretization of the vapor transport equation.

Regarding the termination criterion, it was decided to stop the simulations when the mass flow rate maintained its third significant digit unchanged, rather than terminating iterations based on scaled residuals falling below a fixed value. This computational strategy ensured accurate prediction of potential cavitation within the pump chamber. Approximately 5,000 iterations (2,000 using first-order upwind discretization and 3,000 using second-order upwind discretization) were required to fulfill the convergence criterion without encountering any convergence issues. Table 3 presents the scaled residuals recorded for a simulated inlet reed valve opening $v_{opn_{min}} = 0.1$ mm. Table 4 summarizes the settings employed for the simulations.

4 Numeric Results

In this preliminary analysis, simulations have been conducted for two specific cases:

1. $v_{opn_{min}} = 0.1$ mm, $p_{in} = 20$ bar, $p_{cham} = 0.5$ bar;
2. $v_{opn_{max}} = 0.7$ mm, $p_{in} = 20$ bar, $p_{cham} = 0.5$ bar.

These cases represent extreme possibilities for the behavior of the inlet reed valve during the intake stage, particularly in a critical scenario where the pressure chamber equals 0.5 bar. The simulations were carried out with a relative pressure of 1.01 bar in mind.

The significant difference in the opening values of the inlet reed valve has a notable impact on the velocity and pressure fields within the 2D oil domain under consideration. It is important to note that the cavitation does not occur in the first case, despite the low value of pressure chamber considered. Conversely, the maximum opening of the inlet reed valve leads to a substantial increase in average velocity near the restriction area, with a maximum computed value of 69.7 m/s. This, in turn, causes local pressure values to drop below the vapor pressure, triggering vapor phase formation and subsequently, cavitation. Graphical outcomes from CFD simulations depicting the velocity contour, pressure contour, and volume fraction of vapor oil phase are presented for both cases in Figs. 6 and 7, respectively.

Table 4 Setting employed for the simulations

Equation	Multiphase (Mixture) model
Solver	SIMPLE
Turbulence model	SST k-ω
Cavitation model	Schnerr and Sauer
Discretization of pressure equation	PRESTO!
Discretization of volume fraction	First-order upwind
Discretization of momentum and turbulence	Second-order upwind
Under-relaxation of pressure	0.3
Under-relaxation of volume fraction	0.5
Under-relaxation of momentum	0.7
Under-relaxation of turbulence quantities	0.8
Inlet pressure (p_{in})	20 bar absolute
Chamber Pressure (p_{cham})	0.5 bar
Density and viscosity liquid-oil	866 kg/m^3; 0.027 kg/(m·s)
Density and viscosity vapor-oil	4 kg/m^3; 3 · 10^{-06} kg/(m·s)
Vapor pressure	10 Pa
Bubble number density	Default
Termination criterion	Convergence in mass-flow rate (third digit)

Comparing Figs. 6 and 7, it is evident that the opening size of the inlet reed valve significantly influences both velocity and pressure fields, resulting in different local behaviors near the restriction zone. In general, a larger opening leads to increased oil flow and higher velocity, consequently causing higher pressure drops. However, this pressure drop may occasionally fall below the vapor tension threshold, leading to the problem of cavitation, as shown in Fig. 7.

5 Conclusions

This paper presented an initial investigation into cavitation potential in a piezohydraulic pump developed at the University of Bath (UK). The study, in its early stages, focused on maintaining constant inlet ($p_{in} = 20$ bar) and chamber ($p_{cham} = 0.5$ bar) pressures while varying the opening size of the inlet reed valve ($v_{opn_{min}} = 0.1$ mm and $v_{opn_{max}} = 0.7$ mm). The examination primarily looked at a 2D analysis of velocity and pressure fields in the mid-plane of a slice of the simplified 3D domain of the piezohydraulic pump. The CFD simulation results indicated that wider inlet reed valve openings allowed more oil to pass through the restricted area, resulting in locally elevated velocity values. Specifically, when the velocity exceeded a specified threshold, the cavitation occurred. This phenomenon was identified by plotting the

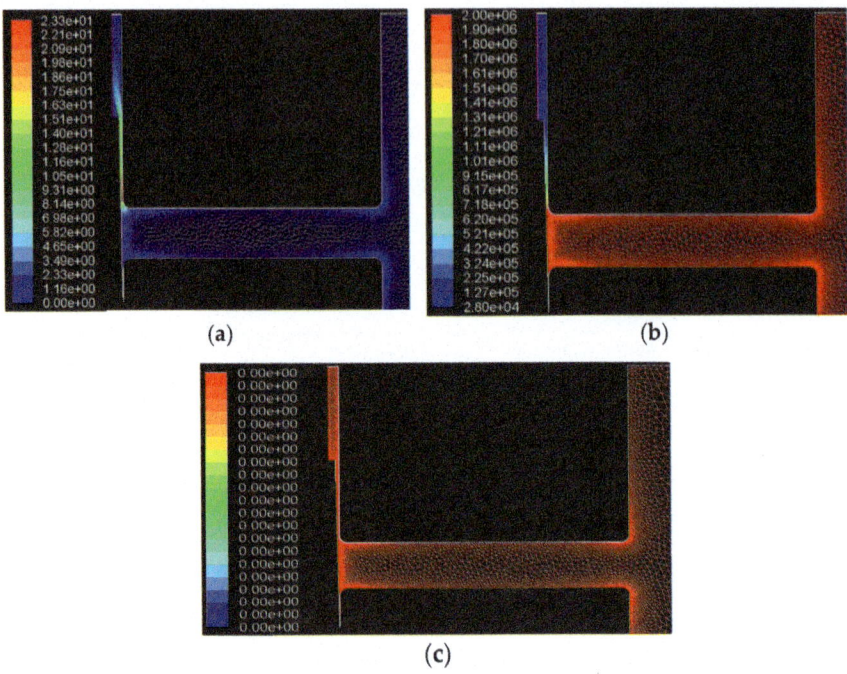

Fig. 6 CFD analysis of the piezopump investigated considering $v_{opn_{min}} = 0.1$ mm; $p_{in} = 20$ bar; $p_{cham} = 0.5$ bar: **a** Velocity contour [m/s]; **b** Pressure contour [Pa]; (**c**) Volume fraction of vapor oil phase [–]

volume fraction of the vapor oil phase, which was nonzero only when the inlet reed valve opening was at its maximum ($v_{opn_{max}} = 0.7$ mm).

Insights gained from this research are crucial for optimizing the performance and longevity of the piezohydraulic pump, particularly regarding key components such as the reed valves and piston. Future investigations will explore a wider range of values between the maximum and minimum inlet reed valve openings, as well as variations in chamber pressure. The proposed 2D model will also be validated by comparing its main flow characteristics with experimental data taken from pump testing.

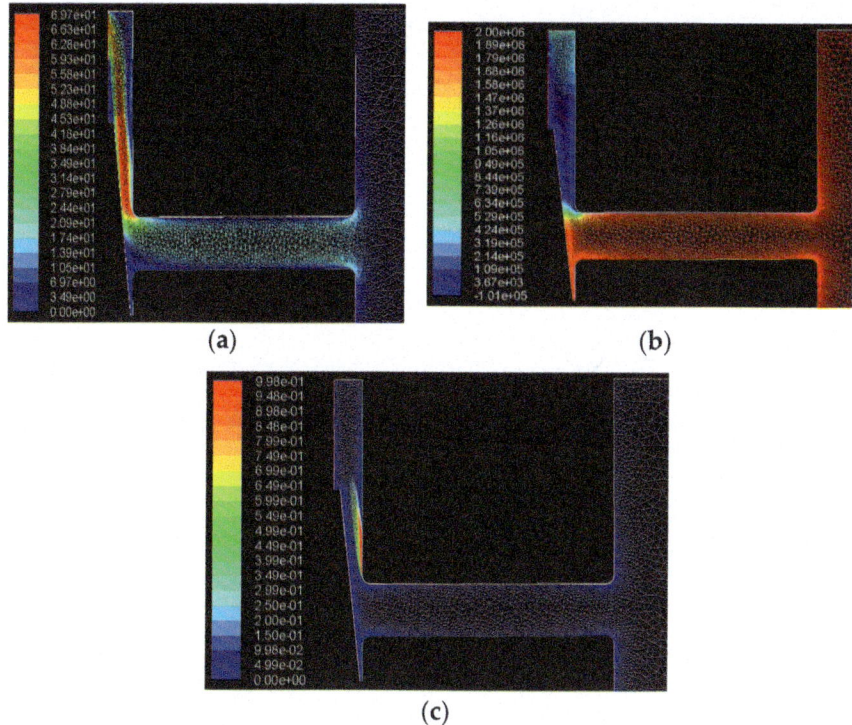

Fig. 7 CFD analysis of the piezopump investigated considering $v_{opn_{max}} = 0.7$ mm; $p_{in} = 20$ bar; $p_{cham} = 0.5$ bar: **a** Velocity contour [m/s]; **b** pressure contour [Pa]; (**c**) Volume fraction of vapor oil phase [–]

References

1. Tamburrano P, Sciatti F, Plummer AR, Distaso E, De Palma P, Amirante R (2021) A review of novel architectures of servovalves driven by piezoelectric actuators. EnergS (Basel) 14(16):4858
2. Tamburrano P, Distaso E, Plummer AR, Sciatti F, De Palma P, Amirante R (2021) Direct drive servovalves actuated by amplified piezo-stacks: Assessment through a detailed numerical analysis. Actuators 10(7). https://doi.org/10.3390/act10070156
3. Tamburrano P, De Palma P, Plummer AR, Distaso E, Sciatti F, Amirante R (2021) Simulation of a high frequency on/off valve actuated by a piezo-ring stack for digital hydraulics. In: E3S Web of Conferences, EDP Sciences, p 05008
4. Tamburrano P, Sciatti F, Distaso E, Amirante R (2023) Comprehensive numerical analysis of a four-way two-position (4/2) high-frequency switching digital hydraulic valve driven by a ring stack actuator. Energies (Basel) 16(21):7355
5. Lindler JE, Anderson EH (2002) Piezoelectric direct drive servovalve. In: Smart structures and materials 2002: industrial and commercial applications of smart structures technologies, SPIE, pp 488–496
6. Zhou M, Gao W, Yang Z, Tian Y (2012) High precise fuzzy control for piezoelectric direct drive electro-hydraulic servo valve. J Adv Mech Des Syst Manuf 6(7):1154–1167

7. Sangiah DK, Plummer AR, Bowen CR, Guerrier P (2013) A novel piezohydraulic aerospace servovalve. Part 1: design and modelling. Proc Inst Mech Eng Part I J Syst Control Eng 227(4):371–389
8. Sciatti F, Tamburrano P, Distaso E, Amirante R (2024) Digital hydraulic valves: Advancements in research, Heliyon
9. Sciatti F, Tamburrano P, Distaso E, Amirante R (2023) Digital hydraulic technology: applications, challenges, and future direction. In: J Phys: Conf Ser, IOP Publishing, 2023, p 012053
10. Park JH, Yoshida K, Yokota S, Seto T, Takagi K (2003) Development of micro machines using improved resonantly-driven piezoelectric micropumps. In: Proceedings of the 4th international symposium on fluid power transmission and control, Wuhan, China, p 536541
11. Nayak S, Muralidhara R (2023) Experimental performance evaluation on piezo-hydraulic pump using flexurally amplified piezoelectric actuators. Mater Today Proc
12. Hu R, He L, Hu D, Hou Y, Cheng G (2023) Recent studies on the application of piezoelectric pump in different fields. Microsyst Technol, 1–20
13. Ma HK, Chen RH, Yu NS, Hsu YH (2016) A miniature circular pump with a piezoelectric bimorph and a disposable chamber for biomedical applications. Sens Actuators A Phys 251:108–118
14. Peng T, Yang Z, Kan J (2009) Investigation on piezo-pump used in water-cooling-system for computer CPU Chip. J Refrig 30(3):30–34
15. Gidde RR, Pawar PM, Dhamgaye VP (2020) Fully coupled modeling and design of a piezoelectric actuation based valveless micropump for drug delivery application. Microsyst Technol 26(2):633–645
16. Wu X, He L, Hou Y, Tian X, Zhao X (2021) Advances in passive check valve piezoelectric pumps. Sens Actuators A Phys 323:112647
17. Huang W et al (2021) Research on a piezoelectric pump with flexible valves. Appl Sci 11(7):2909
18. Spencer WJ, Corbett WT, Dominguez LR, Shafer BD (1978) An electronically controlled piezoelectric insulin pump and valves. IEEE Trans Sonics Ultrason 25(3):153–156
19. Kan J, Tang K, Ren Y, Zhu G, Li P (2009) Study on a piezohydraulic pump for linear actuators. Sens Actuators A Phys 149(2):331–339
20. Sell N et al (2023) Design and testing of a high power piezo pump for hydraulic actuation. J Intell Mater Syst Struct
21. O'Neill C, Burchfield J (2007) 'Kinetic ceramics piezoelectric hydraulic pumps', in industrial and commercial applications of smart structures technologies. SPIE 2007:142–155
22. Sell N, Plummer A, Johnston N, du Bois J (2021) Simulating a high frequency piezo pump with disc reed valves. In Scandinavian international conference on fluid power, pp 274–282
23. Yong YK (2016) Preloading piezoelectric stack actuators in high-speed nanopositioning systems. Front Mech Eng 2:8
24. Fouaidy M, Saki M, Hammoudi N, Simonet L (2007) Electromechanical characterization of piezoelectric actuators subjected to a variable pre-loading force at cryogenic temperature, IPN
25. Claeyssen F, Le Letty R, Barillot F, Sosnicki O (2007) Amplified piezoelectric actuators: static & dynamic applications. Ferroelectr 351(1):3–14
26. Zhao X, Zhao D, Guo Q (2020) A theoretical and experimental study of a piezoelectric pump with two elastic chambers. Sensors 20(20):5867
27. Muralidhara SN, Rao R (2016) Design and simulation of high pressure piezohydraulic pump with active valves. In: 2016 International conference on electrical, electronics, and optimization techniques (ICEEOT), IEEE, pp 1608–1613
28. Lee DG, Or SW, Carman GP (2004) Design of a piezoelectric-hydraulic pump with active valves. J Intell Mater Syst Struct 15(2):107–115
29. Woo J, Sohn DK, Ko HS (2019) Experimental study on pressure pulsation in piezo driven reed valve pump. J Mech Sci Technol 33:661–667
30. Huang J, Zhu Y-C, Shi W-D, Zhang J-H (2018) Theory and experimental verification on cymbal-shaped slotted valve piezoelectric pump. Chin J Mech Eng 31(1):1–8

31. Peng T et al (2019) A high-flow, self-filling piezoelectric pump driven by hybrid connected multiple chambers with umbrella-shaped valves. Sens Actuators B Chem 301:126961
32. Zhang J, Xia Q, Lai D, Akiyoshi O, Hong Z (2004) Discovery and analysis on cavitation in piezoelectric pumps. Chin J Mech Eng (English Edition) 17(4):591–594
33. Opitz K, Schade O, Schlücker E (2011) Cavitation in reciprocating positive displacement pumps. In: Proceedings of the 27th international pump users symposium, Turbomachinery Laboratory, Texas A&M University
34. Ye Y, Chen J, Pan QS, Feng ZH (2019) Suppressing the generation of cavitation by increasing the number of inlet check valves in piezoelectric pumps. Sens Actuators A Phys 293:56–61
35. He X, Deng X, Yang S, Bi Y, Jiang Q (2009) Numerical analysis of cavitation flow in vortex-valve piezoelectric micropump. Drain. Irrig. Mach 27:352–356
36. Accessed on Feb 2024. https://www.piceramic.com/en/products/piezoceramic-actuators/stack-actuators/p-010xxh-p-025xxh-pica-thru-ring-actuators-102800
37. Vanhille C (2020) Numerical simulations of stable cavitation bubble generation and primary Bjerknes forces in a three-dimensional nonlinear phased array focused ultrasound field. Ultrason Sonochem 63:104972
38. Som S (2009) Development and validation of spray models for investigating diesel engine combustion and emissions. University of Illinois at Chicago
39. Yusvika M, Prabowo AR, Tjahjana DDDP, Sohn JM (2020) Cavitation prediction of ship propeller based on temperature and fluid properties of water. J Mar Sci Eng 8(6):465
40. Nezamirad M, Yazdi A, Amirahmadian S, Sabetpour N, Hamedi A (2022) Utilization of Schnerr-Sauer cavitation model for simulation of cavitation inception and super cavitation. Int J Aerosp Mech Eng 16(3):31–35
41. Davidson L (2015) Fluid mechanics, turbulent flow and turbulence modeling. Citeseer
42. Pal E, Kumar I, Joshi JB, Maheshwari NK (2016) CFD simulations of shell-side flow in a shell-and-tube type heat exchanger with and without baffles. Chem Eng Sci 143:314–340
43. Xiao J et al (2022) Assessment of different CFD modeling and solving approaches for a supersonic steam ejector simulation. Atmosphere (Basel) 13(1):144

Open Access This chapter is licensed under the terms of the Creative Commons Attribution 4.0 International License (http://creativecommons.org/licenses/by/4.0/), which permits use, sharing, adaptation, distribution and reproduction in any medium or format, as long as you give appropriate credit to the original author(s) and the source, provide a link to the Creative Commons license and indicate if changes were made.

The images or other third party material in this chapter are included in the chapter's Creative Commons license, unless indicated otherwise in a credit line to the material. If material is not included in the chapter's Creative Commons license and your intended use is not permitted by statutory regulation or exceeds the permitted use, you will need to obtain permission directly from the copyright holder.

Power Analysis of an ePump Applied to the Linear Functions of an Agricultural Planter

Jacob Lengacher, Ryan Jenkins, Peng Li, Michael Conboy, and Andrea Vacca

1 Introduction

In recent years, an increased social and regulatory interest in reducing carbon emissions has driven all sectors to seek methods of cutting down on or eliminating fuel consumption [1]. The agricultural industry is one such field, accounting for 25% of the U.S.'s carbon emissions as reported in [2]. Within this field, in addition to propulsion, tasks such as tilling fields, lifting loads, planting crops, and baling harvested crops are carried out. Hydraulic technology is favored for these solutions due to its robustness, contamination tolerance, and high power density. However, the efficiency of many of these systems can be quite low, with a report by the Oak Ridge National Laboratory placing the efficiency of the average agricultural hydraulic system at just 20% [3]. As a result, methods of reducing the emissions of such systems are of great importance.

One answer to this issue is electrification which allows abatement of local emission by replacing the ICEs with electric motors and a battery system. In 2014, Vauhkonen demonstrated a direct replacement, replacing the engine and fuel tank in a miniature excavator with a battery and electric motor, succeeding in achieving the same dynamic performance, but at the cost of a 75% reduction in run-time before needing to recharge [4]. This highlights the known run-time limitations of EVs in off-road vehicles. A common next step is to seek for increase in the in-vehicle energy transmission efficiency by replacing the hydraulic architecture with a purely electric one. However, this presents challenges particularly pertaining to linear functions. High-power linear electric actuation is usually achieved using a screw/nut mechanism, such as a ball

---

J. Lengacher (✉) · A. Vacca
Purdue University, Lafayette, USA
e-mail: jlengac@purdue.edu

R. Jenkins · P. Li · M. Conboy
CNH Industrial, Chicago, IL, USA

or roller screw to convert rotary motion into linear motion. Such electro-mechanical actuators (EMA) work well for applications with controlled environments and have seen effective application to industrial machinery, as well as in aerospace, where the rigidity of such actuation is desirable [5]. They are less well suited to mobile applications, however, as there is no in-built capacitance or damping; every shock is transmitted directly through the actuator. In spite of this, a number of EMA-only machines have been developed in recent years. In 2017, Volvo presented their first all-electric mini-excavator [6]. In 2023 Bobcat unveiled an all-electric compact loader, as well [7]. Notably, both of these solutions target low-power, speed-controlled applications.

This is not the only option for electrified mobile machinery, however. Joint electric-hydraulic architectures offer many benefits over both EMAs (Capacitance, lower component costs, smaller installed power) and traditional hydraulics (Linear control, load recovery, throttle-less actuation). One of the most popular forms of this is Electro-hydrostatic Actuation (EHA), a form of primary control where pump speed directly controls actuator speed. It was first discussed for aerospace applications, where its simplicity and decentralization made it desirable, as shown by Habibi [8]. Now, its efficiency has led to its expansion into the mobile machinery market. Qu implemented such a system on a miniature skid steer in a distributed architecture, enabling significant power savings and recovery of energy [9, 10]. Minav specifically focused on the load recovery potential of such a layout in [9], applying it to a forklift system. In [10], Pietrzyk proposed an integrated EHA arrangement for a mini-excavator, allowing for only electric lines to be run across the machine, while reducing system losses.

There are many other ways in which these technologies can be paired to obtain better results than either separately. Hao uses an EMA in parallel with a traditional hydraulic cylinder and accumulator to reduce the required power (and thus size) of the EMA [11], while Li uses small electric machines to provide throttle-less control of actuators [14, 15]. On the component side, ePumps, machines consisting of a coupled electric machine and hydraulic machine are being investigated. Assaf proposes a method of optimally sizing such components based on the drive cycle of a machine [12], while Zapaterra proposes a system that nests a small hydraulic machine inside the rotor of a Permanent Magnet Synchronous Machine (PMSM), allowing for an extremely compact installation size [13].

One particularly challenging application for these technologies is force-controlled actuators. Force-controlled actuators are commonly used in industrial applications such as presses, and in agricultural or construction applications where consistent ground contact is desired. Traditionally, force control is attained by setting the pressure in the chambers of a hydraulic cylinder [14]. There are numerous ways this has been implemented efficiently in purely hydraulic systems. Lumkes proposed the multi-pressure rail system as a method for efficient pressure control of actuators, utilizing pressure rails at a variety of levels, with the ability to switch the connections of actuator ports to optimize force output, [15] a concept that was expanded on by Vukovic in [16] and applied to an excavator. This work decoupled the actuator speeds and the engine speed, allowing for engine management and energy recovery.

Guo presents a further development of this idea, proposing a system with dynamically varied pressure rails and applying it to an agricultural system [17]. Expanding further on pressure-supplied systems, Achten proposed the hydraulic transformer system, which uses paired hydraulic machines to achieve throttle-less actuation and load recovery [18]. These works are all built within the framework of ICE-supplied hydraulic systems to reduce power consumption and improve efficiency. This work aims to expand these concepts and apply them to joint electric-hydraulic systems, by proposing a method for efficiently supplying an array of force-controlled actuators using an ePump.

2 Reference Case

This work takes as a reference case the force-controlled linear actuators of an agricultural planter system. These actuators are part of the Down Force (DF) control system and keep the individual row units of the planter in contact with the ground, ensuring a consistent planting depth, and good germination rate. The specific reference machine being considered is a Case IH 2150 early rise planter (a simplified schematic of which is shown in Fig. 1), which uses 18 such cylinders-2 on the wings of the planter, and one on each of the 16 row units. Bidirectional force control is achieved using reducing/relieving valves on both chambers of each cylinder. This strategy allows for a small, modular installed package.

The tight installation space of these cylinders presents a problem, however, as the force required by the row units leads to a much higher pressure being required by the DF cylinders than by the other functions at steady state. Resizing the other functions to accept a lower pressure (by reducing motor volume) would result in unacceptably long startup times, due to the high torque required to start the vacuum and bulk fans spinning. This leads to a load imbalance, and significant throttling loss when the

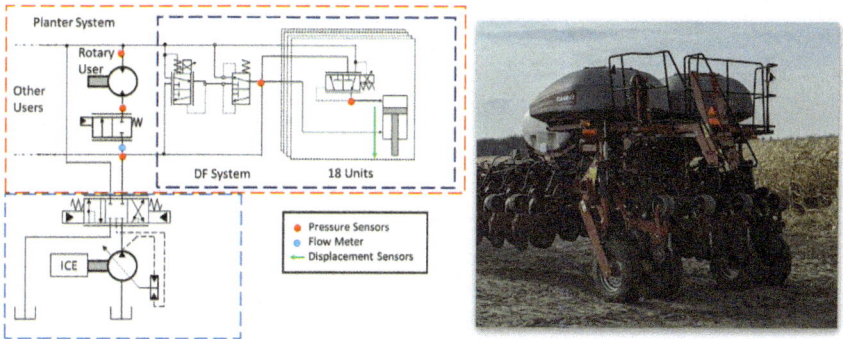

Fig. 1 Sensor placements on reference machine schematic

Fig. 2 Load imbalance problem

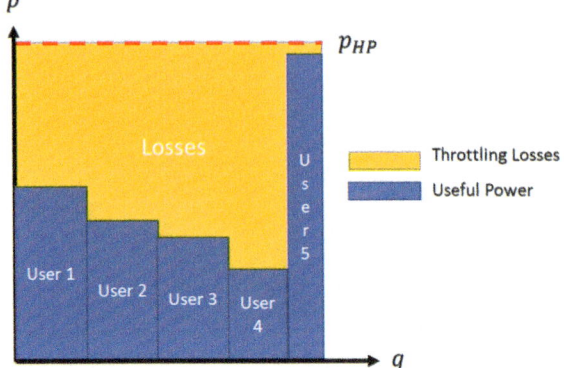

planter is supplied by a traditional LS architecture. This issue is shown qualitatively in Fig. 2.

Removing these functions from the main system and supplying them separately could greatly improve the efficiency of the system by allowing the main system supply to operate at a more efficient point. This work will investigate the possibility of doing this electrically. The duty cycle of this application is fast, as the cylinders ride over the ground, thus maintaining the capacitance of the baseline hydraulic system is crucial. As a result, this system is a good candidate for electric-hydraulic actuation.

3 Baseline System Characterization

To better understand the performance of the system, the baseline reference system was instrumented as shown in Fig. 1. This allowed the total power into and out of the system to be measured, along with the positions of and pressures in several of the planter's cylinders. Data was collected for a range of operating points realistic to those encountered in the field. This information will be used later to validate the modeling method used. These tests were taken replicating the behavior of the baseline system, as shown in Fig. 3 below.

An observation relevant to the design of the system can be made from these experimental results, specifically from the force requirements of the DF cylinders. The peak and average force requirements of the instrumented cylinders, are shown below, normalized for confidentiality and averaged over several tests (Table 1).

As can be seen, the max force requirement is much higher than the average requirement. The downforce cylinders have a short stroke length, and a limited speed largely decouple from net force, leading to their long-term power requirements being dominated by the force requirement. This, in turn, leads to a wide difference between peak and average power for each cylinder. This is not an issue in a system with a centralized prime mover, such as the baseline, but it presents an issue for decentralized

Fig. 3 Field testing procedure

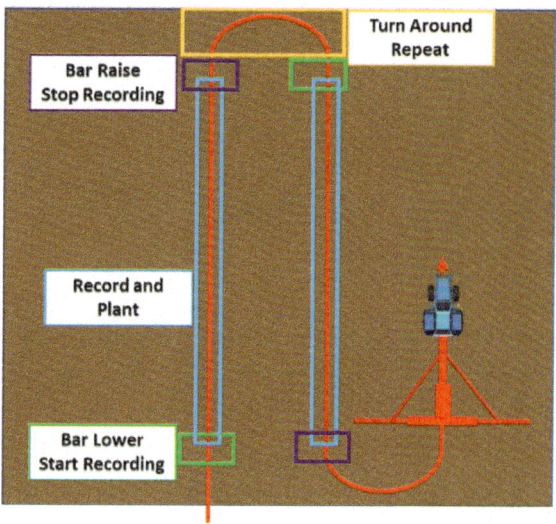

Table 1 Test array and peak and mean force for DF cylinders

Planter settings	Vehicle speed (MPH)	F_mean (%)	F_max (%)
Low speed	6	20	62
High speed	6	63	96

systems such as ones based on EMAs and EHAs. In traditional centralized systems, the prime mover is sized to handle the requirements of the total system; for this case, when one cylinder requires high power, another needs lower power. For decentralized systems, installed power (and thus cost and size) will be higher than for a centralized system as each individual cylinder's motor must be able to supply peak power.

4 Proposed System

4.1 System Working Principle

It is desirable to design a solution that separates the DF cylinders from the rest of the system to reduce throttling losses while maintaining the capacitance provided by hydraulic actuation. The system should use a centralized prime mover to keep the installed power as close as possible to the average required power of the full array of cylinders. The solution proposed here builds on a purely hydraulic architecture proposed by the authors in [19] and aims to further improve the system efficiency by utilizing the strengths of combined electric-hydraulic actuation.

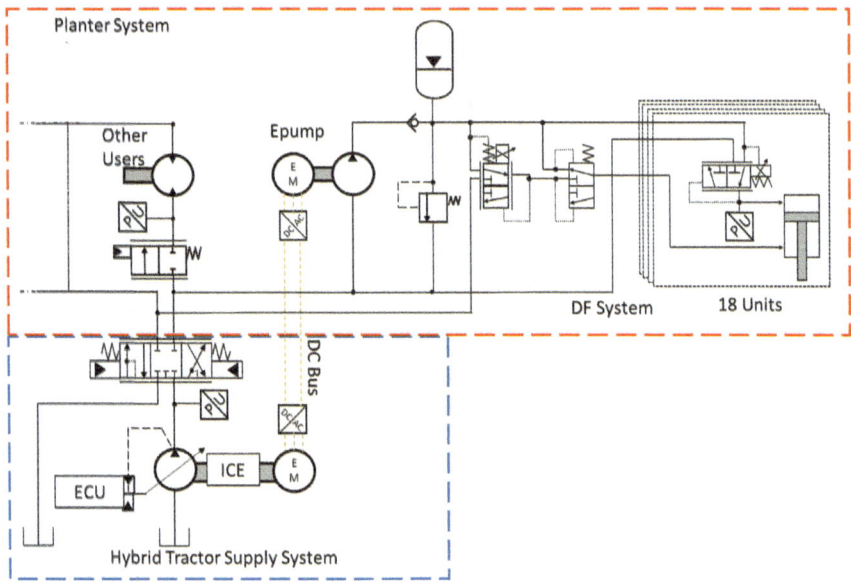

Fig. 4 Simplified schematic of the proposed solution

Figure 4 shows the ISO schematic of the proposed architecture. It uses a pressure supply logic to supply the DF cylinder array. This serves to decouple the flow supplied to the rail from the highly variable flow requirements of the individual cylinders and is achieved using a shared supply rail stabilized by an accumulator. This rail is supplied by a variable speed, fixed displacement ePump, whose speed is varied to regulate the rail pressure, rather than to directly supply the flow needs of the actuators. A gear-pump is used due to its high operating speed and tolerance for a pressurized inlet, paired with a PMSM for its efficient operation and wide operating range enabling good compatibility with the pump. The individual cylinders' pressure commands are then supplied using pressure-reducing valves, as in the baseline system.

4.2 Control Strategies

Two control methods for the ePump are considered. The first prioritizes pump and motor efficiency. Both the gear machine and electric machine will, in general, be more efficient at higher speeds, so it makes sense to run the ePump at this optimal point. The first control logic keeps the pressure between an upper and a lower bound, with the lower bound being the minimum pressure required for the cylinders to function, and the upper bound being the maximum operating pressure of the system, much as presented by the authors for a purely hydraulic system in [19]. As the cylinders operate flow is drained from the accumulator causing the rail pressure to drop. When

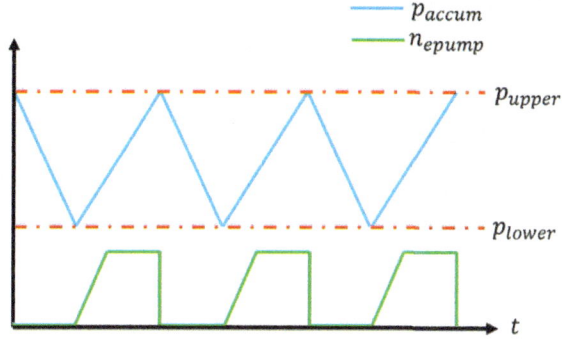

Fig. 5 Behavior of On/Off control strategy

it drops to the lower bound, the ePump is turned on to its ideal speed (near the maximum speed of the gear machine). The ePump supplies flow to the accumulator, causing the pressure to rise, until it eventually reaches the upper bound. At this point, the ePump shuts back off. This control logic has two main advantages: simplicity and ensuring that operation is always at a high-efficiency operating point. This will be referred to as the On/Off control strategy going forward and is demonstrated visually in Fig. 5.

There are also several disadvantages to this strategy. The first is a slight inefficiency due to the pressure of the accumulator raising during charging. This leads to some energy being lost to throttling in the individual row unit regulators, as more than the required pressure is provided. Additionally, it requires frequent start–stop events, which can be detrimental from an endurance stand point.

An alternative Continuous Control Strategy is also proposed, which holds the rail pressure constant, using the expected average flow demand of the DF cylinder array as a feedforward command, then taking the pressure in the accumulator as a feedback term. Figure 6 below shows its block diagram form, as well as its expected behavior.

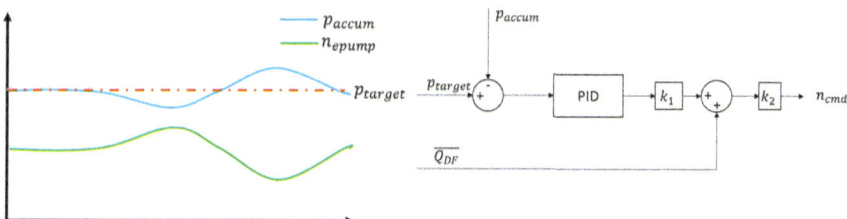

Fig. 6 Behavior of continuous control strategy

4.3 System Sizing

For both control strategies, proper sizing is crucial to good performance, and the method of sizing differs slightly between them. The components that must be sized are the pump, accumulator, and electric motor. First, for the case of the On/Off control strategy, the inputs to the sizing process are laid out in Table 2 below.

Most of these values are straightforwardly given by the system. t_{charge} and $t_{discharge}$ must be selected based on the reference system. They should be selected to be long enough to avoid too fast an on/off cycle, while being short enough to avoid too large a pump and accumulator. For this case 10 and 20 s, respectively, were chosen. Once these values are selected, the working volume differential of the accumulator can be calculated, as shown below:

$$\Delta V_{accum} = t_{discharge} * \overline{q_{user}} \tag{1}$$

Then, the required size of the pump can be calculated using t_{charge}, $n_{pump,max}$ and $\overline{Q_{user}}$, along with an assumed pump efficiency, as shown below.

$$V_{pump} = \frac{\Delta V_{accum}}{n_{pump,max} * \eta_{v,pump}} \tag{2}$$

The accumulator pre-charge pressure is set at 70% of the minimum operating pressure, allowing the accumulator nominal volume to be selected, where k is the polytropic constant:

$$V_0 = \frac{\Delta V_{accum}}{\left(\frac{P_{precharge}}{P_{min}}\right)^{\frac{1}{k}} - \left(\frac{P_{precharge}}{P_{max}}\right)^{\frac{1}{k}}} \tag{3}$$

Then, the maximum operating pressure of the system can be used to calculate the torque requirements for the electric motor.

$$T_{motor} = (p_{max} - p_{supply}) * \frac{V_{pump}}{2\pi \eta_{hm,pump}} \tag{4}$$

Finally, then, the motor must be rated to run at at-least the max speed of the pump.

Table 2 Inputs to On/Off control architecture sizing process

Quantity	Source
$t_{charge,discharge}$	Design choice based on use case
p_{supply}	Based on typical operating pressure of main system
$n_{pump,max}$	Based on pump type selection
$p_{max,min}$	Based on requirements of force-controlled actuator
$\overline{q_{user}}$	Determined experimentally or analytically from user

$$n_{\text{motor,rated}} \geq n_{\text{pump,max}} \tag{5}$$

For the continuous sizing procedure, similar inputs are used, minus the charge and discharge time. The average flow requirements of the cylinders and a working speed selected as a fraction of the maximum pump speed are used to calculate the pump size required. For this case, 1/3 is used, as the flow to the user roughly triples during the initial transient. Thus, during the initial transient, the ePump will run at close to max speed, while, during steady state, it will run at roughly 1/3rd of max speed.

$$V_{\text{pump}} = \frac{\overline{q_{\text{user}}}}{\frac{1}{3} n_{\text{pump,max}} \eta_{v,\text{pump}}} \tag{6}$$

A single pressure target is selected as the average of the maximum and minimum pressures.

$$p_{\text{set}} = \frac{p_{\text{max}} - p_{\text{min}}}{2} \tag{7}$$

The accumulator pre-charge pressure is set to 70% of the set pressure as above. The accumulator must then be sized as a pulsation dampener, using the equivalent capacitance, as shown below:

$$\frac{dp_{\text{ref}}}{dt} = \frac{1}{C_{h,\text{acc}}} \sum q_{\text{ref}} \tag{8}$$

$$C_{h,\text{acc}} = \frac{V_0}{k}\left(\frac{p_0}{p^{1+k}}\right)^{\frac{1}{k}} \tag{9}$$

$\frac{dp_{\text{ref}}}{dt}$ and q_{ref} are the desired flow and pressure response of the accumulator (i.e., for a disturbance of $\pm q_{\text{ref}}$, there is a corresponding disturbance of $\frac{dp_{\text{ref}}}{dt}$). These values can be determined from the behavior of the system and its tolerance to pressure variation. After this, the motor can be sized as for the On/Off system.

This procedure leads to the normalized system sizings shown in Table 3. The pumps for the two cases, after rounding to the closest available manufacturer component, are roughly the same size. The accumulator for the continuous case can be much smaller, however.

Table 3 Relative system sizings

Quantity	On/Off (%)	Continuous (%)
V_{Pump}	100	100
V_{accum}	100	20

5 Simulation Model

To validate the effectiveness of the proposed systems simulation models of both systems and the baseline were assembled. For the hydraulic components, a lumped parameter method following that outlined in [14] was utilized. For the electric components a strategy combining empirical component efficiencies and simple electric governing equations was used. Resistive line losses were modeled simply, using ohms law:

$$I = \frac{\Delta V}{R} \tag{10}$$

Power electronic losses were modeled using a simple efficiency, and voltage conversion ratios:

$$V_{out} = X_{conv} V_{in} \tag{11}$$

$$V_{out} * i_{out} = i_{in} * V_{in} * \eta \tag{12}$$

where η was selected as a conservative, typical efficiency value for such components, and X_{conv} is the voltage conversion ratio.

Electric motor torque was simulated using a first-order dynamic.

$$T_{mot} = \frac{1}{1+\tau s} \min(T_{command}, T_{lim}) \tag{13}$$

where T_{lim} is the maximum torque of the motor. Conservation of energy could then be used to calculate the current draw of the motor

$$i_{mot} = \frac{(T_{mot} * n_{mot})}{V_{sup} * \eta_{mot}} \tag{14}$$

where η is the motor efficiency, pulled from a manufacturer-provided empirical map as shown:

$$\eta_{mot} = f(T_{mot}, n_{mot}) \tag{15}$$

where necessary, algebraic loops were broken by assuming a small node capacitance to ground. This was demonstrated to have no impact on system behavior, but greatly accelerated the simulation solver. Actuator load models and controllers were assembled using the experimental data gathered to ensure that solution systems were truly matching the performance of the baseline. All models were implemented in Siemens Amesim due to its smooth multi-physics integration, and batch run capacity.

6 Baseline Model Validation

To ensure validity of the modeling methods used, the baseline model was validated against the experimental results discussed above. Figure 7 (left) demonstrates the ability of the model to predict the chamber pressures for the measured cylinders (only one of three simulated is shown), as well as the total flow consumption of the DF array.

The primary reason for the deviation in rod side pressure and flow is the method of scaling the system. To reduce computational cost, only 3 of 16 cylinders were simulated fully, and system flow was scaled to match. As a result, flow variation is coupled more closely with those three cylinders than in the true system, leading to higher peaks, but correct average values. Simulating all 16 cylinders would lead to very long simulation times, but likely smoother flow curves. This effect is passed through to the rod pressure, which is influenced by the flow through the pressure regulation valve. The primary goal of this simulation model is comparison of power analysis over the steady state, however, thus these results are acceptable as the variation in the bore pressure (the most unsteady portion) aligns well, and the mean values of the more steady flow and rod pressure match acceptably. Similarly, Fig. 7 (right) demonstrates that the steady state pressures, flows, and power consumption of the other functions track well with the steady state values measured in the baseline tests. One function's validation results are shown, with the others being similarly validated. As can be seen, the simulated steady-state operating points, shown by the bold lines, align well with the relevant steady-state regions in the measured time series data, shown by the thinner line.

7 Results

With the validity of the baseline model established, results can be drawn. First, Fig. 8 above shows the time series behavior of the On/Off ePump system. As can be seen, the system charges in around 15 s, and discharges in around 20. When the accumulator pressure drops to its lower pressure limit, $0.85 p_{max}$ the ePump spins up to its optimal speed providing flow and charging the accumulator. The first charging region is prolonged due to the higher flow requirements of the cylinders in that region. At this point, the pump stabilizes at the target speed, while for the other, later charging periods, the cycle happens fast enough that steady state speed is not reached. When the upper pressure limit is hit, the ePump spins back down to stationary. While this is happening, the main supply pressure is reduced to roughly $0.5 p_{max}$, being set by the vacuum, which is much closer in pressure requirement to the other users, or the compressor, when it intermittently runs (As seen from 30 to 35 s). This arrangement greatly reduces the throttle losses in the system.

Similarly, Fig. 9 shows the time series behavior of the continuous control architecture. Unlike the on/off architecture, the accumulator pressure and ePump speed do

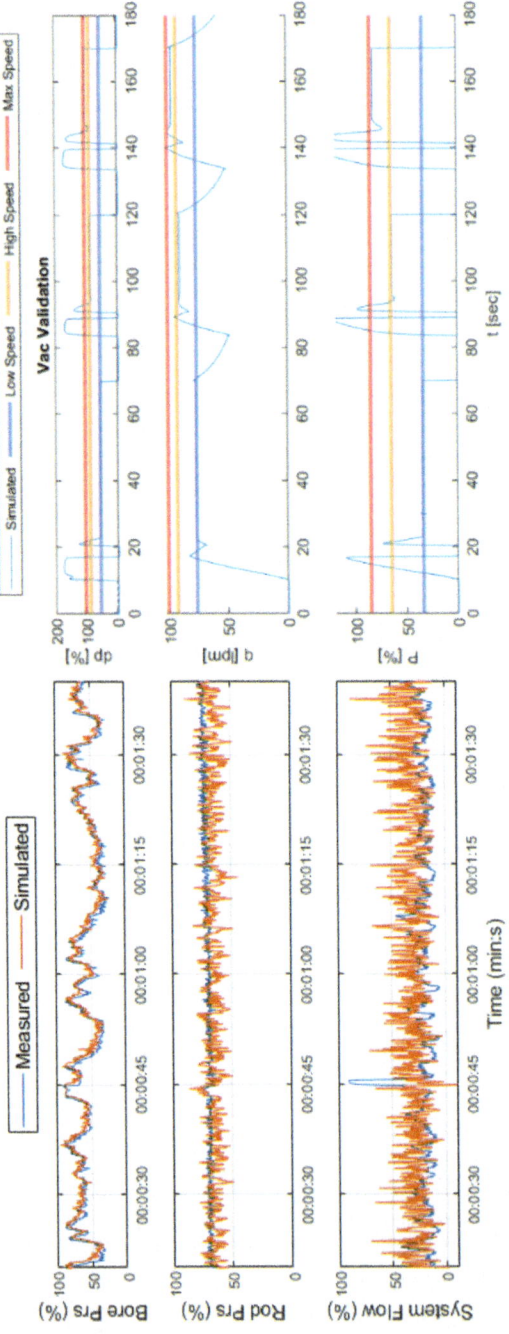

Fig. 7 Baseline versus simulated cylinder pressure and flow behavior (left) and validation of vacuum function behavior (right)

Power Analysis of an ePump Applied to the Linear Functions ...   317

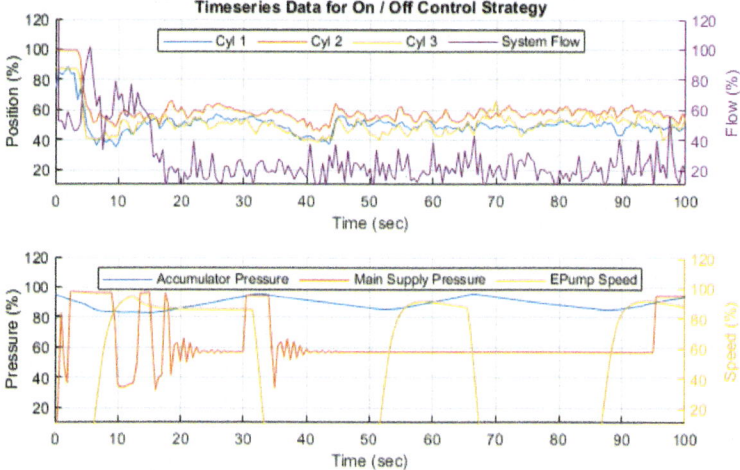

Fig. 8 Time-series behavior of On/Off ePump system

not vary much once steady state is reached. As before, the supply pressure is reduced to match the next highest load user, greatly reducing throttling losses. As expected, during the initial transient, when flow demand is high, the ePump runs close to its maximum speed, before stabilizing out to around 1/3rd of this.

This means, however, that at steady state, the ePump operates at a less efficient speed for the pump and the motor. The effects of this can be seen in Table 4 below where the total efficiency of the ePump motor and pump can be seen during their

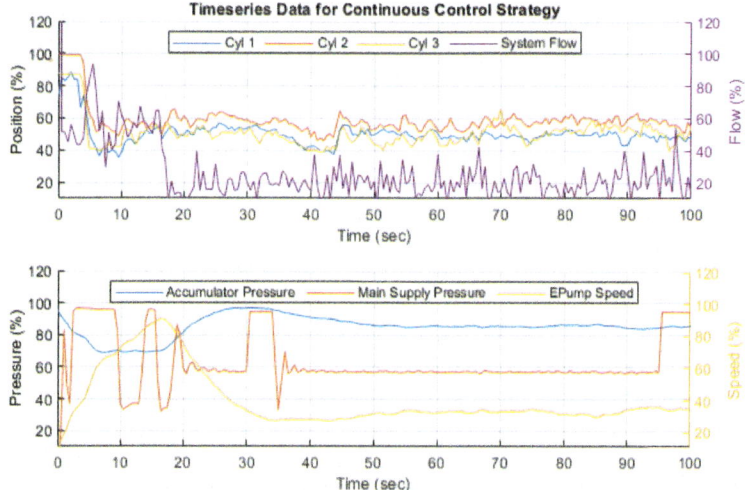

Fig. 9 Time-series data of continuous ePump system

periods of operation (steady state for the continuous architecture, and during the running period for the on/off architecture).

During its operating period, the on / off system operates at a more efficient point than the continuous system. This effect is dominated by the effects of the varying pressure of the accumulator system. As the accumulator stores fluid, its pressure, and thus the power required to charge, increases with the amount of stored fluid. Yet, the cylinders never need more than the lower bound accumulator pressure to operate, so as a result this excess pressure is throttled out by the individual row unit reducing valves. This introduces an inefficiency that outweighs the benefit of operating the ePump at its most efficient point, as can be seen in Table 5, below. The quantities shown are measured at the crossing from the tractor to the planter, to ensure a fair comparison of only planter architectures. Hydraulic power is calculated at the outlet from the tractor remote valves, and electric at the outlet from the tractor DC bus connection.

These results show that both systems display a substantial improvement in efficiency over the baseline system; however, the continuous system performs slightly better than the on/off system, across the board. There are ways to improve the on/off system performance, including decreasing the range between the max and min accumulator pressure. This could be accomplished in two ways: the first is to upsize the accumulator. However, this would lead to a large accumulator, which might be impractical. The second is to decrease the amount of stored flow, increasing the ePump switching frequency. This is also undesirable, as rapid switching might further reduce component life. The sizing simulated above represents a reasonable balance between the 3 considerations of efficiency, accumulator life, and switching speed, but a formal optimization of this trade-off represents a direction for further research.

Table 4 ePump efficiency for both control strategies

Quantity	Motor efficiency (%)	Pump volumetric efficiency (%)	Total (%)
On/Off	82	88	64
Continuous	72	87	56

Table 5 System power consumption and efficiency results

	Reduction in total power over BL (%)		System efficiency (%)			Efficiency improvement over BL	
System	On/Off	Continuous	BL	On/Off	Continuous	On/Off	Continuous
High speed	26	26	23	31	32	34	39
Low speed	40	43	18	30	32	66	77
Average	33	36	20.5	30.5	32	48	56

8 Conclusion

This work set out to demonstrate a method of supplying an array of force-controlled actuators for an agricultural planter which utilized the strengths of both electric and hydraulic actuation to achieve a maximally efficient system. The proposed architecture used a speed-controlled ePump to provide flow to an accumulator, controlling the pressure supplied to the planter DF systems, and freeing up the main hydraulic supply to act at a more efficient point. Two control architectures were then presented, one utilizing an on/off strategy to maximize the efficiency of the ePump, and one utilizing a continuous control strategy to provide a more steady but less efficient operating point to the pump. The baseline planter system was characterized experimentally, and a simulation model of the system was built and validated. The proposed architecture was then applied to the model of the reference system, and tested using both control strategies (with appropriately sized components).

The simulation model demonstrated potential for power savings using both architectures, with the on/off control strategy improving efficiency by 48% of baseline efficiency, while the continuous strategy improved it by 56% of baseline. These results demonstrate that for this application, the continuous architecture is a better option, as the efficiency gains of operating continuously at the required pressure for the system outweigh the efficiency losses of operating at an optimal speed. This is further reinforced by the lack of a need to continuously switch the ePump on and off.

Nomenclatures

Term (Units)	Definition
p bar	Pressure at a given node or component
q lpm	Flow rate through a component
t sec	Time elapsed
n rpm	Rotary speed of a component
V_{pump} cc/rev	Volumetric displacement of a rotary hydraulic machine
V L	Volume of a chamber of component
T Nm	Torque at a shaft
k N/A	Polytropic Constant
C_f N/A	Discharge coefficient of orifice
Ω m^2	Cross-sectional area of an orifice
ρ kg/m^3	Density of oil
C_H m^{4}sec^2/kg	Equivalent Hydraulic capacitance

References

1. Szirniks T (2023) The countries phasing out internal combustion engines, Barrons
2. Kean A, Sawyer R, Harley R (2000) A fuel-based assessment of off-road diesel engine emissions. J Air & Waste Manag Assoc
3. Love L, Alles P (2012) Estimating the impact (Energy, Emissions and Economics) of the US fluid power industry, Oak Ridge National Laboratory, Oak Ride, TN
4. Vauhkonen N, Lilijestrom J, Maharjan D, Mahat C, Sainio P, Kiviluoma P, Kuosmanenm P (2014) Electrification of excavator. In: 9th International DAAAM baltic conference, Tallinn, Estoniaa
5. Qiao G, Liu G, Shi Z, Wang Y, Ma S, Lim T (2018) A review of electromechanical actuators for More/All Electric aircraft systems. In: Proc Inst Mech Eng, Part C: J Mech Eng Sci
6. Allen J (2017) Volvo's EX2 all-electric excavator in action, industrial vehicle technology international
7. Bobcat (2023) T7X & S7X All electric compact track and skid steer loaders. Accessed on Jan 2024. https://www.bobcat.com/na/en/equipment/future-products/t7x-s7x-all-electric-compact-loaders
8. Habibi S (1999) Design of a new high performance electrohydraulic actuator. In: International conference on advanced intelligent mechatronics, Atlanta, USA
9. Qu S, Fassbender D, Vacca A, Busquets E (2021) A high-efficient solution for electro-hydraulic actuators with energy regeneration capability. Energy 216
10. Qu S, Fassbender D, Busquets E, Vacca A (2020) Formulation, design and experimental verification of an open circuit electro-hydraulic actuator. In: Global fluid power symposium
11. Minav T, Immonen P, Laurila L, Vtorov V, Pyrhonen J, Niemela M (2010) Electric energy recovery system for a hydraulic forklift – theoretical and experimental evaluation. IET Electr Power Appl 5(4):377–385
12. Pietrzyk T, Roth D, Schmitz K, Jacobs G (2018) Design study of a high speed power unit for electro hydraulic actuators (EHA) in mobile applications. In: 11th IFK, Aachen
13. Hao YQL, Qiao S, Ge L, Li Z, Zhao B (2023) Energy and operation characteristics of electric excavator with innovative hydraulic-electric dual power drive boom system, IEEE Access, vol 11
14. Li P, Siefert J, Bigelow D (2019) A hybrid hydraulic-electric architecture (hhea) for high power off-road mobile machines. In: Proceedings of the ASME/BATH 2019 symposium on fluid power and motion control, Longboat Key, FL, USA
15. Siefert J, Li P (2021) Optimal control of the energy-saving hybrid hydraulic-electric architecture (HHEA) for Off-highway mobile machines, IEEE transactions on control systems technology
16. Assaf H, Sarode S, Vacca A, Sudhoff S (2023) Electric machine sizing consideration for ePumps in mobile hydraulics. Energy Sci Eng
17. Zappaterra F, Vacca A, Sudhoff S (2022) A compact design for an electric driven hydraulic gear machine capable of multiple quadrant operation. Mech Mach Theory 177
18. Vacca A, Franzoni G (2021) Hydraulic fluid power fundamentals, Applications, and Circuit Designs, Wiley
19. Lumkes J, Andruch J (2011) Hydraulic circuit for reconfigurable and efficient fluid power systems. In: The 12th Scandinavian international conference on fluid power, Tampere, Finland
20. Vukovic M, Leifeld R (2016) Steam–A hydraulic hybrid architecture for excavators. In: 10th International fluid power conference, Dresden, Germany
21. Guo X, Lengacher J, Vacca A (2022) A variable pressure multi-pressure rail system design for agricultural applications. Energies

22. Achten P, Fu Z, Vael G (1997) Transforming future hydraulics: a new design of a hydraulic transformer. In: The 5th scandinavian international conference on fluid power, Linkoping, Sweden
23. Lengacher J, Stump P, Vacca A, Jenkins R, Pintore F, Fiorati S (2023) Application of the hydraulic transformer concept to reduce throttling loss in a multiple-function load sensing system. In: ASME/BATH 2023 Symposium on fluid power and motion control, Sarasota, Florida

Open Access This chapter is licensed under the terms of the Creative Commons Attribution 4.0 International License (http://creativecommons.org/licenses/by/4.0/), which permits use, sharing, adaptation, distribution and reproduction in any medium or format, as long as you give appropriate credit to the original author(s) and the source, provide a link to the Creative Commons license and indicate if changes were made.

The images or other third party material in this chapter are included in the chapter's Creative Commons license, unless indicated otherwise in a credit line to the material. If material is not included in the chapter's Creative Commons license and your intended use is not permitted by statutory regulation or exceeds the permitted use, you will need to obtain permission directly from the copyright holder.

Analysis of Opportunities for Integrated Thermal Management on Battery Powered Mobile Machines

Fabian Lagerstedt, Samuel Kärnell, Marcus Rösth, and Liselott Ericson

1 Introduction

The ongoing electrification of mobile machines faces various challenges regarding system cost, reliability, efficiency, and performance. Battery technology as a solution for energy storage is often considered a key driver behind these challenges, where strict operating constraints define limits and demands of the complete system design. One limiting factor for batteries is operating temperatures, consequently leading to an increased focus on thermal management system designs, that also tend to integrate other components and subsystems with specific cooling and heating demands. Vehicle thermal management system (VTMS) or integrated thermal management system (ITMS) as a holistic approach concept is frequently discussed for passenger cars [4, 7–9]. For mobile machines there are however additional aspects to be addressed such as varying types of operating missions, workloads, as well as thermal impact from systems such as working hydraulics.

F. Lagerstedt (✉)
Huddig AB, Hudiksvall, Sweden
e-mail: fabian.lagerstedt@huddig.se
URL: https://www.huddig.com

F. Lagerstedt · S. Kärnell · L. Ericson
Division of Fluid and Mechatronic Systems, Linköping University, Linköping, Sweden
e-mail: samuel.karnell@liu.se
URL: https://liu.se/en/organisation/liu/iei/flumes

L. Ericson
e-mail: liselott.ericson@liu.se
URL: https://liu.se/en/organisation/liu/iei/flumes

M. Rösth
Hudiksvalls Hydraulikkluster, Hudiksvall, Sweden
e-mail: marcus@rosth.com
URL: https://www.hhk.world

© The Author(s) 2025
L. Ericson and P. Krus (eds.), *Advancements in Fluid Power Technology: Sustainability, Electrification, and Digitalization*, Lecture Notes in Mechanical Engineering,
https://doi.org/10.1007/978-3-031-84505-5_21

This study focuses on an articulated excavator-loader—sometimes called backhoe loader, and in this study exemplified by a Huddig 1370T—the challenges, and possible opportunities with transitioning from an existing series-hybrid to a fully battery electric powertrain from a thermal management perspective. Today's series-hybrid version of the machine is equipped with a 44 kWh battery combined with an internal combustion engine (ICE) and a fuel tank storage corresponding to 1300 kWh of diesel. For a potential fully battery electric powertrain a total capacity of approximately up to 250 kWh is considered reasonable to achieve, with respect to physical space and available battery technology without increasing the size of the excavator-loader. This capacity is also expected to allow for between a half to more than a full working day without charging depending on the work intensity, based on consumption data from the series-hybrid electric version.

Range anxiety is a frequently discussed topic when it comes to battery electric vehicles, which can partly be connected to seasonal effects such as cold batteries during the winter. A study focused on low ambient temperature effects on electric car travel range claims that the reduction impact can be as high as 60% at $-15\,°C$ due to battery and cabin heating demands combined [5]. Travel range reductions can also be seen at high ambient temperatures above $28\,°C$, where a study based on field data reports an increased energy consumption of 2,3 kWh/100 km per additional $5\,°C$—corresponding to an electricity consumption increase of approximately 31% at $38\,°C$ [3]. Mobile machines are typically operated stationary or in lower speeds compared to transport vehicles, which makes them more dependent on cooling devices in lack of forced convection due to vehicle movement. Yang et al. highlight this phenomenon in a simulation study on a fuel cell hybrid car, where a cabin exposed to high ambient temperatures combined with sun radiation results in higher cabin temperatures during low vehicle speeds [10]. The authors explain this as a combination of increased soaked sun radiation and less air flow through the air conditioning (AC) condenser—increasing the thermal load on the cabin itself and decreasing the performance of the AC system.

In other words, temperature effects on energy consumption of electric vehicles are an important topic in the automotive industry, and remains true for electric mobile machines. While passenger vehicles are mainly used for occasional and often short trips such as commuting—with the exception of taxis and buses—mobile machines are built and typically used for continuous full-time operation.

1.1 Thermal Management Trends

To better compete against fossil-fuelled counterparts on the market, a focus on energy efficiency is needed in every energy consuming function of the machine. Thus, operating temperature requirements need to be met with as small negative energy consumption impact as possible.

The current trend within thermal management systems for electric vehicles is not only about managing temperatures, but also about improving the overall vehicle efficiency in order to mitigate challenges such as energy density in batteries and range anxiety. This is achieved by reducing the heat loads on the system, increasing heat exchange efficiencies as well as recycling thermal energies. Integrated system designs allowing redistribution of thermal energy from one component or subsystem to another is a commonly discussed solution [7, 9]. This requires a better understanding of how different cooling and heating demands can vary with operating conditions and applications, in order to create a viable system design. Bennion and Thornton addressed already in 2010 the necessity of developing methods for quantifying thermal loads during different operating cycles, in order to identify possible synergies and integration opportunities between subsystems in hybrid-electric vehicles [1].

Another common trend is the implementation of a heat pump system in order to reduce the energy consumption needed for heating demands, compared to solutions such as resistive heaters. Heat pumps enables heating with a coefficient of performance (COP) beyond 1, unlike resistive heaters [7]. There are a variety of different ways of how heat pumps can be implemented in an electric vehicle system design, which is thoroughly discussed in [4, 7, 9, 11]. However, a challenge such as poor heating performance during cold ambient temperatures is often raised as an issue. This can be solved with recycling of thermal energy from other subsystems which in turn requires an ITMS design [7].

To summarize, system integration is the common denominator in the future development of thermal management systems. State of the art is however mainly focused on passenger cars [8], which leaves a knowledge gap for electric mobile machines.

1.2 Contributions

This paper provides a study of continuous thermal loads on a component and subsystems level, based on measurements and data from an existing series-hybrid excavator-loader. Since performance and capacity even during extreme operating conditions is essential for mobile machines, continuous heating and cooling demands are studied for relevant components and subsystems individually based on realistic worst-case scenarios. In addition, the cooling demand of the working hydraulic system is studied as a potential heat source during cold environment operation. The knowledge from this study can be used to identify how a ITMS for a pure electric excavator-loader can be designed to improve the machine performance and energy efficiency. Furthermore, the study highlights the importance of energy efficiency in working hydraulic systems, in terms of heat dissipation in high ambient temperature scenarios for electric mobile machines.

2 Thermal Load Analysis

In this section, continuous thermal loads of a battery powered excavator-loader are estimated, combined, and analyzed. The studied machine is a multi-purpose tool carrier with a transportation capability reaching speeds up to 50 km/h. The machine is built with an articulated frame structure similar to a wheel loader, and is equipped with a loader unit in the front and an excavating unit in the rear. Enabling a variety of possible use cases also leads to challenges in identifying a typical drive cycle. In this study, different practical high-utility tests are performed to collect data describing some realistic and achievable worst case scenarios for this particular machine. Finally, the results are presented as two different operating modes in cold and hot climate respectively. The data used in the analysis is based on measurements of a diesel-electric series-hybrid version of the machine, which is capable of operating in a fully electric mode. The electric operation mode enables collection of electric power consumption data. Some of the calculations are based on constant efficiency values which in reality are dynamic and dependent on operating points, but which are deemed sufficient for the purpose and the scope of this study.

2.1 Data Collection

Following section is a description of the tests performed in the study, which form the basis for the calculations in the next section. The tests are divided into excavating, transportation, and cabin tests. The basic idea behind is to put each system into a repeatable but realistic high-utility state, and further use recorded data to calculate the mean thermal load from each sub system. The results in this study will be considered as worst case scenarios for this specific machine.

2.1.1 Transportation

To put the electric traction system in a high-utility state a Swedish country road transportation cycle is used, which is performed with a machine in a physical driving test. The cycle begins and ends in Hudiksvall, and covers a route to Delsbo and back. The elevation profile of the transportation cycle is illustrated in Fig. 1. Most of the cycle can be performed at maximum speed request, which in this study is considered a key to find a repeatable and realistic worst case transportation scenario. During the test consumed and regenerated DC current and voltage reported from the inverters are recorded.

Excavating The goal is to determine the thermal load from the working hydraulic system, and the electric motors and inverters powering it. A continuous excavating operation illustrated in Fig. 2 is performed, where gravel is picked up at a low position and dropped at a high position. The cycle is also used in previous work [6], but in

Fig. 1 Elevation profile of the transportation cycle, starting and ending in Hudiksvall at sea level, with the turning point at Delsbo marked at a distance of 35 km and an elevation of 70 meters

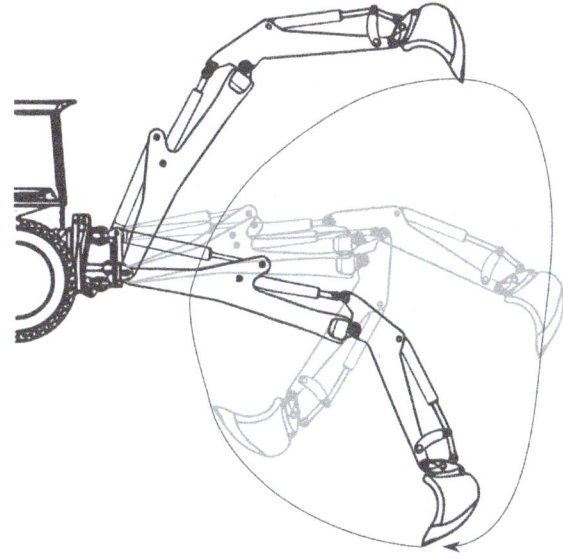

Fig. 2 A 2D illustration of an excavator-loader performing an excavating trajectory, where gravel is picked up at the lowest point and dropped at the highest. The figure originates from [2] and is used with permission from the author

this study the slew function is included as well so that the gravel can be dropped next to the pit.

The first part of the test is a heat balance test, where the working hydraulic system is put into a high-utility state for an extended amount of time by performing the excavating operation. The hydraulic flow as well as the oil temperatures before and after the hydraulic cooler are measured according to Fig. 3. The hydraulic cooling fan is set at maximum speed, and the test is finished when observed temperatures reach a steady state. The thermal load from the hydraulic system is then assumed to be equal to the heat dissipation measured at the hydraulic cooler. The test is performed at an ambient temperature of $-15\,°C$.

The hydraulic system can be powered from either the ICE or the electric motors controlled by a clutch system. Since all the electric motors are cooled with hydraulic oil, the heat balance test is carried out powered by the ICE to simplify isolation of the working hydraulics thermal load. This makes it possible to restrict the hydraulic flow through the motors to avoid that specific passive cooling or heating effect. However, the ICE cooling fan is powered by an individual hydraulic pump and motor circuit,

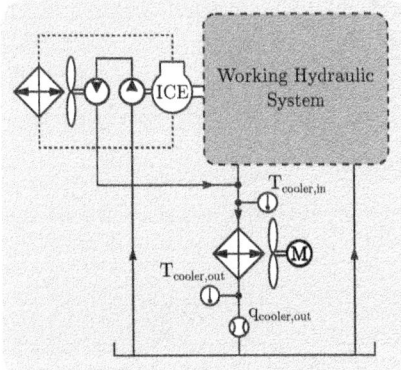 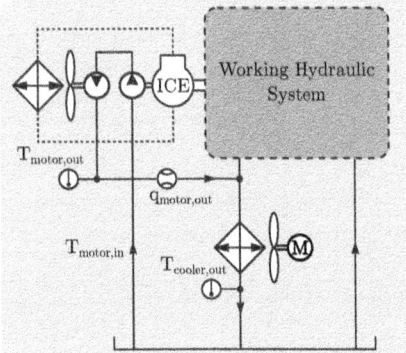

Measure setup for the total hydraulic heat dissipation.

Measure setup for the ICE cooling fan thermal load contribution.

Fig. 3 Two separate diagrams illustrating the measurement setup used to collect data for the working hydraulic thermal load. The left diagram measures the temperature difference and flow across the hydraulic cooler, while the right diagram measures the heat load contribution from the hydraulic motor circuit powering the ICE cooling fan

and contributes with an additional thermal load to the heat balance measurement. To properly isolate the working hydraulic thermal load from the cooling fan circuit contribution, a separate but stationary heat balance tests is performed with only the ICE cooling fan running at full speed. In both tests the ICE is running at the same fixed speed.

The excavating cycle is finally repeated again powered from the electric motors instead of the ICE, at the same fixed speed as the previous heat balance test. This time DC current and voltage reported from the inverters are recorded in order to calculate electric input power.

Cabin Comfort The climate in the cabin is controlled with a heating, ventilation and air conditioning (HVAC) system. Depending on the ambient conditions and input from the driver, this system will heat up or cool down the air inside the cabin. The results from the following tests are used to determine the thermal load on the HVAC system in two different cases—hot climate and cold climate, where maximum possible output power is requested for an extended amount of time. The cabin tests are similar in principle to the heat balance tests performed with the working hydraulic system, but with a determined end point after 30 minutes.

A total of three tests are performed such as a hot climate test, a cold climate test, and an air flow capacity measurement of the HVAC unit. In this study, the hot climate as well as the air flow calculations relies on data from previous tests performed externally, but are described below. The cold climate test, however, aims to replicate the main principles of the hot climate test procedure but in an outdoor environment during the winter.

The hot climate test is a stationary test in a controlled climate chamber with an ambient temperature and humidity set to 38 °C and 50% relative humidity (RH), combined with a simulated solar load consisting of an array of light bulbs corresponding to 950 W/m². The first step of the test is a soaking period of 30 min, where the machine is parked in the chamber with the predefined conditions. The second step is activation of the HVAC system, where lowest possible temperature setpoint and maximum possible blower speed is requested. In this test the recirculation function is activated by closing the fresh air throttle valve, restricting but not preventing fresh air from entering the cabin. Air temperature before and after the HVAC unit are measured according to Fig. 4. The test is finished after 30 min of cooling.

The cold climate test is performed stationary outdoors during the winter, with an ambient temperature of approximately −15 °C. The ICE is used as a heat source for the HVAC system. The test begins with a cold vehicle. The first step is preheating the ICE to operating temperature and adding an external load. In this case a hydraulic circuit is used to create a waste heat margin and a steady ICE operating temperature. The second step is activation of the HVAC system, requesting highest possible temperature setpoint and maximum possible blower speed. In this test the recirculation function is deactivated, allowing more fresh air to enter the cabin without mechanically restricting recirculation of cabin air.

The amount of air flowing into the cabin is measured separately from the cooling and heating tests. The test is performed with full blower speed with opened and closed fresh air throttle valve as well as opened and closed cabin door.

2.2 Data Analysis

Based on the measurement data collected in the previous section, calculations are performed in order to estimate the continuous thermal loads. The data from the excavating test is used to calculate the thermal load contribution both from the working hydraulic system as well as from the power source in the form of electric motor and inverter. The transportation test is used to calculate the thermal load contribution from the traction motors and inverters, and the cabin tests to calculate the cabin comfort thermal load. Lastly, the thermal load from battery cooling is derived from the input power calculations.

Working Hydraulics The total hydraulic heat dissipation $\overline{\dot{Q}}_{tot}$ is calculated according to equation (1) and (2), where the temperatures are the stabilized temperatures in the end of the heat balance test. The hydraulic flow $\overline{q}_{cooler,out}$ is the calculated mean flow of the whole heat balance test. The oil density ρ_{oil} is assumed to be 850 kg/m³, and the specific heat capacity of oil $C_{p,oil}$ 2093 J/kgK. The mean thermal load from the working hydraulic system $\overline{\dot{Q}}_{wh}$ is further isolated from the ICE fan thermal load, and calculated according to equations (3–6).

$$\overline{\dot{Q}}_{tot} = \overline{\dot{m}}_{tot}\, C_{p,oil}\, (T_{cooler,out} - T_{cooler,in}) \qquad (1)$$

$$\dot{m}_{tot} = \overline{q}_{cooler,out}\, \rho_{oil} \tag{2}$$

$$\overline{\dot{Q}}_{fan} = \dot{m}_{fan}\, C_{p,oil}\, (T_{motor,out} - T_{motor,in}) \tag{3}$$

In this case $T_{motor,in}$ is practically difficult to measure and the temperature according to equation (4) is assumed instead.

$$T_{motor,in} \approx T_{cooler,out} \tag{4}$$

$$\dot{m}_{fan} = \overline{q}_{motor,out}\, \rho_{oil} \tag{5}$$

$$\overline{\dot{Q}}_{wh} = |\overline{\dot{Q}}_{tot}| - \overline{\dot{Q}}_{fan} \tag{6}$$

Electric Motor and Inverter Input DC voltage, current as well as motor speed reported by the inverters are recorded during the excavating cycle and the transportation cycle. The total power during each cycle is calculated according to equation (7). The mean heat losses are further calculated according to the equations (8–9).

$$P_{DC} = U_{DC}\, I_{DC} \tag{7}$$

$$\overline{\dot{Q}}_{inverter} = |\overline{P}_{DC}|\, (1 - \eta_{inverter}) \tag{8}$$

$$\overline{\dot{Q}}_{motor} = |\overline{P}_{DC}|\, \eta_{inverter}\, \cos\phi\, (1 - \eta_{motor}) \tag{9}$$

The power factor $\cos\phi$ and the motor efficiency η_{motor} are roughly approximated as a constant from the efficiency map of the motor, by considering the recorded mean motor speed and calculated total mean power $|\overline{P}_{DC}|$ during the full test cycle. The inverter efficiency $\eta_{inverter}$ is approximated as a constant 0,98.

Cabin Comfort The thermal load within the cabin is divided into sensible heat and latent heat. Sensible heat is the amount of energy required to change the temperature of the air, given that no phase change is taking place and is calculated according to the equations (10–11).

$$\dot{Q}_{sens,air} = \dot{m}_{air,tot}\, C_{p,air}\, (T_{hvac,out} - T_{hvac,in}) \tag{10}$$

$$\dot{m}_{air,tot} = q_{air,amb,open}\, \rho_{air} \tag{11}$$

The specific heat capacity of dry air at constant pressure $C_{p,air}$ is in Eq. (10) assumed as 1005 J/kgK. The total air mass flow at full blower speed $\dot{m}_{air,tot}$ in Eq. (11) is approximated by measuring the air flow $q_{air,amb}$ with both the fresh air throttle and the cabin door fully opened simultaneously. It is assumed that the recirculated air flow in this case is negligible. The density of the air ρ_{air} is temperature dependent and approximated at the temperature between $T_{hvac,out}$ and $T_{hvac,in}$.

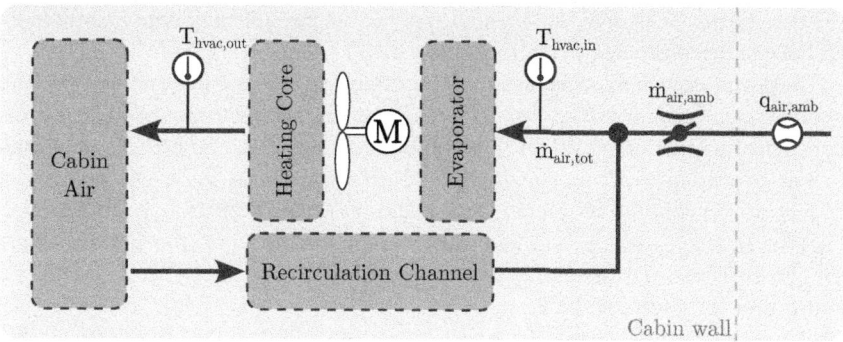

Fig. 4 Flow diagram of the cabin HVAC system, consisting of the evaporator, heating core, recirculation channel, and fresh air throttle valve, illustrating the measurement setup used to collect data for cabin comfort tests. The fresh air flow rate is measured before entering the cabin, and the temperature difference across the HVAC unit is measured after the recirculation channel junction

The latent heat is the energy required or released when a substance is changing its phase, e.g., through dehumidification of moist air. During a cooling process both of these effects happen simultaneously, when the air is cooled below its current dew point. The latent heat is calculated according to the equations (12–13)

$$\dot{Q}_{lat,vapor} = \dot{m}_{air,amb} \left(H_{amb} - H_{hvac,out} \right) h_{vapor} \quad (12)$$

$$\dot{m}_{air,amb} = q_{air,amb,closed}\, \rho_{air} \quad (13)$$

where:

- $\dot{m}_{air,amb}$ is the calculated fresh air mass flow with closed fresh air throttle valve.
- $H_{air,amb}$ is the absolute humidity in the ambient air in kg of water per kg of dry air.
- $H_{hvac,out}$ is the absolute humidity in the air exiting the HVAC system, assuming 100% RH at $T_{hvac,out}$.
- h_{vapor} is the specific enthalpy of water vaporization, in this case $-2257\,\text{kJ/kg}$.

The total heat load during cabin cooling is finally calculated according to equation (14).

$$\dot{Q}_{cabin,cool} = \dot{Q}_{sens,air} + \dot{Q}_{lat,vapor} \quad (14)$$

Cabin heating however does only consist of sensible heat $\dot{Q}_{sens,air}$ since the absolute humidity is assumed to remain constant over the heating core unit and is calculated according to equation (15).

$$\dot{Q}_{cabin,heat} = \dot{Q}_{sens,air} \quad (15)$$

The sensible heat calculation does also apply to the absolute amount of vapor in the air, but can be considered negligible in this case.

The power needed to achieve required cooling performance in an air conditioning system depends on the COP. This coefficient is more commonly used as a performance measure on a systems level, and is derived from the ratio between the actual cooling or heating power and input power. The COP depends on several factors such as ambient conditions, refrigerant types, and compressor efficiency. In this study energy consumers such as fans and pumps are not included, so the COP value can only be derived from the compressor itself. In this case the COP of the electric compressor is assumed to be 2.

$$P_{cab,cool} = \frac{\dot{Q}_{cabin,cool}}{COP} \qquad (16)$$

To heat the cabin, a low efficient solution is through a resistive heater. Without considering any heat exchange losses, the power needed to supply the cabin with sufficient heat is assumed to be equal to the calculated thermal load according to equation (17).

$$P_{cab,heat} = \dot{Q}_{cabin,heat} \qquad (17)$$

Battery The battery thermal load during discharge is calculated according to equation (18), where the total battery power P_{bat} is represented by one of five different cases described by Eqs. (19a–19d). The DC power consumed by the pump motors and the wheel motors $P_{DC,pumps}$ and $P_{DC,wheels}$ are calculated according to equation (7). The battery efficiency $\eta_{battery}$ is assumed as a constant 0,9.

$$\dot{Q}_{bat} = P_{bat}(1 - \eta_{battery}) \qquad (18)$$

$$P_{bat,exc,cold} = P_{cab,heat} + P_{DC,pumps} \qquad (19a)$$
$$P_{bat,exc,hot} = P_{cab,cool} + P_{DC,pumps} \qquad (19b)$$
$$P_{bat,travel,cold} = P_{cab,heat} + P_{DC,wheels} \qquad (19c)$$
$$P_{bat,travel,hot} = P_{cab,cool} + P_{DC,wheels} \qquad (19d)$$

3 Results

The highlighted operating cases in this study are excavating and transportation during both cold and hot ambient conditions. Figure 5a shows the total mean thermal load for each case as well as the distribution between the subsystems. The difference seen between hot and cold conditions is the thermal load from cabin heating compared to

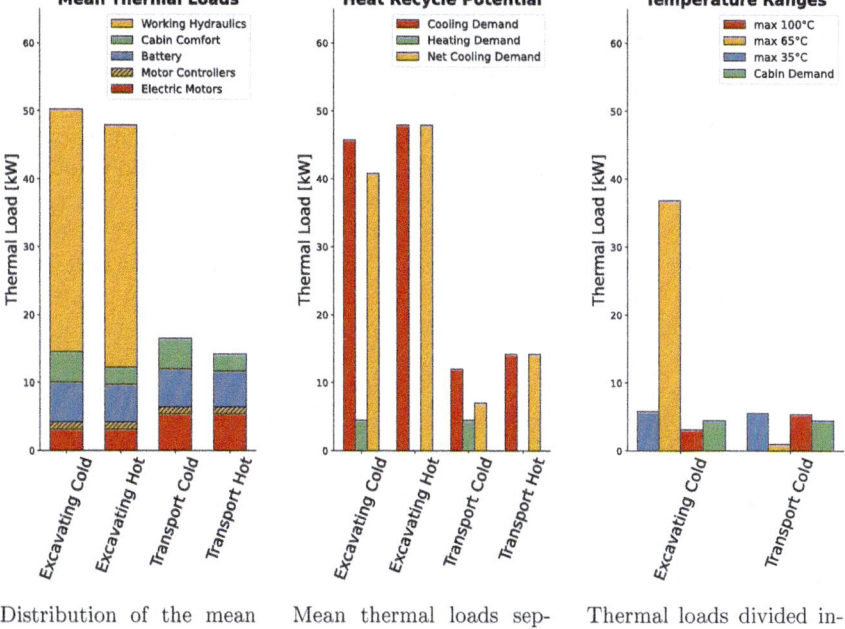

Fig. 5 Resulting mean thermal loads for the studied high-utility cases—excavating and transportation under cold and hot ambient conditions—illustrated from different perspectives: on the left, total thermal load with distribution among each system; in the middle, separated into heating and cooling demand; and on the right, divided by temperature ranges

cabin cooling. The estimated thermal load from the battery differs slightly due to the estimated power consumption difference required for heating and cooling the cabin.

The major difference between excavating and transport is related to the working hydraulic system, which corresponds to 71–75% of the total thermal load during the excavating cycle. The working hydraulics seen in Fig. 5a is the isolated thermal load from Eq. (6). The contribution from the ICE fan is negligible compared to the working hydraulics for the excavating cycle, in this case corresponding to 2% of the total measured hydraulic heat dissipation.

Figure 5b shows the thermal loads separated into cooling and heating demands. It also illustrates the potential of reducing the net cooling demand by recycling thermal energy for cabin heating. In this case by assuming zero heat exchange losses and direct recycling. In the transportation case during cold ambient conditions the cooling demand is potentially reduced by 38%, while the reduction during excavating is 10%.

Figure 5c shows the cooling demands of each subsystem during cold ambient conditions separated into approximate maximum temperatures, in relation to the cabin heating demand. The purpose is to identify possible ways of recycling thermal energies, depending on the specific temperature range of intended heat source. In this case 35 °C is represented by the battery, 65 °C by the working hydraulics together with motor controllers and 100 °C by the electric motors. During transportation it can be seen that the thermal load with a maximum temperature of 100 °C slightly exceeds the cabin heating demand with 0,9 kW. For excavating however, the cabin heating demand exceeds the potential 100 °C thermal load with 1,4 kW. On the other hand, there is instead a thermal load with a potential temperature maximum temperature of 65 °C that well exceeds the cabin demands.

4 Discussion

The idea behind this paper is to perform and present a thermal load overview of an electric excavator-loader, in order to better understand the cooling and heating demands as well as provide input to how a thermal management system can be designed in the future. As previously mentioned, excavator-loaders as many other mobile machines are versatile machines which makes the actual daily usage difficult to fully represent based in only a few single tests. This study can be seen as an example approach on exploring high-utility cases, which is an important consideration from a mobile machine design perspective.

It is quite obvious that the working hydraulic efficiency is a major concern in mobile machine electrification. Not only in terms of useful work per spent kWh of battery capacity, but also as additional load on the thermal management system. Even if some of the working hydraulic thermal energy could be recycled in order to achieve sufficient comfort in the cabin, there is still a substantial amount that potentially needs to be dumped. It should be noted that the working hydraulic thermal load in this study is derived from a heat balance test performed at -15 °C. The hydraulic system in this case benefits from passive cooling to the surroundings through all the exposed surfaces, which is not quantified. This means that the thermal load contribution from the working hydraulic system in reality is higher in the hot ambient case than presented here. It is however uncertain how much this impacts the results. The determination of the ICE fan circuit contribution to the total measured heat dissipation is also slightly underestimated due to the assumption made in Eq. (4). Theoretically this contribution can reach between 3 and 4% instead of 2% with a mechanical efficiency assumption of around 0,9 for the internal gear pump and motor. Regardless, energy efficiency improvements of the working hydraulic system is considered a topic of interest from a thermal management perspective in order to reduce the total heat load.

Regarding recycling of thermal energy for cabin heating, the highest potential impact on the total cooling demand is seen in the transportation case. In the excavating case, it can clearly be seen that the working hydraulic system provides good opportunities for heat recycling. However, it should be mentioned that the HVAC system needs a certain heat exchange temperature difference in order to heat the cabin in a satisfactory way. During the cabin heating test the ICE provides the HVAC heating core with a steady water supply temperature of approximately 80 °C. Therefore, specific temperatures of potential reusable energies are of interest in terms of designing an integrated system that allows heat recycling opportunities from subsystems. The heat can be either recycled directly by routing the coolant from other systems into the HVAC heating core, but also indirectly through the evaporator and condenser of a heat pump system. In both presented cases there is enough heat loads to cover the cabin heating demand in terms of power. Depending on the temperature distribution and the specific thermal loads in each temperature range, one technical solution might be more realistic or effective than the other. However, it is uncertain how far away from a daily use case the presented worst cases in this study are.

5 Conclusions

The working hydraulic efficiency is the largest potential contributor to the thermal heat load of a battery powered excavator-loader, in this study corresponding to 71–75%. Hydraulic system efficiency improvements in combination with heat recycling methods are considered necessary measures in order to maximize the overall machine efficiency. The knowledge from this study enables future sizing of cooling components in order to efficiently mitigate the thermal loads. Furthermore, it gives specific insight into potential sources for heat recycling in an ITMS, where the total cooling demand potentially can be reduced by up to 10% during excavating and 38% during transportation.

Acknowledgements This research was funded by the Swedish Energy Agency (Energimyndigheten) with grant number P2023-00596.

Nomenclature

Designation	Denotation	Unit
$\dot{Q}$	Thermal load	[W]
$\dot{m}$	Mass flow	[kg/s]
C_p	Specific heat capacity	[J/kgK]
T	Temperature	[K]
q	Flow	[m^3/s]
ρ	Density	[kg/m^3]
P	Power	[W]
U	Voltage	[V]
I	Current	[A]
η	Efficiency	[−]
$cos\phi$	Power factor	[−]
H	Absolute humidity	[kg/kg]
h	Specific enthalpy	[J/kg]
COP	Coefficient of performance	[−]

References

1. Bennion K, Thornton M (2010) Integrated vehicle thermal management for advanced vehicle propulsion technologies. In: SAE 2010 world congress and exhibition. SAE International. https://doi.org/10.4271/2010-01-0836
2. Fernlund E (2020) A novel pump-controlled asymmetric cylinder with electric regeneration: implementation and evaluation of a closed hydraulic system on a backhoe. Master's thesis
3. Hao X, Wang H, Lin Z, Ouyang M (2020) Seasonal effects on electric vehicle energy consumption and driving range: a case study on personal, taxi, and ridesharing vehicles. J Clean Prod 249:119–403 (2020). https://doi.org/10.1016/j.jclepro.2019.119403
4. He L, Jing H, Zhang Y, Li P, Gu Z (2023) Review of thermal management system for battery electric vehicle. J Energy Stor 59:106–443. https://doi.org/10.1016/j.est.2022.106443
5. Iora P, Tribioli L (2019) Effect of ambient temperature on electric vehicles' energy consumption and range: model definition and sensitivity analysis based on Nissan leaf data. World Electr Veh J 10(1) (2019). https://doi.org/10.3390/wevj10010002
6. Kärnell S, Fernlund E, Lagerstedt F, Ericson L (2021) Pump-controlled actuators with dump valves. In: Proceedings of the 17th scandinavian international conference on fluid power. SICFP'21. https://doi.org/10.3384/ecp182p150
7. Lei S, Xin S, Liu S (2022) Separate and integrated thermal management solutions for electric vehicles: a review. J Power Sour 550:232,133. https://doi.org/10.1016/j.jpowsour.2022.232133
8. Leoncini G, Mothier R, Michel B, Clausse M (2024) A review on challenges concerning thermal management system design for medium duty electric vehicles. Appl Therm Eng 236:121,464. https://doi.org/10.1016/j.applthermaleng.2023.121464
9. Liang K, Wang M, Gao C, Dong B, Feng C, Zhou X, Liu J (2021) Advances and challenges of integrated thermal management technologies for pure electric vehicles. Sustain Energy Technol Assess 46:101,319. https://doi.org/10.1016/j.seta.2021.101319
10. Yang Q, Zeng T, Zhang C, Zhou W, Xu L, Zhou J, Jiang P, Jiang S (2023) Modeling and simulation of vehicle integrated thermal management system for a fuel cell hybrid vehicle. Energy Convers Manag 278:116,745. https://doi.org/10.1016/j.enconman.2023.116745

11. Zhang Z, Wang J, Feng X, Chang L, Chen Y, Wang X (2018) The solutions to electric vehicle air conditioning systems: a review. Renew Sustain Energy Rev 91:443–463. https://doi.org/10.1016/j.rser.2018.04.005

Open Access This chapter is licensed under the terms of the Creative Commons Attribution 4.0 International License (http://creativecommons.org/licenses/by/4.0/), which permits use, sharing, adaptation, distribution and reproduction in any medium or format, as long as you give appropriate credit to the original author(s) and the source, provide a link to the Creative Commons license and indicate if changes were made.

The images or other third party material in this chapter are included in the chapter's Creative Commons license, unless indicated otherwise in a credit line to the material. If material is not included in the chapter's Creative Commons license and your intended use is not permitted by statutory regulation or exceeds the permitted use, you will need to obtain permission directly from the copyright holder.

A Hydraulic Architecture Based on Multi-common Pressure Rail Principle Using Multi-chamber Cylinders for Excavators

Zihao Xu, Mateus Bertolin, Andrea Vacca, and Jan Nilsson

1 Introduction

The recent trend towards developing sustainable off-road vehicles is pushing towards solutions for the hydraulic actuation system that can increase the overall energy efficiency and reduce fuel consumption. Considering that most of the energy consumption in the off-road vehicle sectors concentrates on agricultural tractors, loaders, and excavators [1], there is high interest towards the development of high-efficient solutions suitable for multi-actuator systems like in excavators. In excavators, the state-of-the-art technology for the actuation system is dominated by centralized systems based on either load sensing solutions or advanced open-center solutions [2]. Research towards more efficient actuation systems for such vehicles has focused on solutions based on higher efficient components, without altering the basic actuation architecture. A significant example is the application of the digital pump concept by Danfoss [3]. However, most of the R&D effort presented in literature pertains to the research of alternative layout architectures. An omni-comprehensive review of promising solutions, some of these also translated to real-market applications, can be found in the review papers [4, 5]. The most energy-efficient hydraulic control architectures reduce throttling losses and allow for energy recuperation during instances of overrunning loads. Among these, there are distributed hydraulic systems (such as displacement control architectures [6]) which increase the number of pumps, as well as centralized architectures based on the concept of common pressure rails, which can retain the cost-effectiveness of current state-of-the-art technology.

Z. Xu (✉) · M. Bertolin · A. Vacca
Maha Fluid Power Research Center, Purdue University, West Lafayette, USA
e-mail: xu1376@purdue.edu

J. Nilsson
Wipro Infrastructure Engineering, Skellefteå, Västerbotten, Sweden

In an MPR system, flow to multiple actuators is provided by several common pressure lines kept at different pressure levels. The flow to each chamber of an actuator is controlled by proportional throttling from the pressure level nearest to the required function pressure, such that the throttling losses can be reduced when multiple actuators need different pressures. Lumkes and Andruch [7] first proposed the idea and used two main supply lines powered by variable displacement pumps and separate meter-in and meter-out valves. By adding accumulators to the pressure rails and integrating optimized engine operations, Vukovic, Sgro, and Murrenhoff [8] improved the architecture and validated the system on a prototype 18-ton excavator [9], which claimed 23.2% reduction in fuel consumption in an air grading cycle comparing to the baseline load sensing (LS) system, showing the outstanding capability of such systems in energy recuperation under overrunning loads frequently seen by construction machines.

The use of multi-chamber cylinder can further increase the flexibility in configuring the connection between pressure rails and cylinder chambers, which uniquely define discrete actuator operation modes. As the actuator has more options to choose as the most efficient mode within certain operating conditions, the throttling losses can be better minimized. Among many studies conducted on the use of multi-chamber cylinders, the most relevant one is the system developed by Heybroek and Sahlman [10], which used four-chamber cylinders with two pressure rails (high and low pressures), which was implemented on a 30-ton excavator and achieved up 50% fuel efficiency improvement. The integration of MPR systems and multi-chamber cylinders showed noticeable potential regarding improved efficiency by introducing flexibility in the system configurations during operation even prior to the above-cited work. Such flexibility also leads to additional design considerations. Huova, Laamanen, and Linjama [11] showed that proper area selection is needed when multi-chamber cylinders were used to increase system efficiency. In the study carried out by Dell'Amico et al. [12] on controlling a four-chamber cylinder with three-pressure rails as an excavator arm, the trade-off between cylinder accuracy, smoothness, and switching frequency was found by testing different control strategies. The number of pressure rails and cylinder chambers also plays a key role in determining the energy-saving potential. The study by Guo et al. [13], in an agricultural application, shows the efficiency gains per added pressure rail significantly decrease after three rails. Bertolin and Vacca [14] also compared the efficiency differences potentials among different possible configurations of cylinder chambers and pressure rails, considering the case of a mid-size excavator performing a digging cycle. Their findings, which also inspired the present work, can be summarized in Table 1. Essentially, a solution of three-pressure rails and three-chamber cylinders seems to be, at least efficiency-wise, more promising than the case of two pressure rails and four chamber cylinders as considered in prior literature [10]. The three-pressure rail system with three-chamber cylinders consumes 21.5% less energy by including one additional chamber and three extra valves per cylinder, alongside implementing a controller evaluating 16 more operation modes.

Table 1 Analysis on energy consumption differences conducted by Bertolin and Vacca [14]

# Pressure rails	# Cylinder chambers	# Valves per cylinder	# Operation modes	Percent difference [%]
2	2	4	4	+57.6
2	3	6	8	+9.9
3	2	6	9	Baseline (STEAM [9])
2	4	8	16	−8
3	3	9	27	−21.5
3	4	12	81	−29.3

With one more pressure rail but one less cylinder chamber, such a configuration does not require significantly more hardware compared to the successful implementation by Heybroek and Sahlman [10]. Furthermore, it exhibits promising efficiency improvement at a reasonable cost of increasing controller complexity.

Driven by the above considerations, this paper aims extensively to analyze the design of a three-pressure rail and three-chamber cylinder system, to overcome the possible limitation of the previous study [14]. As a matter of fact, [14] considered the power consumed only by the linear actuators. It also neglected the sizing of the power supply system, so that the analysis could be agnostic with respect to the number of pressure rails. In the presented study, both the engine and the hydraulic pump operation features are taken into consideration, to construct a full simulation model of the overall excavator system. To achieve the best efficiency, a compound controller which estimates the required force or torque and then determines the most efficient operation mode is developed for both the linear and rotary actuators. A pump controller is purposely designed to reduce the pump losses. The overall controller also ensures the engine operation is such that the Brake-Specific Fuel Consumption (BSFC) is minimized. Such model provides simulation results with realistic estimates of the fuel consumption reduction of the proposed solution with respect to a reference baseline where measurement data were available. The simulation results also allow assessing the feasibility of the proposed controllers.

2 Baseline Machine and Duty Cycle

A 22-ton excavator equipped with a negative-flow open center system [2] is taken as reference in the study. The reference cycles considered in this study involve only the four actuators as shown in Fig. 1, although other functions (such as the tracks) are present in the vehicle.

The system essentially separates the four actuators into two circuits to limit load interference, although it includes merging features for the boom and arm systems, where both circuits can supply the actuators. The two supply pumps are usually with

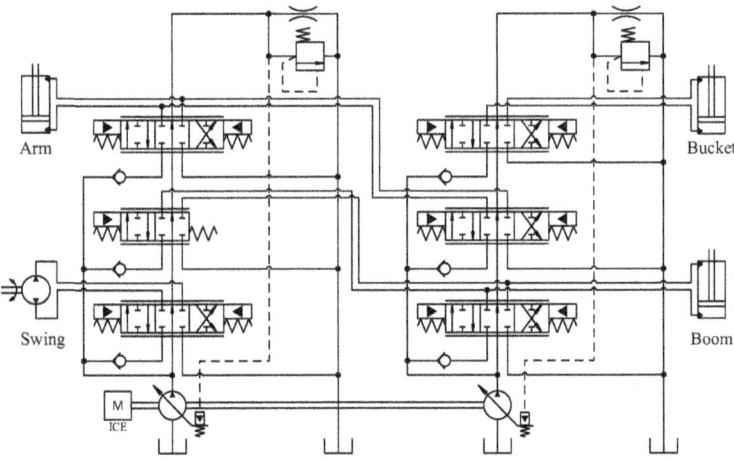

Fig. 1 Simplified schematic of the baseline open center system

the same displacement, to allow meeting the equal flow requirements of the two tracks (not shown in the figure). To evaluate the performance of the simulated system in different conditions, two different duty cycles are studied. The first cycle, named as the "digging cycle", is a 90° truck loading cycle where the machine rotates 90° from where it starts digging while lifting the boom, and dumps the bucket content into a load receiver. This duty cycle is similar to the one used in [10, 14]. In the second cycle, named as the "grading cycle", the excavator is used to level and grade the land with empty buckets. The measurements from duty cycles performed by professional operators include cylinder positions, swing velocities, actuator pressures, actuator pilot pressures, engine torque and speed, and pump delivery pressure. These measurements are used as inputs and references for evaluating energy consumption and tracking performance.

3 Hardware Configuration and Simulation Model

The conceptual schematic of the proposed system is presented in Fig. 2. The two variable-displacement pumps in the baseline machine are used to charge one high-pressure (HP) line and one medium-pressure (MP) line, respectively. An additional small fixed-displacement pump is used to charge the low-pressure (LP) line through the integration of a pressure relief valve.

The two variable displacement pumps have equal size as the baseline system, to meet the same high flow demands on both tracks (not shown also in Fig. 2). Each cylinder control manifold consists of nine two-way bi-directional proportional valves which are also capable of load holding. Such a control approach is also applied to the swing motor so that the connections between the motor and the pressure rails can

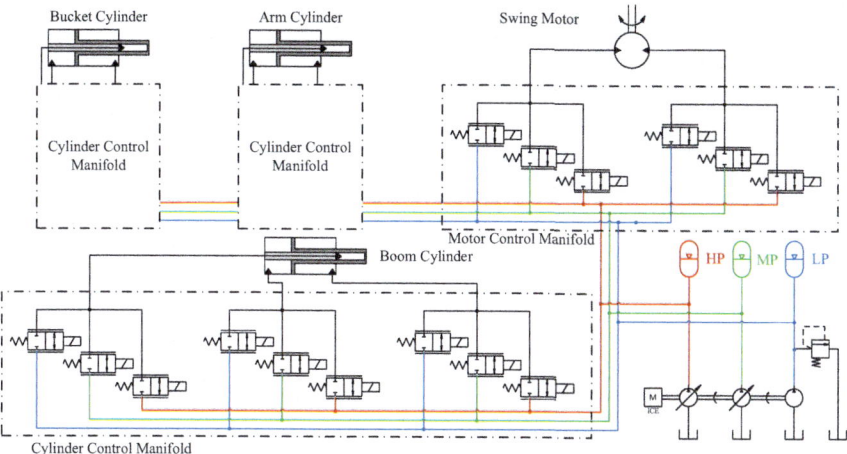

Fig. 2 Conceptual schematic of the proposed system

also be configured independently. The three-chamber cylinders and rail pressures are optimized in a similar manner as proposed in [14].

With the main purpose of evaluating the total power consumption of the proposed system and verifying the feasibility of the designed controllers, a simulation model including the engine operation and all the primary hydraulic components is developed. As shown in Fig. 3, the full simulation model consists of four main sub-system models, including hydraulic system model, engine model, machine dynamics model, and controller. This section describes the first three models which define the dynamics of the excavator, and the controller would be introduced in the next section. All simulation models are developed in MATLAB Simulink.

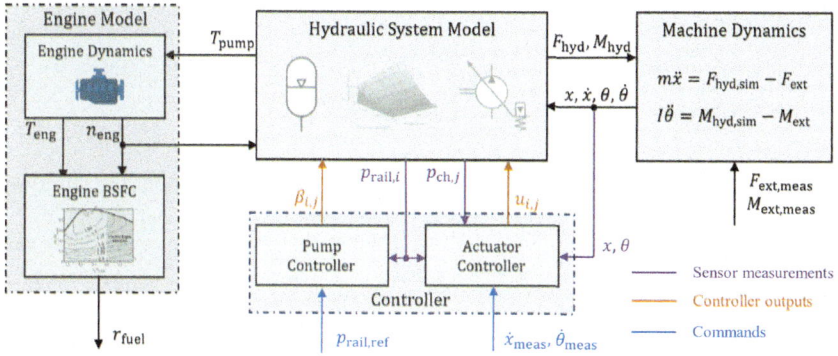

Fig. 3 Conceptual schematic of the simulation model

3.1 Hydraulic System Model

The hydraulic system model is responsible for accounting for all the losses associated with the hydraulic architecture while providing the force or torque generated by the actuators. The primary components in the hydraulic system are pumps, accumulators, cylinders, swing motor, and proportional valves.

3.2 Hydraulic Pumps

To account for the influence of energy losses on the flow rate and the torque at the shaft, the pumps are modeled with empirically derived volumetric efficiencies and hydromechanical efficiencies that vary with pressure drop, displacement, and shaft speed.

The dynamics of the pump normalized displacement is also taken into consideration by a first-order transfer function as

$$\dot{\beta} = \frac{1}{\tau_P}\beta_{cmd} - \frac{1}{\tau_P}\beta \qquad (1)$$

3.3 Accumulators

With the assumption of ideal gas and adiabatic processes and the oil pressure forced to be equal to the gas pressure, the resulting hydraulic capacitance of the accumulator is

$$C_{acc} = \frac{V_g}{\gamma}\left(\frac{p_g}{p_{oil}^{1+\gamma}}\right)^{\frac{1}{\gamma}} + \frac{V_{oil}}{B_T} \qquad (2)$$

The calculated hydraulic capacitance is then applied to the pressure build-up equation to model the accumulator pressure dynamics.

3.4 Cylinders

Ignoring the internal leakage between chambers, the pressure build-up equation, as derived in [2], is applied to evaluate the pressure in each chamber of the cylinders.

With the pressures in each chamber defined, the hydraulic force generated by the cylinder can be calculated using cross-sectional areas.

3.5 Swing Motor

The swing motor is modeled alongside the pressure dynamics of the hydraulic lines around it by pressure build-up equations. Since the motor can operate in either positive or negative speeds with either positive or negative pressure differential, the logic to use the volumetric efficiency and hydromechanical efficiency of hydrostatic units needs to be modified according to operation quadrants.

3.6 Proportional Valves

To predict the throttling losses as accurately as possible, the pressure-flow characteristics of the proportional valves are implemented based on steady-state performance curves provided by the manufacturer. Thus,

$$Q_{prop} = \text{sgn}(\Delta p) \cdot f\left(y_{prop}, |\Delta p|\right) \tag{3}$$

The dynamics of the proportional valves relative to an electronic control signal are modeled by a first-order system as

$$\dot{y}_{prop} = \frac{1}{\tau_{prop}} u_{prop} - \frac{1}{\tau_{prop}} y_{prop} \tag{4}$$

The response time of such valves is assumed to be 100 ms, which is reasonable for such proportional valves available in the market.

3.7 Engine Model

The engine model is used to evaluate the fuel consumption based on engine torque and engine speed in operations, as well as to account for the speed dynamics and torque limitations. The engine speed dynamics is obtained with torque balance in the engine shaft, described as

$$I_{eng}\dot{\omega}_{eng} = M_{eng} - M_L - b_{eng}\omega_{eng} \tag{5}$$

To model the dynamics of the generated engine torque, an approach similar to the one proposed by Dekraker, Stuhldreher, and Kim [15] is used. A PI controller tracking a reference is used to model the engine governor and to output the engine torque command, which is then saturated by the maximum the engine can provide at the given engine speed. Subsequently, two first-order transfer functions are applied to the saturated toque to model the delay in engine torque generation.

To validate the engine model, the measured pump pressures, flow rates, and engine shaft speeds are input into the pump model described in the previous section, which calculates the effective load torque at the engine. The engine model parameters are then optimized in such a way that the simulated speed profile fits the measured engine speed with the calculated load torque.

The fuel consumptions are evaluated with the optimized model and BSFC data in steady state for both the baseline system and the proposed system. The BSFC is given by

$$\text{BSFC} = \frac{r}{P} = \frac{r}{M_{\text{eng}} \cdot n_{\text{eng}}} \tag{6}$$

The BSFC map with respect to engine torque and speed is provided by the engine manufacturer which is covered by confidentiality.

3.8 Machine Dynamics Model

The machine dynamics model is developed to simulate the movement of the actuators and to check the position/velocity tracking performances. Instead of using the machine kinematic information to predict the equivalent mass/inertia with actuator positions and estimating the friction forces based on empirical evaluations, the total external force/torque acting on each actuator is approximated directly through measurements. Take the case of linear actuators as an example, the following equation holds during measurements:

$$m\ddot{x}_{\text{cyl,meas}} = F_{\text{hyd,meas}} - F_{\text{L}} - F_{\text{f}}(\dot{x}_{\text{cyl}}). \tag{7}$$

The sum of the load force and the friction force can be calculated as

$$F_{\text{ext}} = F_{\text{load}} + F_{\text{f}}(\dot{x}_{\text{cyl}}) = -F_{\text{hyd,meas}} - m\ddot{x}_{\text{cyl,meas}}. \tag{8}$$

Then the calculated total external force can be input to the equation of motion of the cylinder piston in simulation:

$$m\ddot{x} = F_{\text{hyd,sim}} - F_{\text{ext}}. \tag{9}$$

The key idea of such approximation is to ask the generated cylinder force to follow the hydraulic force in measured duty cycles, while also keeping well track of the actuator positions and velocities. In this way, a relatively fair energy comparison is attained by forcing the power output by actuators as close as possible. On one hand, such an approach can be used for energy evaluations of a given duty cycle with measurements available. On the other hand, it cannot be used to predict the performance of any arbitrary duty cycles due to lack of a kinematic model. It also needs to be mentioned that this approach assumes the same friction force for different cylinder designs.

4 Controllers

Since charging of pressure rails and controlling actuators through throttling from pressure rails are almost fully decoupled, it is possible to design the pump displacement controllers and the actuator controllers so that both the pumps and the actuators operate with the best efficiency. On the one hand, the design objective of the pump controller is to minimize fuel consumption while ensuring power availability to the actuators. On the other hand, the design objective of the actuator controller is to minimize the total power taken from the pressure rails while maintaining the tracking performance.

4.1 Pump Controller and Modified Engine Operation

To minimize the total fuel consumed for charging the pressure rails, both reducing pump losses and increasing engine fuel efficiencies are taken into account. As pump efficiencies are maximized with full displacements, it is preferred to set the pumps either in charging mode with maximal available displacements or standby mode with minimum displacements to account for the leakage due to pressure drop across the pump. To maintain a fairly constant power demand and to avoid frequent part loading which leads to inefficient use of the engine, the engine is set to operate in the highest efficiency range with a lower reference shaft speed. Thus, the pumps tend to charge longer with the lower shaft speed and unload the engine less frequently. Moreover, a torque limiter is needed to avoid engine stalling because the torque of operating both pumps at full displacements may exceed the maximum torque that is available from the engine. Since the HP rail ensures power availability to the system, the displacement of MP pump would be saturated according to available engine torque.

4.2 Actuator Controller

The actuator controller consists of four main components: a velocity controller, a mode selector, a pressure controller, and a flow allocator. An overview of the controller concepts is presented in Fig. 4.

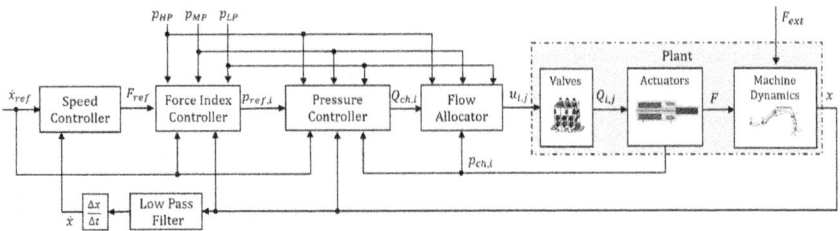

Fig. 4 Overview of the actuator controller structure

4.3 Speed Controller

Similar to the concept of secondary control, closed-loop velocity control is utilized for the actuators. Different from the controller developed by Heybroek and Sahlman [10] which only uses the machine model to generate the reference force, an additional adaptive speed controller is implemented to assist speed tracking under changing loads depending on the actuator positions. A similar secondary control approach has been successfully implemented by Busquets and Ivantysynova [6] on rotary actuators. As the focus of the paper is to evaluate the overall performance of the architecture on the full machine level, the details of the derivation of the adaptive controller are out of scope and are not included.

4.4 Force Index Controller

Once the reference force is determined, the best operation mode can be selected based on certain criteria evaluating each mode in terms of energy efficiency and force tracking. An approach similar to the Force Index Controller (FIC) proposed by Heybroek and Sjöberg [16] is used but with a different cost function. In addition to the first term evaluating the throttling losses of each mode

$$J_1 = k_1 \cdot \left| v_{\text{cyl}} \cdot (F_{\text{ref}} - F_{\text{mode}}) \right|, \tag{10}$$

and the second term evaluating the compressibility losses due to switching

$$J_2 = k_2 \cdot \sum_{i=1}^{3} p_i \cdot \underbrace{-(p_{i,\text{new}} - p_i) \cdot \frac{V_i}{B}}_{\Delta V_i} \tag{11}$$

a term penalizing the number of chambers whose connection with the rails are to be changed

$$J_3 = \begin{cases} k_3 \cdot \#sw^4, & \#sw \leq 2 \\ \infty, & \#sw = 3 \end{cases}, \tag{12}$$

and a term evaluating the switching loss due to cross leakage between pressure rails

$$J_4 = k_4 \cdot \sum_{i=1}^{3} \left(p_{i,\text{rail,new}} - p_{i,\text{rail}}\right)^{3/2}, \tag{13}$$

are added to better optimize the mode selection and tradeoff between minimizing throttling losses and reducing switching frequency. The coefficients are manually optimized based on the performance in the two given duty cycles.

4.5 Pressure Controller

With the given desired pressure in each chamber, an adaptive model-predictive controller is used to find the optimal pressure trajectories during transient to minimize the force tracking error, thus smoothing the dynamic performances during switching. The desired flow rate Q_{ref} for each chamber is determined by minimizing

$$J_{\text{MPC}} = \sum_{t=1}^{N} W^F \left(F_{\text{ref}} - A^T \cdot \varphi \cdot p_{\text{ch},t}\right)^2 + \left(p_{\text{ch,ref}} - p_{\text{ch},t}\right)^T W^p \left(p_{\text{ch,ref}} - p_{\text{ch},t}\right)$$
$$+ \left(Q_{\text{ref},k} - Q_{\text{ref},k-1}\right)^T W^{\Delta Q} \left(Q_{\text{ref},k} - Q_{\text{ref},k-1}\right), \tag{14}$$

where the first term penalizes force tracking error, the second term penalizes pressure tracking error, and the last term penalizes large changes in desired flow rates as control inputs. Upper bounds and lower bounds on the flow rates are implemented to ensure that the commanded flow rates are available through the proportional valves. The MATLAB MPC toolbox is used for implementing the controller.

4.6 Flow Allocator

Based on the flow rates commanded by the pressure controller and chamber pressures, the flow allocator calculates valve opening commands to get enough flow from pressure rails. While it is always possible to get enough flow from HP line or to send enough flow to LP line, the logic is implemented such that the flow from or going to nearest pressure lines is always preferred. Such logic is primarily used in transient states when changing operating modes, while at steady state the pressure source for each chamber would be determined by the pressure distribution optimized by the Force Index Controller. An inverse map provided by the valve manufacturer is used to derive the required normalized opening commands for a given flow rate request and pressure differential.

5 Simulation Results

The model and controller described above are used to perform the simulation of the two reference duty cycles. It is essential to ensure similar delivered power from actuators when energy consumption is in interest. Instead of inputting the measured operator commands and compare the resulting performances, which might lead to different actuator position profiles and delivered power, the controllers are utilized to back calculate the valve commands for tracking the measured positions under external forces calculated from measurements.

5.1 Cycle Tracking

The simulated results showing the position tracking performances in the grading cycle are presented in Fig. 5. Despite the tracking errors due to estimating equivalent loads for each actuator based on tracking performances, the simulated trajectories show a good match with the measurements.

Force tracking and transient performances during mode switches are shown in Fig. 6, where the simulated force follows well the reference commanded by the velocity controller. The short spikes during changing forces modes are considered to be acceptable given the large inertia of cylinder pistons. The result shows the controller is able to change operations modes for reduced throttling losses when external forces change.

Verifying the cycle tracking performance not only confirms the feasibility of the developed controller but also backs fair energy consumption comparisons.

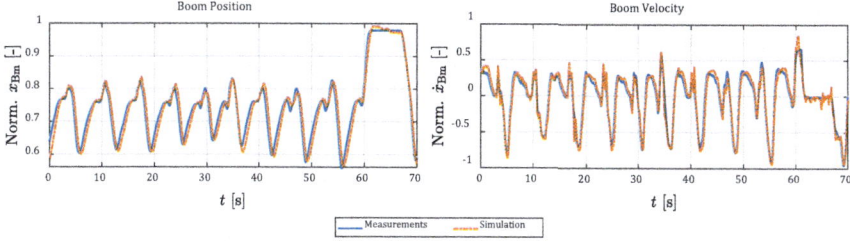

Fig. 5 Position and velocity tracking with normalized properties in the grading cycle

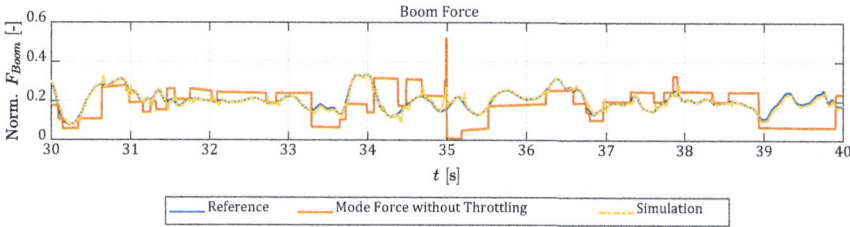

Fig. 6 Boom force tracking performance with normalized properties in the grading cycle

5.2 Power Supply Performance

The pressure dynamics in accumulators and corresponding pump operations are presented in Fig. 7. While the pumps are controlled simply to maintain pressures within a certain range, there are moments when actuators are taking more flow than a single pump can provide. Nevertheless, low pressure due to the excessive flow command only occurs when both the arm and boom cylinders are pushed to maximum position rapidly, after which maintaining high pressures in the high-pressure rail for potential high power output is no longer necessary.

The pump displacements show that the HP pump is frequently charging the pressure rail and the MP pump is always trying to charge the accumulator under the torque limits. This means the system is operating close to its maximum capacity. Another important thing to be noticed is the similar accumulator pressures in the beginning and the end of the duty cycles, which reflects that the energy consumption evaluation does not take any advantage of the power stored in the accumulator.

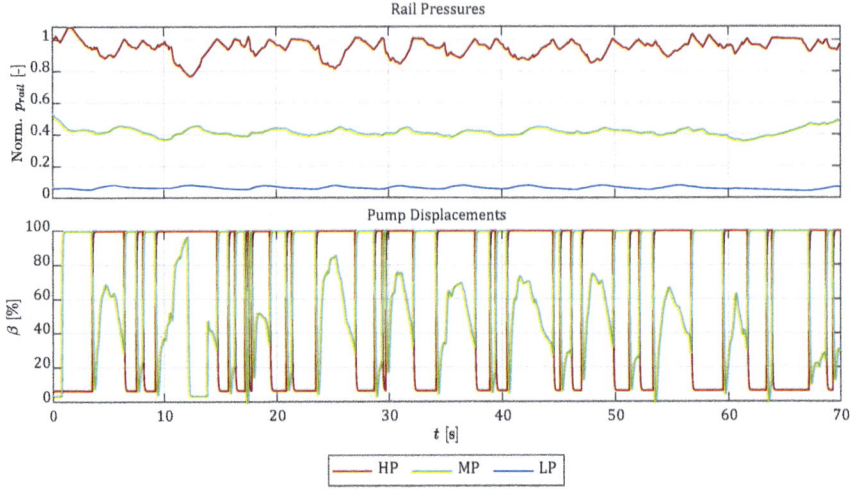

Fig. 7 Power supply performances with normalized properties in the grading cycle

5.3 Engine Operation

The high charging frequency leads to less partial load operation of the engine, as shown in Fig. 8. The contours represent the fuel efficiencies evaluated by BSFC within maximum torque. The time spent at each operation point is indicated by the histogram. With modified speed optimized for the given two duty cycles, the engine can operate in the desired region with the best fuel efficiency throughout the majority

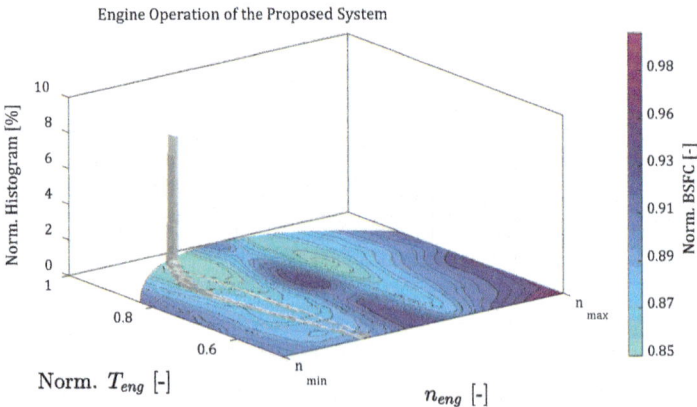

Fig. 8 Engine operations of the proposed system with normalized properties in the grading cycle

Table 2 Percentage difference of the proposed system in average power, BSFC, and total fuel consumption when compared to the baseline system

Duty cycle	Average Power	Average BSFC	Fuel Consumption
Digging	−18.8%	−5.1%	−22.3%
Grading	−34.3%	−2.2%	−36.0%

of the cycle duration. This leads to higher fuel efficiencies than the baseline engine operations, where a higher engine speed is maintained with frequent partial loading.

5.4 Fuel Consumption Evaluations

With the previous sections showing the evidence of fair energy consumption comparisons and intuitively more efficient engine operations, Table 2 summarizes the comparisons between the proposed system and the baseline system in average power, average BSFC, and total fuel consumption in the two duty cycles.

The results indicate that the proposed structure can reduce the fuel consumption by 22.3% in the digging cycle and 36.0% in the grading cycle when compared to the baseline machine. Due to the improved engine efficiency, the final fuel savings are even slightly higher.

It is interesting to observe the differences in efficiency gains between two duty cycles. The differences reveal the different loss distributions in the two duty cycles. Depending on operator behaviors, the energy that can be restored and the throttle losses that can be reduced are distinct in various cycles.

6 Conclusions

A hydraulic architecture based on the concept of MPR systems and multi-chamber cylinders, with the specific configuration of three-pressure rails and three-chamber cylinders, was proposed and simulated in the case of a mid-size excavator. The system uses two variable displacement pumps and one small, fixed displacement pump to charge three accumulators, which are viewed as pressure rails and connected to all primary actuators. Parallel installed proportional valves enable flexible connections between pressure rails and actuators. A compound controller aiming at tracking given actuator speed command with maximized efficiency was developed. The pump controller and engine operations were also modified to reduce pump losses and partial loading conditions.

Two different duty cycles were simulated to evaluate the fuel consumption and to assess the controllability. With competent cycle tracking performances as well as capability of changing operation modes for optimized efficiency, results showed a fuel consumption reduction up to 36%, depending on the cycles. The reduction is mainly due to reduced power requirements from the hydraulic system, while improved engine efficiency also contributes.

The next step would be to further reason the different efficiency gains in different cycles for a better understanding of whether better performances are possible given the work cycle and hardware configurations.

Abbreviations

Nomenclature

β	Normalized pump displacement
τ	Time constant
C	Equivalent hydraulic capacitance
p	Pressure
V	Volume
B_T	Isothermal bulk modulus
γ	Polytropic index
Q	Flow rate
Δp	Pressure differential
y	Position of valve poppet
u	Opening area of valves
I	Equivalent rotational inertia
n/ω	Shaft speed
M	Shaft Torque
b	Empirical friction coefficient
r	Fuel consumption rate
P	Produced power by the engine
m	Mass
x	Position
F	Force
J	Penalty term
k	Weight of penalty term
v	Velocity
#sw	Number of switches
k	Weight of penalty term
W	Weight matrix
φ	Force direction coefficient
A	Cross-sectional area

Subscripts

P	Pump
cmd	Command
acc	Accumulator
mech	Hydromechanical
g	Pre-charged gas in accumulators
oil	Oil in accumulators
prop	Proportional valve
eng	Engine
L	Load
cyl	Cylinder
meas	Measurement
hyd	Hydraulic
f	Friction
ext	Sum of all external forces
sim	Simulation
ref	Reference
mode	Operation mode without throttling
rail	Pressure rail
i	Chamber index
j	Pressure rail index
HP	High-pressure rail
MP	Medium-pressure rail
LP	Low-pressure rail

References

1. Lynch L, Zigler B (2017) Estimating energy consumption of mobile fluid power in the United States. United States. https://doi.org/10.2172/1408087
2. Vacca A, Franzoni G (2021) Hydraulic Fluid Power: Fundamentals, Applications, and Circuit Design. Wiley, United Kingdom
3. Pellegri M, Green M, Macpherson J, Mckay C, Caldwell N (2020) Applying a multi-service digital displacement® pump to an excavator to reduce valve losses. In: The 12th international fluid power conference, Dresden, Germany, 12–14. https://doi.org/10.25368/2020.7
4. Beltrami D, Iora P, Tribioli L, Uberti S (2021) Electrification of compact off-highway vehicles—overview of the current state of the art and trends. Energies 14(17):5565. https://doi.org/10.3390/en14175565
5. Quan Z, Ge L, Wei Z, Li YW, Quan L (2021) A survey of powertrain technologies for energy-efficient heavy-duty machinery. Proc IEEE 109(3):279–308. https://doi.org/10.1109/JPROC.2021.3051555
6. Busquets E, Ivantysynova M (2015) Adaptive robust motion control of an excavator hydraulic hybrid swing drive. SAE Int J Commer Veh 8(2):568 582. https://doi.org/10.4271/2015-01-2853

7. Lumkes J, Andruch J (2011) Hydraulic circuit for reconfigurable and efficient fluid power systems. In: The 12th scandinavian international conference on fluid power, Tampere, Finland, 18–20
8. Vukovic M, Sgro S, Murrenhoff H (2013) STEAM: A mobile hydraulic system with engine integration. In: Proceedings of the ASME/BATH 2013 symposium on fluid power and motion control. ASME/BATH 2013 Symposium on Fluid Power and Motion Control, Sarasota, Florida, USA. 6–9. https://doi.org/10.1115/FPMC2013-4408
9. Vukovic M, Leifeld R, Murrenhoff H (2016) STEAM–A hydraulic hybrid architecture for excavators. In: Proceedings of the 10th international fluid power conference, pp 151–162. https://nbn-resolving.org/urn:nbn:de:bsz:14-qucosa-200445
10. Heybroek K, Sahlman M (2018) A hydraulic hybrid excavator based on multi-chamber cylinders and secondary control – design and experimental validation. Int J Fluid Power 19(2):91–105. https://doi.org/10.1080/14399776.2018.1447065
11. Huova M, Laamanen A, Linjama M (2010) Energy efficiency of three-chamber cylinder with digital valve system. Int J Fluid Power 11(3):15–22. https://doi.org/10.1080/14399776.2010.10781011
12. Dell'Amico A, Carlsson M, Norlin E, Sethson M (2013) Investigation of a digital hydraulic actuation system on an excavator arm. In: Proceedings from the 13th scandinavian international conference on fluid power, pp 505–511. https://doi.org/10.3384/ecp1392a50
13. Guo X, Madau R, Lengacher J, Vacca A, Cardoso R (2022) Multi-pressure rail system design with variable pressure control strategy. In: Proceedings of the 13th international fluid power conference, Aachen, Germany. https://doi.org/10.18154/RWTH-2023-04609
14. Bertolin M, Vacca A (2022) A parametric study on architectures using common-pressure rail systems and multi-chamber cylinders. In: Proceedings of the 2022 global fluid power society PhD symposium, Naples, Italy. https://doi.org/10.13052/rp-9788770047975
15. Dekraker P, Stuhldreher M, Kim Y (2017) Characterizing factors influencing si engine transient fuel consumption for vehicle simulation in alpha. SAE Int J Engines 10(2):529–540. https://doi.org/10.4271/2017-01-0533
16. Heybroek K, Sjöberg J (2018) Model predictive control of a hydraulic multichamber actuator: a feasibility study. IEEE/ASME Trans Mechatron 23(3):1393–1403. https://doi.org/10.1109/TMECH.2018.2823695

Open Access This chapter is licensed under the terms of the Creative Commons Attribution 4.0 International License (http://creativecommons.org/licenses/by/4.0/), which permits use, sharing, adaptation, distribution and reproduction in any medium or format, as long as you give appropriate credit to the original author(s) and the source, provide a link to the Creative Commons license and indicate if changes were made.

The images or other third party material in this chapter are included in the chapter's Creative Commons license, unless indicated otherwise in a credit line to the material. If material is not included in the chapter's Creative Commons license and your intended use is not permitted by statutory regulation or exceeds the permitted use, you will need to obtain permission directly from the copyright holder.

Harmonic Characterisation of Electrically Driven Pumps

Thomas Heeger, Martin West, and Liselott Ericson

1 Introduction

Conventional mobile fluid power systems often have low efficiency. The low efficiency of these systems is due to their design and operational priorities, which favour factors such as low investment cost, increased productivity and high robustness over achieving high efficiency. However, these priorities are shifting with the increasing importance of avoiding greenhouse gas emissions as well as increasing energy costs. Research on many different solutions to increase energy efficiency is ongoing, including energy recovery and reducing throttling losses [1]. Electrification of mobile machinery is a substantial part of this trend. As a result, there is a lot of research into the electric drive of pumps, and many different integrated combinations of pumps and electric motors are being investigated [2]. There is great potential for energy savings by using such machines (e.g. 65% in the work of Qu et al. [3]).

Low noise emission is important to protect the health of humans in the proximity of the machines and to increase customer acceptance. Although electric motors emit significantly less noise than internal combustion engines, noise is a major challenge for electrification because the noise of other components is no longer masked by the engine's noise. This is critical, as hydraulic pumps are known to emit unpleasant noise with distinct frequencies. While electric motors are significantly less noisy than hydraulic pumps the noise spectrum is also characterised by narrow-band emissions at distinct frequencies. Besides minimising the noise contributions of each component, the interaction between the components should be such that noise is minimised. However, the interaction between hydraulic pumps and electric motors in terms of noise generation is not well-researched yet.

T. Heeger (✉) · L. Ericson
Linköping University, Linköping, Sweden
e-mail: thomas.heeger@liu.se

M. West
Tin Arm Engineering, Göteborg, Sweden

© The Author(s) 2025
L. Ericson and P. Krus (eds.), *Advancements in Fluid Power Technology: Sustainability, Electrification, and Digitalization*, Lecture Notes in Mechanical Engineering,
https://doi.org/10.1007/978-3-031-84505-5_23

This paper provides background on the origin of harmonics present in piston pumps and PMSMs and shows the design parameters that affect the harmonic orders. Analytical equations for predicting harmonic orders are shown and simulation results are analysed using torque ripple as an example. The effects of the combination of the numbers of pistons and poles on the noise of an electrically driven pump are discussed.

This paper is structured as follows: Sect. 2 discusses the harmonics in hydraulic pumps that contribute to noise emission. Section 3 discusses the harmonics in PMSM, with a special focus on fractional slot machines with 12 slots and 10 poles. Section 4 calculates the harmonic orders for the example of a pump with ten respectively eleven pistons in combination with a 12-slot/10-pole PMSM. FEM and lumped parameter simulation results of the torque ripples are combined and analysed. Section 5 discusses the choice of piston number and the introduction of a gearbox into the system. Section 6 summarises the conclusions.

Other noise sources such as cavitation are not considered in this paper. The scope for the electric machine is limited to a machine type particularly suited for this kind of application [4].

2 Harmonics in Hydraulic Machines

Accurate noise prediction is generally an extremely challenging task, as the radiated acoustic power is only a fraction of the machine's power. Very many parameters are needed to model noise behaviour, and laboratory testing is always required [5]. In hydraulic machines, several characteristics contribute to noise generation: flow ripples in the discharge port and inlet port, piston forces, bending moments and drive shaft torque [6, 7]. These contributors cause fluid-borne and structure-borne noise, which eventually become air-borne (and thus potentially audible) noise. The system in which the hydraulic machine is used strongly affects how much noise is emitted by each of the contributors, so there is no general trade-off between the different contributors, making the search for an ideal design challenging. However, there are usually positive correlations between minimising individual contributors such as flow ripple or cylinder pressure rate and the other contributors [6].

Ivantysyin and Ivantysynova [8], and Manring [9] have published books on pump design, which describe the origin of flow, force, bending moment and drive ripples in detail.

For the sake of simplicity, the following sections (except Sect. 2.1 on compressible flow ripples) assume that the displacement chambers are pressurised as soon as they are in contact with a port and that the examples are based on zero-lapped valve plates (i.e. as soon as a chamber leaves one port, it connects to the other port). Commutation effects are neglected (except for compressible flow pulsations), although they introduce additional harmonics with frequencies kzn in all machines.

2.1 Flow Ripples

There are kinematic and compressible flow ripples, and their combination is the overall flow ripple.

Kinematic Flow Ripples In axial piston machines, the piston velocity and therefore the fluid velocity that one piston displaces follows a sine shape. A pump is connected to the delivery port during the positive part of the sine curve. The total delivered flow is the sum of the positive flow of all chambers. This sum is not constant, and thus kinematic flow ripples occur. Typical shapes of kinematic flow ripples are shown in Fig. 1.

A distinction must be made between machines with odd and even numbers of pistons z. When the number of pistons is odd, the events of a displacement chamber connecting to and disconnecting from the delivery port occur alternately. However, when the number of pistons is even, the chambers connect to and disconnect from the ports in pairs. The peak-to-peak kinematic flow ripple is significantly lower for pumps with an odd number of pistons [8]. Commonly, odd numbers of pistons are used [10].

The harmonic frequencies f_{kinFlow} of the kinematic flow ripples are given in Eq. 1, with an integer number $k = 1, 2, 3, \ldots$ and the mechanical speed n.

$$f_{\text{kinFlow}} = \begin{cases} kzn & \text{for even piston numbers} \\ 2kzn & \text{for odd piston numbers} \end{cases} \quad (1)$$

Compressible Flow Ripples During commutation, the fluid's compressibility, in combination with a pressure mismatch between the displacement chamber and the port, causes compressible flow ripples. The combination of compressible and kinematic flow ripple creates the overall flow ripple. Typically, the compressible flow ripples dominate over the kinematic flow ripples in mobile working machines [10].

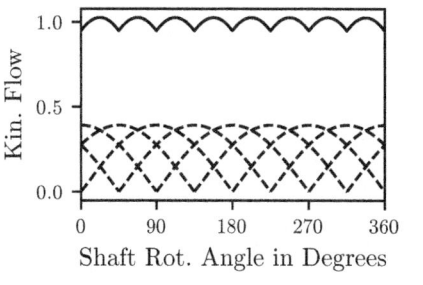
Machine with 8 pistons.

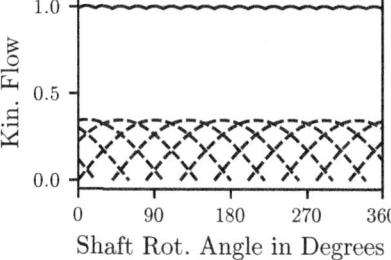
Machine with 9 pistons.

Fig. 1 Examples of kinematic flow ripples. The dashed lines show the contributions of the individual pistons, and the solid line shows the total kinematic flow. The figures are normalised with the average total kinematic flow

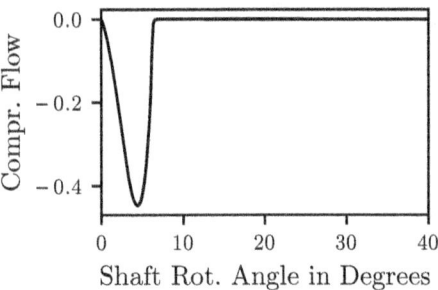

Fig. 2 Typical shape of compressible high-pressure flow ripples for one displacement chamber of a pump without commutation features. This ripple occurs once per pump period. The figure is normalised with the average total kinematic flow

Figure 2 visualises the compressible flow ripples for a valve plate without any commutation features, also known as a zero-lapped valve plate. Reviews of various commutation features can be found in Johansson [10] and Heeger et al. [11].

Assuming the same displacement, compressible flow ripples increase for a smaller number of pistons, as both the stroke and dead volume of each piston increase, and resulting in a larger volume that needs to be compressed (or decompressed). Pre- and de-compression occur z times per revolution, so the harmonic frequencies f_{compFlow} of the compressible flow ripples in the high-pressure port and low-pressure port are as stated in Eq. 2.

$$f_{\text{compFlow}} = kzn \qquad (2)$$

2.2 Force, Moment and Torque Ripples

Axial Force Ripples Each piston alternates between high-pressure and low-pressure. The sum of all piston forces varies, if the number of pressurised pistons varies over time. This is the case for machines with odd piston numbers. For even piston numbers, the resulting force is constant, except for the time during commutation. Therefore, even piston number pumps exhibit significantly lower harmonic contents for force ripples and are not considered in Eq. 3.

The alternating piston forces are the main source of structure-borne noise [10]. The harmonics of the piston force ripples occur at the frequencies $f_{\text{pistonForce}}$ given in Eq. 3.

$$f_{\text{pistonForce}} = \begin{cases} 0 & \text{for even piston numbers} \\ k_{\text{odd}} zn & \text{for odd piston numbers} \end{cases} \qquad (3)$$

Typical shapes of piston force ripples are visualised in Fig. 3.

Valve plate design can modify the shape of the force transition for both pump types. Johansson [10] recommends keeping the pressurisation and depressurisation rates of the displacement chambers as low as possible. The smoother the pressure profile, the lower the harmonic content of the force pulsations (and especially the lower the high-frequency content) and thus the emitted noise.

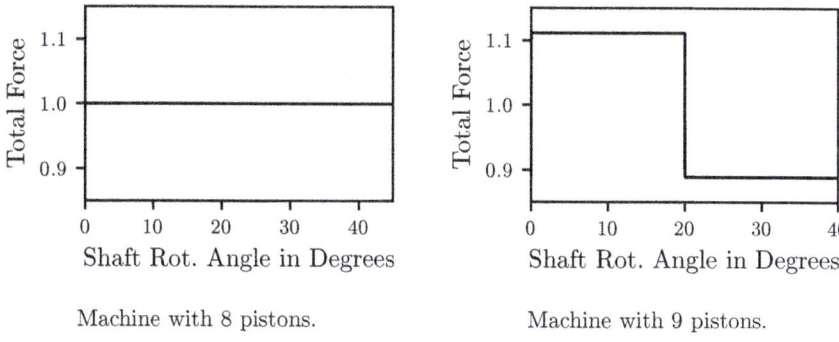

Fig. 3 Examples of normalised piston force ripples

Bending Moment and Drive Torque Ripples The varying number of pressurised pistons, and the varying piston positions (thus varying lever) cause bending moments, which contribute to structure-borne noise. Typical positions of the location of the force resultants are shown in Fig. 4.

Bending moments can be analysed around the y-axis (refer to Fig. 4 for the axis definition), and around the x-axis. The moment around the z-axis (caused by the swash-angle in swashplate machines) is the drive torque. Typical shapes of bending moment ripples and drive torque ripples are visualised in Fig. 5.

Even piston numbers create much larger peak-to-peak moments than odd piston numbers, as cylinders enter and leave each port in pairs. Furthermore, bending moments around the x-axis have a larger ripple than those around the y-axis, as the pistons that enter and leave a kidney have a larger lever around the x-axis. Also, the moment around the x-axis changes sign, and can thus cause noise by releasing geometric plays. Therefore, the bending moment ripple around the x-axis is responsible for more noise than the bending moment ripple around the y-axis and the drive torque ripple [10].

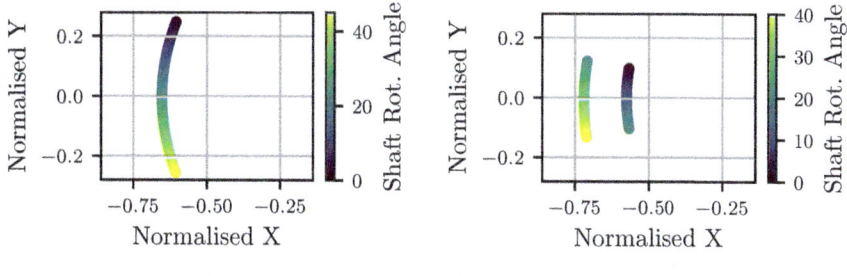

Fig. 4 Normalised location of resultant of piston forces (normalised with the radius of cylinder barrel bore location)

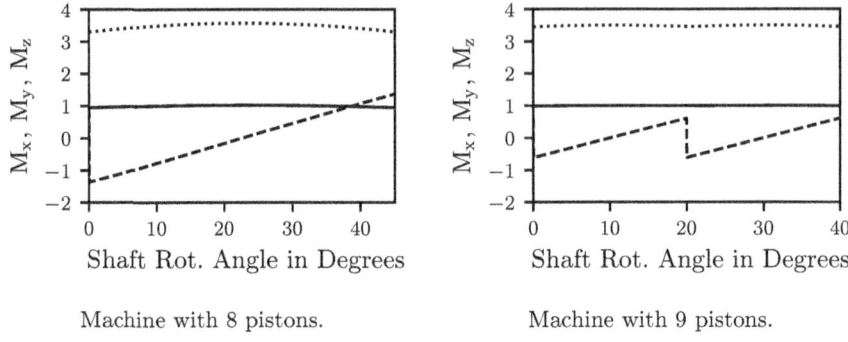

Fig. 5 Examples of bending moment ripples (M_x—dashed, M_y—dotted) and drive torque ripples (M_z—solid line). They are normalised in relation to the average drive torque

The harmonics of the bending moment ripples and the drive torque ripples occur at the frequencies f_{moments} given in Eq. 4.

$$f_{\text{moments}} = \begin{cases} kzn & \text{for even piston numbers} \\ 2kzn & \text{for odd piston numbers} \end{cases} \quad (4)$$

3 Harmonics in Permanent Magnet Synchronous Machines

The noise and vibration behaviour differs between different electric machine types. This paper focuses on PMSMs. The use of permanent magnets enables high efficiency (due to the elimination of conductor losses in the rotor), which reduces operating costs. Other advantages include high compactness, a high power factor [12] and ease of control. However, permanent magnets are relatively expensive, and there are concerns about the environmental impact of their production.

This paper focuses on fractional slot concentrated winding PMSMs, using a 12-slot/10-pole machine as sketched in Fig. 6 as an example. These machines gain popularity due to their simple structure, short end windings (thus reduced copper material and copper loss), high efficiency, low cogging torque, and high torque production at low speeds. However, fractional slot design introduces low-order harmonics, resulting in increased vibration levels [13, 14].

The magnetic flux in the air gap produces not only torque but also a radial magnetic force ripple, resulting in vibrations and noise (see Sect. 3.2). Typically, the stator housing is the primary radiator of electrical machine noise. Besides the structure-borne noise, electrical machines also exhibit a torque ripple (see Sect. 3.3) [5].

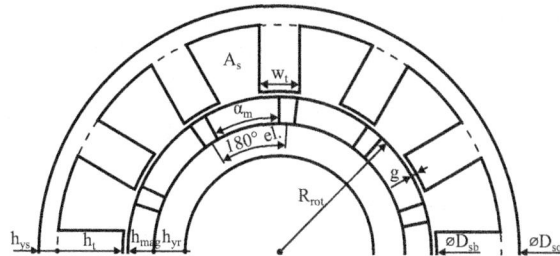

Fig. 6 Sketch of a PMSM with 12 slots and 5 pole pairs

3.1 Electromagnetic Harmonics

This paper follows the simplifications suggested by Hannon et al. [15] for the study of PMSM. Therefore, end effects are neglected, a balanced current with an odd number of phases is assumed, windings are distributed with the star-of-slot technique (this technique is described, e.g. by Bianchi et al. [16]), saturation effects are neglected, and synchronous operation is assumed.

For a single mechanical revolution, the electric field must change p times, where p represents the number of pole pairs. The fundamental electrical frequency $f_{\text{electrical}}$ (which is also the frequency of the stator current), is given in Eq. 5 [5, 15]:

$$f_{\text{electrical}} = pn \tag{5}$$

For an easier comparison of the electric motor harmonics to the pump harmonic orders, this paper will refer to the harmonics of both machine types in relation to shaft speed/frequency.

Hannon et al. [15] show that a rotating magnetic field is caused by the combination of time and spatial harmonics. The speed of this rotating field depends on the ratio of the orders of the time harmonics and the spatial harmonics.

Rotor Magnetic Field Harmonics As stated above, the harmonic orders of the rotor's magnetic field are multiples of p. Furthermore, if the magnets are symmetrical in the angular direction, each half of the fundamental period of the magnetic field is symmetrical. Consequently, the magnetic field of one fundamental period is only represented by odd harmonics, and even harmonics are not present [15].

The rotor harmonic frequencies $f_{\text{B,Rotor}}$ are then given with Eq. 6.

$$f_{\text{B,Rotor}} = k_{\text{odd}} pn \tag{6}$$

Stator Current Time Harmonics The stator current changes with time, thus introducing time harmonics. Based on the definition of the fundamental stator current frequency in Eq. 5, stator current harmonic orders must be a multiple of p.

The assumption of a balanced system means that the sum of current over the m different phases is always zero. As each phase provides the same current, but shifted by $2\pi/m$ electrical radians, any current with a frequency of $k_{\text{time}} m f_{\text{electrical}}$

experiences negative interference with the current of the other phases, and thus those orders are not present. Therefore, only k_{time} which are not a multiple of m are present [5, 15]. Additionally, as the stator current can be assumed to be symmetric over half a fundamental electrical period, only odd k_{time} are considered [15].

For higher harmonics, the two conditions (k_{time} is odd, and not a multiple of m) can be summarised for three-phase systems, stating that $k = 1 + 6c$, with c being an integer taking positive and negative values. Waves with positive k rotate in the same direction as the rotor, and waves with negative k rotate in the opposite direction [5]. The above-given considerations are summarised in Eq. 7.

$$f_{\text{B,statorCurrent}} = kpn, \quad \text{where } k \text{ is odd and not a multiple of } m \tag{7}$$

Stator Current Spatial Harmonics due to Winding Distribution The star-of-slot technique is commonly used to assign phases to the individual coils. The number of slots N_s must be a multiple of the phases m to have the same number of coils for each phase. The stator slots introduce spatial harmonics.

The machine's periodicity τ is defined as the greatest common divisor of p and N_s in Eq. 8 [15, 16].

$$\tau = \gcd(p, N_s) \tag{8}$$

The term *slot group* summarises slots that are connected to the same zone (positive or negative) of the same phase to a group. Under synchronous operation, the mechanical shift between two slot groups equals the time shift of their current densities. The number of slots per group N_g differs between two cases (see Eq. 9) [15].

$$N_g = \begin{cases} \frac{N_s}{2m\tau} & \text{for } \frac{N_s}{\tau} \text{ even} \\ \frac{N_s}{m\tau} & \text{for } \frac{N_s}{\tau} \text{ odd} \end{cases} \tag{9}$$

The time required for this mechanical shift is equal to the time shift between two current zones. After this shift, the rotor experiences the same stator current density again. For the stator, the armature reaction to the rotor's magnetic field is identical but shifted by $\varphi_{\text{mechShift}}$ mechanical radians. Therefore, the winding distribution introduces spatial harmonics to the magnetic field. The spatial harmonics h_{spatial} must satisfy Eq. 10 [15], along with the previously introduced time harmonics h_{time}. Note that the integer c comes in pairs of positive and negative values.

$$h_{\text{spatial}} - h_{\text{time}} = c \frac{N_s}{N_g} \tag{10}$$

Magnetic Field Spatial Harmonics due to Stator Slots The stator slots introduce a variation of the magnetic permeance over the machine's rotational angle. Assuming identical slots, the rotor will experience the same permeance N_s times per revolution. Therefore, the spatial harmonic orders that satisfy Eq. 11 are introduced [15].

Table 1 Harmonic frequencies in the magnetic field of PMSM

	Rotor field	Stator current	Winding distribution	Stator slot permeance
Harmonic order (in reference to one mechanical revolution)	$h_{\text{Rotor}} = kp$	$h_{\text{sc}} = kp$	$h_{\text{winding}} =$ $h_{\text{sc}} + c\frac{N_s}{N_g}$	$h_{\text{slot}} =$ $h_{\text{Rotor}} + cN_s$
Condition for k in a three-phase system	Odd	$k = 1 + 6c$	included in N_g (Eq. 9)	–
Orders in a 12 slot, 10 pole, 3-phase PMSM	5, 15, 25, 35, 45, ...	5, (−)25, 35, (−)55, 65, ...	For first time order: 5, (−)1, 11, (−)7, 17, ...	For first time order: 5, (−)7, 17, (−)19, 29, ...

$$h_{\text{spatial}} - h_{\text{time}} = cN_s \qquad (11)$$

Pyrhönen et al. [17] state that the largest flux harmonics are caused by slotting. Slot harmonics always occur in pairs (i.e. positive and negative values of c occur together). Pyrhönen et al. only consider the fundamental time harmonic when calculating the slot harmonics. For $c = \pm 1$, they are called the first slot harmonics [17]. The slotting effect introduces harmonics of lower orders. A summary of the harmonics due to slotting is given by Zhu et al. [18].

Summary of Main Electromagnetic Harmonics The orders of the electromagnetic harmonics are summarised in Table 1.

Other Effects Causing Electromagnetic Harmonics In addition to the previously described harmonics, eccentricity (harmonic orders $h_{\text{eccentricity}} = p \pm 1$), magnetic saturation (harmonic orders $h_{\text{saturation}} = 2kp$) and rotor saliency affect the magnetic field in PMSM [5, 15].

3.2 Radial Force Ripples and Structural Vibration

Section 3.1 summarised the most important harmonics in the electromagnetic fields of PMSM. The magnetic flux density in the airgap is the sum of the stator's and rotor's magnetic flux densities (see Eq. 12) [5].

$$B_{\text{airgap}} = B_{\text{stator}} + B_{\text{rotor}} \qquad (12)$$

Radial Forces The *magnetic pressure* p_r (i.e. the radial magnetic force per unit area) is proportional to the square of the magnetic flux density as shown in Eq. 13 [5, 19].

$$p_{\rm r} \approx \frac{B^2}{2\mu_0} = \frac{B_{\rm stator}^2 + 2B_{\rm stator}B_{\rm rotor} + B_{\rm rotor}^2}{2\mu_0} \qquad (13)$$

The trigonometric identities as shown in Eqs. 14 and 18 are useful for analysing the frequencies of radial magnetic forces. Equation 14 shows that each harmonic in the stator or rotor field produces a force harmonic with twice its frequency. Therefore, a first-order harmonic of the magnetic flux density causes a second-order harmonic of the radial force. Furthermore, Eq. 18 in Appendix 7.1 shows that the interaction of stator and rotor harmonics causes harmonics at the sum and difference of their frequencies.

$$\sin^2(\omega t) = \frac{1 - \cos(2\omega t)}{2} \qquad (14)$$

An overview of the spatial order, angular frequency and magnitude of the radial force harmonics is given by Wang et al. [13], who show that radial force harmonics of all even orders $2kp$ occur in fractional slot machines. Zhu et al. [18] point out that in a 12-slot/10-pole machine, even lower harmonics occur due to several interactions between the rotor and stator field, e.g. the rotor fifth harmonic (fundamental rotor harmonic) interacts with the seventh stator harmonic (due to slotting) to produce a harmonic of second order, and so does the 15th rotor harmonic with the 17th stator harmonic.

In the work of Ballo et al. [20], the radial forces for two 12-slot/10-pole PMSMs are shown. The force spectrum is dominated by forces of second time order, and second as well as tenth space order. Ballo et al. state that this is typical for 12-slot/10-pole PMSM. Islam and Husain [19] also find second- and tenth-order harmonics, as well as their subharmonics. They find that the low-order harmonics are amplified at full load.

Stator Vibration The radial forces excite vibrations of the stator. Large resonances occur when the excitation frequency matches a natural frequency, and when the excitation wave number matches the modal shape of the stator [21]. Ballo et al. [20] model the stator of two 12-slot/10-pole PMSMs as a curved beam and conclude that the second-order harmonics dominate. They state that this is typical for 12-slot/10-pole PMSM.

The high relevance of the harmonics of low spatial order is in line with the literature, which states that the vibration is inversely proportional to the fourth power of the spatial order [22]. Islam and Husain [19] state that the low-order radial force harmonics are the main source of noise in 12-slot/10-pole PMSM.

3.3 Torque Ripples

Torque ripples in PMSM originate from the cogging torque, the electromagnetic field torque, reluctance torque, magnetic saturation, current ripples (e.g. caused by

PWM), and phase current commutation [5, 12]. The focus of this paper is on cogging torque and electromagnetic torque as main contributors from electric motor design.

Cogging Torque Cogging torque is created by the interaction of the rotor's magnetic flux with the angular variation of the stator's magnetic reluctance [23]. In simpler terms, it is caused by the poles trying to align with the stator teeth. Cogging torque is also present when there is no stator current, and it can disturb a "smooth" running at low speeds. Cogging torque is proportional to the change of reluctance with the rotational angle [24], and its harmonic frequencies f_{cogging} are given in Eq. 15 [25], with the cogging periodicity τ_{cogging}.

$$f_{\text{cogging}} = kN_s\tau_{\text{cogging}}n \quad (15)$$

Fractional slot machines possess a low cogging torque at a high frequency and low amplitude, as their number of slots is not a multiple of the number of poles [23]. Within the rotation of one slot pitch, the cogging torque waveform is repeated τ_{cogging} times (see Eq. 16) [25].

$$\tau_{\text{cogging}} = \frac{2p}{\gcd(N_s, 2p)} \quad (16)$$

Field-Harmonic Electromagnetic Torque The harmonics introduced to the magnetic field have been discussed in Sect. 3.1. The electromagnetic torque is proportional to the product of the phase electro-motive force (EMF) and the phase current, see Eq. 17. Therefore, the orders of the electromagnetic torque ripple are caused by the interactions of the EMF and the phase currents [14]. Equation 18 in Appendix 7.1 describes the multiplication of two waves in general.

$$T_{\text{electromagnetic}} = \frac{1}{2\pi n}(e_A i_A + e_B i_B + e_C i_C) \quad (17)$$

An example: when the electric current harmonic and the EMF harmonic have the same order, their interaction produces a constant torque (of order 0). This is the desired drive torque [14, 16]. But the interaction of the first-order electric harmonic (of order p in reference to a mechanical revolution) and the seventh-order EMF (of order $7p$ in reference to a mechanical revolution) produces a sixth-order torque harmonic (of order $6p$ in reference to a mechanical revolution). The same applies to the fifth-order EMF, as it rotates the opposite way. More details on this example can be found in Huang et al. [24].

Reluctance Torque Reluctance torque is created by the interaction of the stator current's magneto-motive force with the variation of the rotor's magnetic reluctance. The reluctance torque of surface-mounted PMSM is typically very low [23], as the magnets have a similar reluctance as air.

Minimisation of Torque Ripple The torque ripple of electrical machines can be minimised by their design, and/or their control. Concerning the design, there are several options such as skewing of slots or magnets, selection of the slot number, and increasing the number of phases [5, 23].

4 Example for Harmonics in Electro-Hydraulic Machines

In this section, the harmonics in exemplary electrically driven pumps are discussed. Two pumps are studied, one with an even piston number ($z = 10$), and one with an odd piston number ($z = 11$). Both pumps are studied in combination with a PMSM with 12 slots and 10 poles (i.e. $p = 5$). As a simplification, the focus in this paper is only on the harmonics described in Sects. 2 and 3. Other noise contributors such as rotor unbalance, shaft misalignment, bearing noise, fan noise, and couplings are not considered. Furthermore, a direct drive is assumed (no gearbox is used).

4.1 Analytical Harmonic Orders

Table 2 and Fig. 7 summarise the drive torque harmonics in the exemplary electrically driven pumps. It can be seen that the drive torque harmonics of the PMSM are of higher order than those of the pumps. For the pump with $z = 10$, the third fundamental pump harmonic coincides with the fundamental electromagnetic torque harmonic.

Table 3 and Fig. 8 summarise the force harmonics. For the pump with $z = 10$, there are no force harmonics and the moment harmonics coincide with the fundamental radial force harmonic of the PMSM. For the pump with $z = 11$, the fundamental piston force harmonic is close to the fundamental force harmonic of the PMSM's radial force. Furthermore, the fundamental bending moment harmonic is close to the second main harmonic of the PMSM's radial force.

Flow harmonics are not considered here, as they are less relevant for the interaction between the pump and the PMSM.

Table 2 Torque harmonics in exemplary electrically driven pumps

	Pump drive torque	Cogging torque	Electromagnetic torque
Specific equation	$h_{\text{moments, 10}} = 10k$, $h_{\text{moments, 11}} = 22k$	$h_{\text{cogging}} = 60k$	$h_{\text{T,em}} = 30k$
Harmonics	10, 20, 30, ... resp. 22, 44, 66, ...	60, 120, 180, ...	30, 60, 90, ...

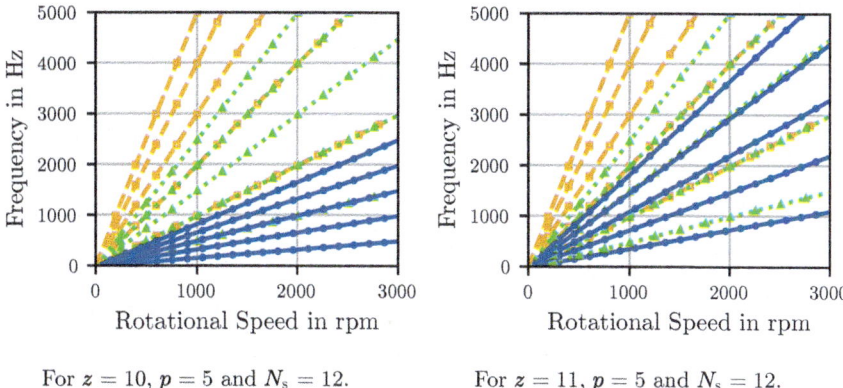

For $z = 10$, $p = 5$ and $N_s = 12$. For $z = 11$, $p = 5$ and $N_s = 12$.

Fig. 7 Campbell diagram for first 5 orders of the torque harmonics shown in Table 2. Pump drive torque is blue, cogging torque is amber, and electromagnetic torque is green

Table 3 Force harmonics in exemplary electrically driven pumps

	Piston forces	Pump bending moments	E-machine radial forces	E-machine radial forces—subharmonics
Specific equation	$h_{F,piston,10} = 0$, $h_{F,piston,11} = 11k_{odd}$	$h_{moments,10} = 10k$, $h_{moments,11} = 22k$	$h_{F,rad} = 10k$	$h_{F,rad,sub} 2k$
Harmonics	0 resp. 11, 33, ...	10, 20, 30, ... resp. 22, 44, 66, ...	10, 20, 30, ...	2, 4, 6, ...

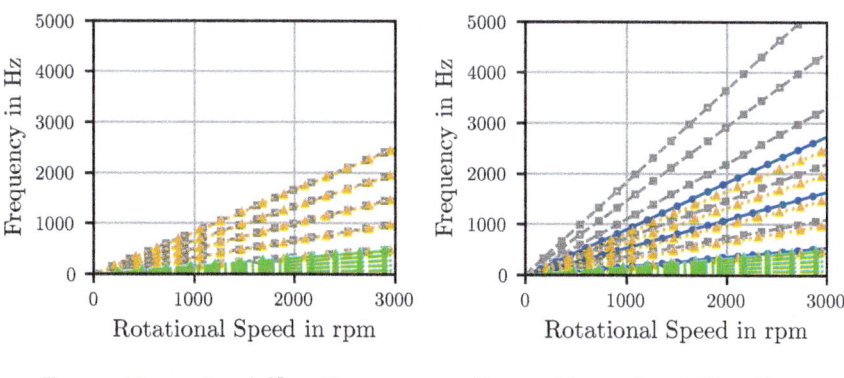

For $z = 10$, $p = 5$ and $N_s = 12$ For $z = 11$, $p = 5$ and $N_s = 12$

Fig. 8 Campbell diagram for first 5 orders of the force harmonics shown in Table 3. Pump piston forces are blue, pump bending moments are grey, the main harmonics of the electromagnetic forces are amber, and the subharmonics of the electromagnetic forces are green

4.2 Simulated Harmonic Orders

A FEM simulation for a 12-slot/10-pole PMSM was done using Elmer FEM [26], and an axial piston pump of swash-plate type was simulated using a lumped parameter model in Hopsan [27]. For both simulations, the torque ripple was extracted. Assuming an infinitely stiff connection between the PMSM and the pump, the torque ripples can be added to compute the combined torque ripple. To quantify the combined effect of pump and electric motor harmonics on the structural vibration would be significantly more challenging, due to the complexity of the components and their interfaces, as well as influences from mounting and the load side.

Figure 9 shows the torque ripples in the time domain, which have been normalised with respect to the average torque. Figure 9a, b both show the torque ripples for a PMSM in combination with a pump with ten pistons, but the phase angle between the magnet poles and the pistons is modified. This affects the peak-to-peak value of the torque ripple. As the number of poles and pistons are the same, the same pattern is repeated $z = 2p$ times for one revolution. Figure 9c shows the torque ripple for a PMSM in combination with a pump with eleven pistons. Aligning with Sect. 2.2, the torque ripple for an odd number of piston machine is lower.

Figure 10 shows the torque ripples in the frequency domain. The harmonic orders align with the expectations from Sect. 4.1. For the pump with ten pistons, the tenth harmonic is of large magnitude. A comparison between Fig. 10c, d shows that the interaction between the 30th harmonic of the PMSM and the pump is affected by the phase angle between the magnet poles and the pistons. Thus, this phase angle affects the combined torque ripple. Figure 10e, f show that odd numbers of pistons lead to decreased magnitudes, and the PMSM and pump harmonics do not align with each other.

5 Discussion

5.1 Choice of Piston Number

Section 4 shows that for $z = 2p$, some harmonics coincide. For the drive torque ripple, the third pump fundamental coincides with the first electromagnetic fundamental. However, these orders are quite high, and therefore the magnitudes are low. For the force harmonics, the bending moments of the pump coincide with the fundamental electromagnetic harmonics. This can increase the amplitude of structural vibrations caused by the combination of those harmonics if they are in phase and thus interact constructively, but conversely, there is also the possibility of creating destructive interaction. This could be achieved by controlling the phase and amplitude of the electric machine's harmonics, through design or by using the inverter to inject additional harmonics into the stator currents.

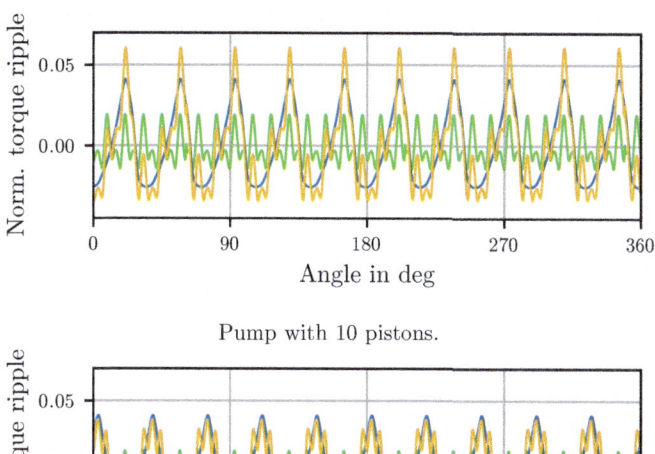

Pump with 10 pistons.

Pump with 10 pistons and modified phase angle between poles and pistons.

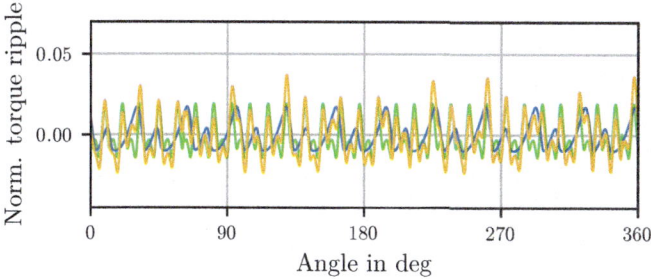

Pump with 11 pistons.

Fig. 9 Normalised torque ripple of a PMSM (in green), a ten- or eleven-piston pump (in blue), and the combined ripple (in amber). The values were computed for operation at medium pressure and speed (1500 rpm, 175 bar)

For the piston number $z = 2p + 1$, the harmonics do not coincide. However, their proximity to one another will cause the modulation of low-order waves (see the *beating effect* in Appendix 7.2), which can be critical for noise perception. For example, the interaction between the first piston force harmonic with order z and the first fundamental of the electromagnet forces with order $2p$ produces a wave of order $z - 2p = 1$. When the number of poles and pistons are the same, the same torque pattern is repeated $z = 2p$ times for one revolution, and the phase angle between poles and magnets affects the torque ripple, and it can be used to optimise

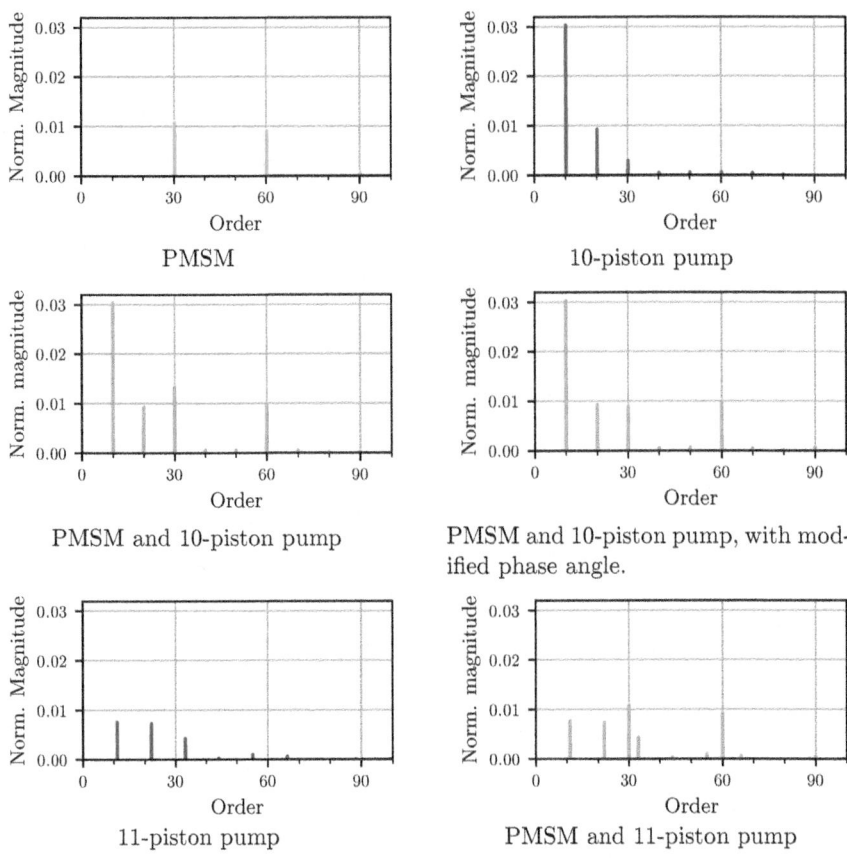

Fig. 10 Torque ripples in the frequency domain

the combined torque ripple. However, due to the low torque ripple of the considered PMSM, the combined torque ripple in this study remains higher when using even piston numbers than when using odd piston numbers.

Note that flow pulsations remain an important noise source. Section 2.1 shows that flow pulsations are decreased when using odd piston numbers in comparison to even piston numbers, and generally decreased by larger piston numbers.

5.2 Effect of Using a Gearbox

The introduction of a gearbox introduces additional harmonics into the system, which makes it more challenging to limit interactions between the harmonics.

Gearboxes reduce the torque requirement of the PMSM, at the expense of higher rotational speeds. With higher speeds, PMSMs typically are designed with fewer pole pairs p to limit frequency-dependent iron losses. As p directly affects the harmonics, all harmonics of the PMSM will occur at different frequencies.

6 Conclusion and Future Work

The harmonic characterisation of electrically driven pumps has been presented, and the design parameters which affect harmonic orders have been discussed. The combination of piston number and pole number is an important parameter for the noise of a pump combined with a PMSM. If the piston number and pole number are the same, many harmonics coincide. This can lead to increased excitation forces, but the interaction could also be used to reduce them. The phase angle between the poles and the pistons affects the magnitude of the coincidental harmonics, either constructively or destructively. The combination of piston and pole number $z = 2p \pm 1$ reduces the torque ripple, but less than expected when just considering the pump. When considering only the pump, the torque ripple of the ten-piston pump is 145% larger than that of the eleven-piston pump. However, when looking at the combined torque ripple of the pump and motor, the torque ripple of the combination with ten pistons (where $z = 2p$) is only 28% larger than for the combination with eleven pistons (where $z = 2p \pm 1$). The choice of odd numbers of pistons in electrically driven pumps is therefore less obvious. Also, piston numbers $z = 2p \pm 1$ lead to the modulation of first-order force harmonics, which can increase noise perception. Further work is needed to quantify the trade-offs due to the combinations of piston and pole numbers in electrically driven pumps.

Acknowledgements This research was funded by the Swedish Electromobility Centre with grant number 13070 and the Swedish Energy Agency (Energimyndigheten) with grant number P2023-00594.

7 Appendix: Combining Harmonics

7.1 Product of Two Harmonics

In the case of electrical machines, the product of the magnetic flux densities excited by the rotor and by the stator is proportional to the magnetic stress wave in the airgap (see Eq. 13). Equation 18 shows that the product of the waves has an amplitude $A_r = 0.5 B_1 B_2$, the angular frequencies $\omega_r = \omega_1 \pm \omega_2$, the orders $k = k_1 \pm k_2$, and the phase $\varphi_r = \varphi_1 \pm \varphi_2$ [5].

$$B_1 \cos(\omega_1 t + k_1 \alpha + \varphi_1) \cdot B_2 \cos(\omega_2 t + k_2 \alpha + \varphi_2)$$
$$= 0.5 B_1 B_2 \cos[(\omega_1 + \omega_2)t + (k_1 + k_2)\alpha + (\varphi_1 + \varphi_2)] \qquad (18)$$
$$+ 0.5 B_1 B_2 \cos[(\omega_1 - \omega_2)t + (k_1 - k_2)\alpha + (\varphi_1 - \varphi_2)]$$

7.2 Beating Effect

Another phenomenon is the so-called *beating effect*. The beating effect occurs when harmonics with similar frequencies appear. These two sine waves together create a slowly modulated high-frequency wave as shown by the trigonometric identity in Eq. 19 [28]. In practice, this implies that harmonics close to one another should be avoided to avoid the modulation of low-frequency waves.

$$\sin(\omega_1 t) + \sin(\omega_2 t) = 2 \sin\left(\frac{\omega_1 + \omega_2}{2}t\right) \cos\left(\frac{\omega_1 - \omega_2}{2}t\right) \qquad (19)$$

Nomenclature

Designation	Denotation	Unit
c	An integer number	–
e	Stator phase EMF	V
f	Frequency	1/s
h	Harmonic order in reference to one mechanical revolution	–
i	Stator phase current	A
k	An integer number	–
m	Number of phases	–
n	Mechanical speed	1/s
p	Number of pole pairs	–
p_r	Radial magnetic pressure	N/m^2
t	Time	s
z	Number of displacement chambers (e.g. pistons)	–
B	Magnetic flux density	kg/(s^2 A)
N_g	Number of slots per group	–
N_s	Number of slots	–
T	Torque	Nm
μ_0	Magnetic constant	N/A^2
τ	Periodicity	–
ω	Angular velocity	rad/s

References

1. Padovani D, Dimitriou P, Minav T (2024) Challenges and solutions for designing energy-efficient and low-pollutant machines in off-road hydraulics. energy conversion and management: X, 21 Dec 2023, 100–526. https://doi.org/10.1016/j.ecmx.2024.100526
2. Heeger T (2023) Design of electro-hydraulic energy converters : with focus on integrated designs and valve plate rotation, Linköping studies in science and technology. Licentiate thesis, vol 1971. Linköping University Electronic Press, Linköping. https://doi.org/10.3384/9789180752442. https://urn.kb.se/resolve?urn=urn:nbn:se:liu:diva-194262
3. Qu S, Zappaterra F, Vacca A, Busquets E (2023) An electrified boom actuation system with energy regeneration capability driven by a novel electro-hydraulic unit. Energy Conv Manag 293(May):117–443. https://doi.org/10.1016/j.enconman.2023.117443
4. Heeger T, West M, Heybroek K, Ericson L (2023) Methodology for dimensioning of integrated electro-hydraulic machines. In: Proceedings of the 18:th scandinavian international conference on fluid power, SICFP23 (2023). https://liu.diva-portal.org/smash/record.jsf?pid=diva2%3A1820972&dswid=-8547
5. Gieras JF, Wang C, Lai JC (2018) Noise of polyphase electric motors. CRC Press. https://doi.org/10.1201/9781420027730. https://www.taylorfrancis.com/books/9781420027730
6. Ericson L, Ölvander J, Palmberg JO (2008) On optimal design of hydrostatic machines. In: Proceedings of the 6th international fluid power conference, 6. IFK, pp 273–286
7. Harrison AM (1997) Reduction of axial piston pump pressure ripple. PhD thesis, University of Bath
8. Ivantysyn J, Ivantysynova M (2001) Hydrostatic pumps and motors: principles, design, performance, modelling, analysis, control, and testing. Akademia Books International, New Delhi. ISBN: 8185522162
9. Manring ND (2013) Fluid power pumps and motors: analysis. Design and Control, McGraw Hill Book CO. ISBN: 0071812202
10. Johansson A (2005) Design principles for noise reduction in hydraulic piston pumps. PhD thesis, Linköping University
11. Heeger T, Kärnell S, Ericson L (2024) Challenges for multi-quadrant hydraulic piston machines. Energy Conv Manag: X 100578. https://doi.org/10.1016/j.ecmx.2024.100578. https://www.sciencedirect.com/science/article/pii/S2590174524000564
12. Vaez-Zadeh S (2018) Control of permanent magnet synchronous motors. Oxford University Press, Oxford. https://doi.org/10.1093/oso/9780198742968.001.0001. https://academic.oup.com/book/26764
13. Wang J, Xia ZP, Howe D, Long SA (2006) Vibration characteristics of modular permanent magnet brushless AC machines. Conf Rec—IAS Ann Meet (IEEE Industry Applications Society) 3(c):1501–1506. https://doi.org/10.1109/IAS.2006.256728
14. Zhang J, Zhang B, Feng G (2021) Influence of pole and slot combination on torque characteristics and radial force of fractional slot permanent magnet machines. IEEJ Trans Electr Electron Eng 16(8):1055–1066. https://doi.org/10.1002/tee.23402
15. Hannon B, Sergeant P, Dupré L (2017) Time- and spatial-harmonic content in synchronous electrical machines. IEEE Trans Magn 53(3). https://doi.org/10.1109/TMAG.2016.2637316
16. Bianchi N, Bolognani S, Pré MD, Grezzani G (2006) Design considerations for fractional-slot winding configurations of synchronous machines. IEEE Trans Ind Appl 42(4):997–1006. https://doi.org/10.1109/TIA.2006.876070
17. Pyrhönen J, Jokinen T, Hrabovcová V (2013) Design of rotating electrical machines. Wiley (2013). https://doi.org/10.1002/9781118701591. https://onlinelibrary.wiley.com/doi/book/10.1002/9781118701591
18. Zhu ZQ, Xia ZP, Wu LJ, Jewell GW (2010) Analytical modeling and finite-element computation of radial vibration force in fractional-slot permanent-magnet brushless machines. IEEE Trans Ind Appl 46(5):1908–1918. https://doi.org/10.1109/TIA.2010.2058078

19. Islam R, Husain I (2010) Analytical model for predicting noise and vibration in permanent-magnet synchronous motors. IEEE Trans Ind Appl 46(6):2346–2354. https://doi.org/10.1109/TIA.2010.2070473
20. Ballo F, Gobbi M, Mastinu G, Palazzetti R (2023) Noise and vibration of permanent magnet synchronous electric motors: a simplified analytical model. IEEE Trans Transp Electrification 9(2):2486–2496. https://doi.org/10.1109/TTE.2022.3209917
21. Devillers E, Degrendele K, Hecquet M, Lecointe JP, Besnerais JL, Cousin G (2019) Open-access testbench data for NVH benchmarking of e-machines under electromagnetic excitations. SAE Tech Paper (2019)
22. Deng W, Zuo S (2019) Electromagnetic vibration and noise of the permanent-magnet synchronous motors for electric vehicles: an overview. IEEE Trans Transp Electrification 5(1):59–70. https://doi.org/10.1109/TTE.2018.2875481
23. Jahns TM, Soong WL (1996) Pulsating torque minimization techniques for permanent magnet AC motor drives—a review. IEEE Trans Ind Electron 43(2):321–330. https://doi.org/10.1109/41.491356
24. Huang S, Aydin M, Lipo TA (2001) Torque quality assessment and sizing optimization for surface mounted permanent magnet machines. Conf Rec—IAS Ann Meet (IEEE Industry Applications Society) 3(1):1603–1610. https://doi.org/10.1109/ias.2001.955749
25. Wang X, Yang Y, Fu D (2003) Study of cogging torque in surface-mounted permanent magnet motors with energy method. J Magn Magn Mater 267:80–85. https://doi.org/10.1016/S0304-8853(03)00324-X
26. CSC—IT Center For Science Ltd: Elmer. https://www.csc.fi/web/elmer
27. Linköping University (Division of Fluid and Mechatronic Systems): Hopsan. https://liu.se/en/research/hopsan
28. Vold H, Herlufsen H, Mains M, Corwin-Renner D (1997) Multi axle order tracking with the Vold-Kalman tracking filter. Sound Vib 31(5):30–34

Open Access This chapter is licensed under the terms of the Creative Commons Attribution 4.0 International License (http://creativecommons.org/licenses/by/4.0/), which permits use, sharing, adaptation, distribution and reproduction in any medium or format, as long as you give appropriate credit to the original author(s) and the source, provide a link to the Creative Commons license and indicate if changes were made.

The images or other third party material in this chapter are included in the chapter's Creative Commons license, unless indicated otherwise in a credit line to the material. If material is not included in the chapter's Creative Commons license and your intended use is not permitted by statutory regulation or exceeds the permitted use, you will need to obtain permission directly from the copyright holder.

The manufacturer's authorised representative in the EU is Springer Nature Customer Service Centre GmbH, Europaplatz 3, 69115 Heidelberg, Germany. If you have any concerns regarding our products, please contact ProductSafety@springernature.com

Printed and bound by CPI Group (UK) Ltd, Croydon, CR0 4YY
26/03/2026
02078989-0001